Springer Proceedings in Mathematics & Statistics

Volume 510

This book series features volumes composed of selected contributions from workshops and conferences in all areas of current research in mathematics and statistics, including data science, operations research and optimization. In addition to an overall evaluation of the interest, scientific quality, and timeliness of each proposal at the hands of the publisher, individual contributions are all refereed to the high quality standards of leading journals in the field. Thus, this series provides the research community with well-edited, authoritative reports on developments in the most exciting areas of mathematical and statistical research today.

Adibah Shuib · Yu-Hong Dai ·
Iskandar Shah Mohd Zawawi
Editors

Mathematical Modeling and Numerical Analysis

UICMS 2024, Kuala Lumpur, Malaysia, October 12–13

Springer

Editors
Adibah Shuib
Faculty of Computer and Mathematical Sciences
Universiti Teknologi MARA
Shah Alam, Selangor, Malaysia

Yu-Hong Dai
Academy of Mathematics and Systems Science
Chinese Academy of Sciences
Beijing, China

Iskandar Shah Mohd Zawawi
Faculty of Computer and Mathematical Sciences
Universiti Teknologi MARA
Shah Alam, Selangor, Malaysia

ISSN 2194-1009 ISSN 2194-1017 (electronic)
Springer Proceedings in Mathematics & Statistics
ISBN 978-981-96-9349-8 ISBN 978-981-96-9350-4 (eBook)
https://doi.org/10.1007/978-981-96-9350-4

Mathematics Subject Classification: 92C37, 90C29, 53A10, 54E15, 65L05, 90B50, 68T05, 65N30, 65L20, 90C10

This Springer imprint is published by the registered company Springer Nature Singapore Pte Ltd.
The registered company address is: 152 Beach Road, #21-01/04 Gateway East, Singapore 189721, Singapore

Preface

This book comprises 20 selected original and unpublished papers presented at the "UiTM International Conference of Mathematical Sciences 2024 (UICMS 2024)" organized by the College of Computing, Informatics and Mathematics, Universiti Teknologi MARA (UiTM) Shah Alam, Selangor, Malaysia, held from October 12 to 13, 2024, at the Perdana Kuala Lumpur City Centre (KLCC), Malaysia. The UICMS 2024 with the theme "Mathematics in the Age of Intelligence and Analytics: Bridging Theory and Applications" aims to bring together a diverse group of thought leaders, innovators and practitioners from around the world. The UICMS 2024 focusses on the critical role that mathematics plays in the current artificial intelligence and analytics age, emphasizing the bridge between theoretical advancements and practical applications. Mathematics undeniably fosters a robust theoretical foundation and provides feasible real-world solutions across various fields. This consequently drives advancements in both artificial intelligence (AI) and the mathematical sciences. The UICMS 2024 was supported by Malaysian Mathematical Sciences Society.

As part of the conference event, three workshops on mathematical imaging techniques for digital image processing, decision-making and fuzzy system and integrating mathematics and machine learning were organized. The UICMS 2024 features four key tracks: Differential Equations and Numerical Analysis (DENA), Applied and Computational Mathematics (ACM), Algebra and Analysis (AA) and Statistics, Analytics, Actuarial Science and Decision Science (SAAD), each promoting significant contributions in their respective fields.

Talks were held in parallel sessions and 81 papers were presented, of which 20 were accepted for publication in this volume with acceptance rate 25%. Each paper had undergone peer review by using a double-blind reviewing process by at least two reviewers with strong academic background. Reviewers are asked to evaluate the relevance of the topic, significance, originality and contribution of the work. Volume editors assess comments and make necessary recommendations to respective publishers and contributing authors. These papers represent the collective effort and dedication of the authors, whose remarkable research and findings continue to push the boundaries of mathematical sciences and applications.

The contributed scientific papers published in this volume cover a wide range of topics within mathematical sciences, presenting novel findings and promoting sustainable developments in mathematical sciences in addressing contemporary challenges in the era of AI and analytics. The first-seven chapters cover a range of advanced numerical methods applied to various real-world problems. These include the numerical analysis of rod behavior in both linear and nonlinear formulations, stability analysis of numerical method for fractional order differential equations and the piecewise constant argument method for nonlinear systems. This book also addresses the numerical solution of the susceptible-infected-recovery model for Coronavirus disease transmission with vaccination, validation of numerical method in chronic lymphocytic leukemia therapy studies and modeling tuberculosis transmission with vaccination strategies. Additionally, it explores computational methods for solving second-order nonlinear differential equations by using piecewise constant argument techniques, highlighting the importance of these methods in engineering, biology and medical research.

The next-four chapters explore various optimization and modeling techniques applied across different fields. The performance measurement of Malaysian international airports uses data envelopment analysis. Ranking success factors of open distance learning applies the interval neutrosophic measurement of alternatives and ranking according to compromise solution method. Optimizing workforce scheduling for a veterinary clinic uses a binary programming model to enhance staff allocation and improve operational efficiency. Finally, the susceptible-exposed-infected-recovery model for disease transmission is solved using a multi-step block method to simulate the spread of infectious diseases. These approaches reflect the broad applicability of optimization and modeling in enhancing decision-making across diverse sectors.

The following chapters explore diverse applications of advanced analytical and computational techniques across various fields. They include using hybrid machine learning models for credit score prediction, applying Bayesian Poisson generalized linear mixed models for simulating dengue incidence and investigating the odd Lomax–Fréchet distribution for risk modeling. The chapters also focus on optimizing travel itineraries for Islamic landmarks in Kuala Lumpur by using the Miller–Tucker–Zimmer traveling salesmen problem model, evaluating the efficiency of water service providers in Malaysia with the scale directional distance functions method and developing a green mixed integer goal programming model for optimizing vegetable freshness in supply chains. Additionally, the impact of road accidents during Coronavirus disease in Malaysia is analyzed statistically. Together, these chapters highlight the role of computational modeling, statistical analysis and optimization in solving real-world problems in health, transportation, finance and environmental management.

The last-two chapters delve into advanced topics in geometry and fuzzy topology. The first explores the recovery of convex surfaces in semi-hyperbolic spaces, a non-Euclidean geometric framework, focusing on methods to reconstruct convex shapes for applications in fields like computer graphics and optimization. The last chapter introduces generalized separation axioms in fuzzy bitopological spaces, merging fuzzy set theory with bitopology to study imprecise information. This work advances theoretical understanding of topological properties in uncertain contexts, with potential applications in fuzzy logic, decision-making and information theory. Both chapters contribute to the development of mathematical models for complex systems involving geometry and uncertainty.

Organized in a hybrid mode, the conference had 42 face-to-face participants and 40 online participants. Professor Dr. Rakhmatillo Aloev (National University of Uzbekistan, Uzbekistan), Prof. Dr. Mukhiddin Muminov (Samarkand State University, Uzbekistan), Dr. Ts. Dr. Adibah Shuib (UiTM Shah Alam, Selangor, Malaysia) and Prof. Dr. M. Nirmala (Sathyabama Institute of Science and Technology, India) were invited as keynote speakers. Professor Dr. Mirzoali Khudoyberganov (National University of Uzbekistan, Uzbekistan), Prof. Dr. Abdullo Rakhmonovich Hayotov (Institute of Mathematics, Uzbekistan Academy of Sciences, Uzbekistan), Prof. Dr. Rasulov Tulkin Husenovich (Bukhara State University, Uzbekistan), Dr. Ural Bekbaev (Turin Polytechnic University in Tashkent, Uzbekistan), Prof. Dr. La Zakaria (Universitas Lampung, Indonesia), Dr. Nanang Susyanto (Universitas Gadjah Mada, Indonesia), Dr. Priya Shirley Muller (Sathyabama Institute of Science and Technology, India), Assoc. Prof. Dr. J. Jayapriya (Sathyabama Institute of Science and Technology, India) and Dr. M. Prasanna Jeyanthi (Sathyabama Institute of Science and Technology, India) delivered invited speeches.

We express our thanks and appreciation to the keynote and invited speakers, session chairs, reviewers and UICMS 2024 committee for their commitments and efforts in ensuring the success of this conference. The conference organizer extends gratitude to Springer for accepting to publish these papers in a proceedings form.

Shah Alam, Selangor, Malaysia — Adibah Shuib
Beijing, China — Yu-Hong Dai
Shah Alam, Selangor, Malaysia — Iskandar Shah Mohd Zawawi

Contents

Numerical Analysis of the Calculation of Rods in Linear and Geometrically Nonlinear Formulations

Anarova Shahzoda, Karshieva Soliha, Samidov Muhriddin, Uralova Iroda, and Amonova Oftoboy

Abstract The article is devoted to the numerical analysis of the calculation of rods in linear and geometrically nonlinear formulations. The article presents graphs of the results of longitudinal displacement along the length of the rod at different time points. The transverse vibration propagation along the length of the rod is shown for boundary conditions of rigid fixing at two ends in a linear formulation at different time points. The sections of the rod where the largest excitation is observed are determined, and it is shown that the results differ at different times at different points along the length of the rod. Several practical problems can be solved setting specific values of parameters (Young's modulus, Poisson's ratio (for steel), rod length, cross-sections under consideration, surface loads). Considering the theoretical foundations, a software package was created using finite difference methods, the Bubnov–Galerkin method, the V. L. Rvachev *R*-functions method (RFM), and the method of successive approximations. The method of successive approximations is used in the article, by virtue of which linear and nonlinear problems are solved at each iteration step. The results obtained are presented in the form of graphs. The article highlights the importance of numerical methods in the analysis of bar structures and provides valuable information for engineers and researchers involved in this field.

Keywords Bubnov–Galerkin · Numerical analysis · Poisson's ratio · Rod · *R*-function · Young's modulus

---

A. Shahzoda · K. Soliha (✉) · S. Muhriddin · U. Iroda · A. Oftoboy
Tashkent University of Information Technologies, Tashkent, Uzbekistan
e-mail: solihakarshieva@gmail.com; s.qarshiyeva@tuit.uz

A. Shahzoda
e-mail: anorova@tuit.uz

S. Muhriddin
e-mail: maker96bek@gmail.com

U. Iroda
e-mail: i.abduvaliyevna@tuit.uz

A. Oftoboy
e-mail: amonovasitora22@gmail.com

A. Shuib et al. (eds.), *Mathematical Modeling and Numerical Analysis*,
Springer Proceedings in Mathematics & Statistics 510,
https://doi.org/10.1007/978-981-96-9350-4_1

1 Introduction

It is known that in mechanical engineering, structural elements in the form of prismatic bodies of arbitrary cross-sections (sprockets, polygons, etc.) with cavities of various shapes to lighten the weight without reducing their strength are used. When designing such structural parts, it is necessary to know what state they are in: elastic or elastoplastic. Details of any engineering structure, regardless of its purpose, must be strong, rigid, and lightweight at the lowest cost of materials. Therefore, one of the main design tasks is the study of elastic and elastoplastic states in machine parts or structural elements of a given shape, and, based on the research, the creation of new, more rational structural forms with a given strength. In this regard, the optimal design of prismatic structural parts of arbitrary cross-sections in various branches of mechanical engineering and the study of their strength quality are urgent problems that require the implementation of modern calculation methods that would make it possible to more fully consider the actual operating conditions of the configuration of a given part, accounting for the properties of the material. Therefore, it is advisable to develop new algorithms based on existing problem-solving methods that allow for a wide-ranging study without any difficulty.

In this regard, in the Cartesian coordinate frame (x, y, z), prismatic bodies of arbitrary cross-sections with physical and mechanical characteristics are considered. Generally, they are described by a linear or nonlinear system of partial differential equations with appropriate boundary and initial conditions. In this form, the problem presented in the differential formulation is not always solved; therefore, to eliminate the difficulties of this formulation, variational principles are used. The combined use of the variational principle and classical methods such as Ritz, Bubnov–Galerkin, and Vlasov–Kantorovich methods does not always lead to the desired results. Since the domain under consideration (a section) has a complex shape, these methods are related to the construction of a system of coordinate functions that satisfy the boundary conditions. Until recently, constructing coordinate functions for complex domains was very difficult. Therefore, a new approach is required that, at an analytical level, could consider the geometry of the domain and the pattern of conditions at the edges. This problem can be solved using R-function method [1, 2] or finite element method [3] using which a system of coordinate functions can be constructed.

2 Methods, Results, and Discussions

Numerical analysis of the vibrations of spatially loaded rods is carried out within the framework of linear formulations to determine their dynamic responses under various loading conditions.

In [4–9], mathematical models and a computational algorithm for spatially loaded rods are presented in linear and geometrically nonlinear formulations under the following hypothesis:

$$\left.\begin{array}{l} u_1(x,y,z,t) = u(x,t) - z\alpha_1(x,t) - y\alpha_2(x,t), \\ u_2(x,y,z,t) = v(x,t) + z\theta(x,t), \qquad u_3(x,y,z,t) = w(x,t) - y\theta(x,t). \end{array}\right\}$$

To calculate the bars, the following parameter values are used: Young's modulus $E = 2 \times 10^5$ Pa, Poisson's ratio $\nu_1 = 0.3$ (for steel), the length $l = 10$ m, cross-sections under consideration $a = 0.02$ m, $b = 0.02$ m, surface loads $q_1 = 0.015$ N, $q_2 = 0.01$ N, $q_3 = 0.02$ N, $M = 0.012$ N m.

The problems posed are solved when the following relations are given [10, p. 60]:

$$a_1 = \psi_1 - z, \quad a_2 = \psi_2 - y,$$

$$\psi_1 = \frac{3z}{2k}\left(1 - \frac{4z^2}{3h^2}\right), \quad \psi_2 = \frac{3y}{2k}\left(1 - \frac{4y^2}{3b^2}\right),$$

$$a_1 = a\left(\frac{5}{4}z - \frac{5}{3}z^3 - z\right), \quad a_2 = b\left(\frac{5}{4}y - \frac{5}{3}y^3 - y\right).$$

Here, as an example, consider a rod with fixed edges, longitudinal force q_1, and transverse forces q_2 and q_3.

Based on the studies conducted in [11–16], a complex of programs (CP) was created for solving boundary value problems of systems of linear and nonlinear partial differential equations, aimed to calculate the stress–strain state of elastic prismatic bodies of arbitrary cross-sections with a cavity [17–21]. Considering the theoretical foundations, a software package was created using finite difference methods, Bubnov–Galerkin, V. L. Rvachev *R*-functions (RFM) methods, and the method of successive approximations. Here, the method of successive approximations is used; therefore, linear and nonlinear problems are solved at each iteration step [22–27].

Figure 1 shows graphs of the results obtained of longitudinal displacements along the length of the rod at times $t = 20$, 40, 80 s. As seen from the graphs, at different values of t, at different points along the length of the rod, different results were obtained: increasing sinusoidal amplitudes appeared closer to the points of the center of the rod.

In Fig. 1, it can be seen that for longitudinal displacement u, the largest excitation is formed in the middle of the rod. Further, the displacement values decrease along the ends of the rod. At $t = 20$ s, the longitudinal displacement u has the smallest value and gradually increases at $t = 40$ s, and at $t = 80$ s, it has the largest value.

Figure 2 shows the propagation of transverse vibrations v along the length of the rod for boundary conditions of rigid fixing at two ends in a linear formulation at different time points. It can be seen in the figure that for transverse vibrations v, the largest excitation appears at the ends of the rod and gradually decreases in the middle of the rod. At $t = 20$ s, the transverse vibration has the smallest value and gradually increases at $t = 40$ s, and at $t = 80$ s, it has the largest value. With increasing time t, the transverse vibration v at the edge points of the rod takes on the largest value and symmetrically moves to the maximum value relative to the center of the length of the rod.

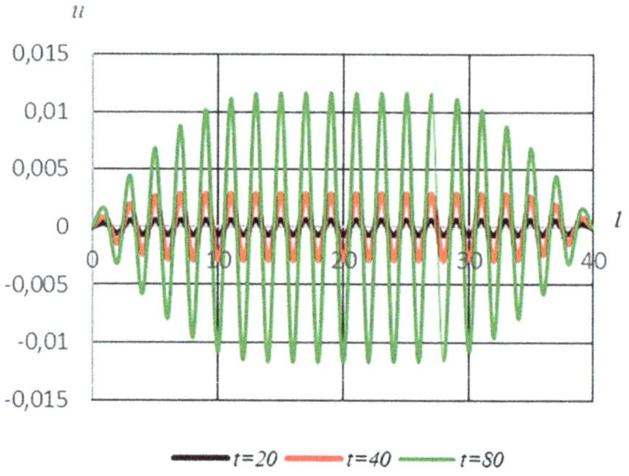

Fig. 1 Propagation of longitudinal vibrations $u \cdot 10^6$ along the length of the rod for $l = 10$ m at different time points

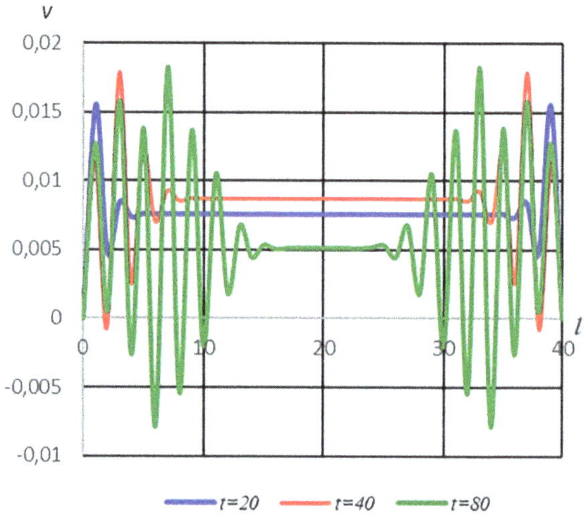

Fig. 2 Propagation of transverse vibrations along the length of the rod at different values of $t = 20$, 40, 80 s

Figure 3 shows the propagation of transverse vibration along the length of the rod for the boundary conditions of rigid fixing of two ends in a linear formulation at different time points. It can be seen in the figure that for transverse vibration, the largest excitation is observed at the ends of the rod, which gradually decreases as it approaches the middle of the rod. At $t = 20$ s, the transverse vibration has the smallest value and gradually increases at $t = 40$ s, and at $t = 80$ s, it takes on the largest value. With increasing time t, the transverse vibration w at the edge points of the rod takes on the largest value and symmetrically moves to the maximum value relative to the center of the length of the rod. The value of transverse vibrations w corresponding to the largest amplitude is observed at the initial time t and, at $t =$ 20 s near the boundary points, with increasing time t, for example, at $t = 80$ s, it is observed close to the center and has a large value.

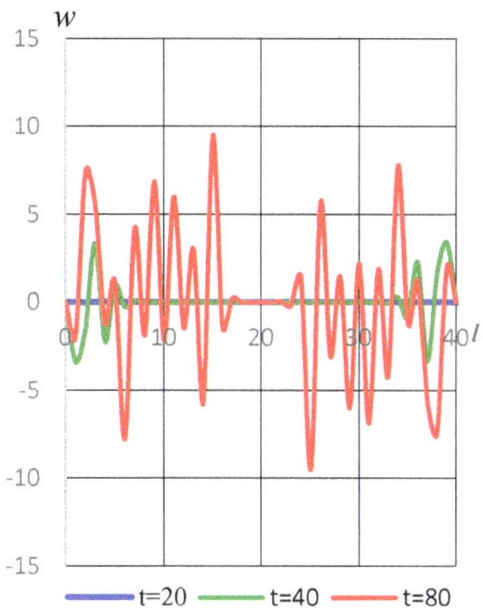

Fig. 3 Propagation of transverse vibrations $w \cdot 10^2$ along the length of the rod at $t = 20$, 40 and 80 s

Figure 4 shows graphs of the vibration propagation of the angle of inclination α_1 along the length of the rod at different times ($t = 20$, 40, and 80 s). According to these graphs, the largest excitation of the inclination angle α_1 appears at the ends of the rod, and it gradually decreases toward the middle of the rod. At $t = 20$ s, the inclination angle α_1 has the smallest value, which gradually increases at $t = 40$ s, and at $t = 80$ s, it has the largest value.

Figure 5 shows graphs of the vibration propagation of the angle of inclination α_2 along the length of the rod at different times ($t = 20$, 40, and 80 s). According to the graphs, the angle of inclination α_2 has the greatest excitation at the ends of the rod and gradually decreases toward the middle of the rod. At $t = 20$ s, the inclination

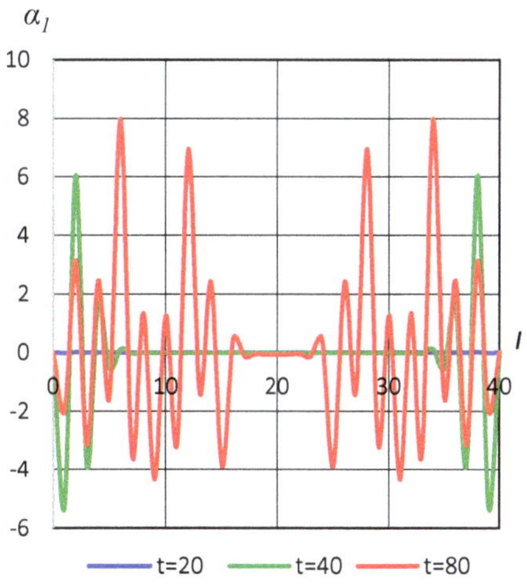

Fig. 4 Propagation of inclination angle $\alpha_1 \cdot 10^5$ along the length of the rod at $t = 20$, 40, 80 s

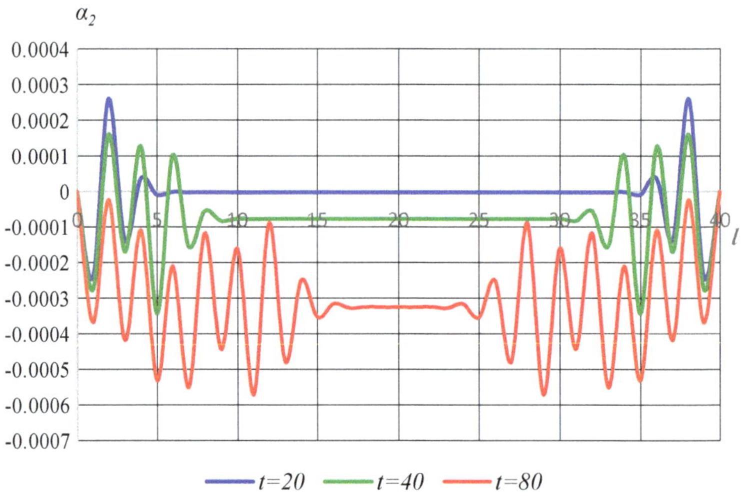

Fig. 5 Vibration propagation of the inclination angle $\alpha_2 \cdot 10^2$ along the length of the rod at different time points

angle α_2 has the smallest value and gradually increases at $t = 40$ s and, at $t = 80$ s, it takes on the largest value. According to these graphs, with an increase in time at two rigidly fixed edges, the vibration begins to increase toward the center of the rod and different values of vibration are observed at different points of the rod. At $t = 20$ s, the amplitude of the vibrations is small and only at the boundary points takes on different values that differ from zero; at $t = 40$ s and 80 s, the amplitude of the vibrations increases, and the largest values are achieved at points close to the center. With an increase in time, the inclination angle α_2 has a negative value.

Figure 6 shows graphs of the vibration propagation of the torsion angle θ along the length of the rod at times $t = 20$, 40 and 80 s. From these graphs, it follows that the torsion angle θ has the greatest excitation at the ends of the rod, and gradually decreases toward the middle of the rod. As time increases, the large value of the excitation gradually increases. At small values of time in a rod with fixed edges, the fundamental vibrations occur in the boundary region. In the middle parts, a constant value is set. Over time, the excitations are transferred closer to the center, and the middle part at $t = 40$ s and 80 s has a constant and non-zero value. If the average part of the constant torsion at $t = 20$ s had a positive value, then over time, it takes on a negative value.

Figure 7 shows graphs of the vibration propagation of the angle of linear twist ϑ along the length of the rod at $t = 20, 40$, and 80 s. The pattern of vibrations repeats the result of the torsion angle: over time, the middle part of the constant value decreases gradually. Large vibrations gradually move from the edges to the middle of the rod. The difference from Fig. 6 is that the constant longitudinal torsion in the middle part of the rod increases gradually and remains positive at all times.

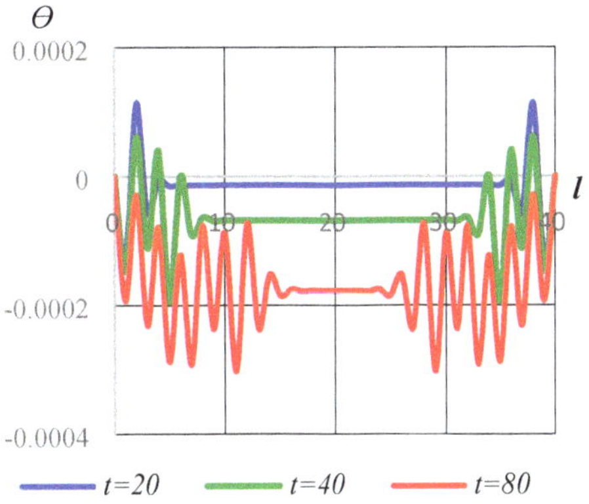

Fig. 6 Vibration propagation of the torsion angle $\theta \cdot 10^3$ along the length of the rod at different times

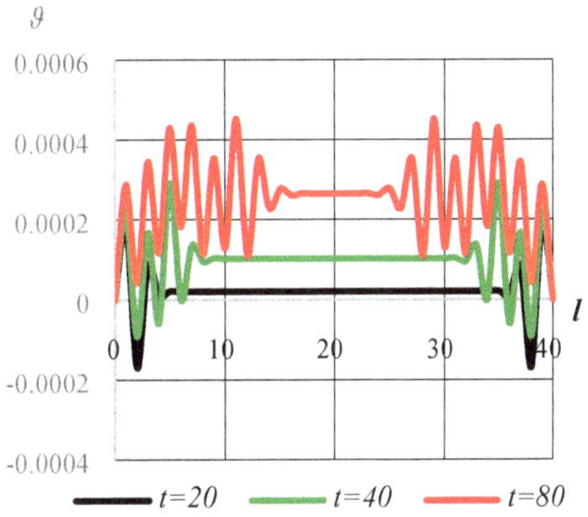

Fig. 7 Vibration propagation of the angle of linear twist $\vartheta \cdot 10^2$ along the length of the rod at different times

The graphs of the vibration propagation of the shear angles β_1 and β_2 along the length of the rod at different time points qualitatively repeat the graphs in Fig. 6. The interval of change in $10^2 \cdot \beta_1$ and $10^2 \cdot \beta_2$ is (– 0.0005; 0.0015).

Figures 8, 9, 10, 11, 12, and 13 show longitudinal, transverse, and torsional vibrations of the rod with the functions of longitudinal displacement, transverse displacement of inclination angles, and torsion angle for a rod, the two ends of which are rigidly fixed. The results were obtained for linear and nonlinear formulations at different time points.

Figure 8 shows graphs of the propagation of longitudinal vibrations along the length of the rod $l = 10$ m at $t = 40$ and 80 s. According to Fig. 8, longitudinal displacement $u(x, t)$ has the largest excitation in the middle of the rod. As moving to the ends of the rod, the excitations attenuate. The amplitude of the vibrations increases as time goes on. The number of iterations here is $k = 9$.

Figures 9 and 10 show transverse vibrations $v(x, t)$ and $w(x, t)$ of the rod in linear and nonlinear formulations for boundary conditions, when the rod is rigidly fixed at both ends at time values of $t = 40$ s and 80 s. As seen in the figures, the largest

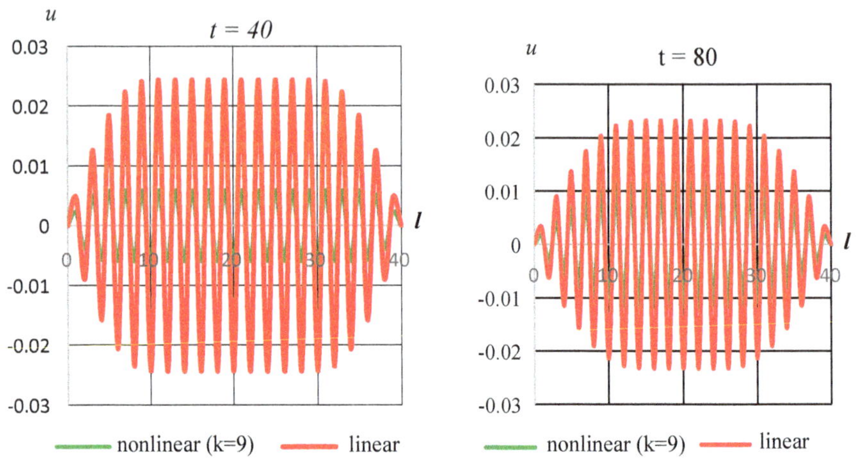

Fig. 8 Propagation of longitudinal vibrations $u \cdot 10^6$ along the length of the rod at different times, $l = 10$ m

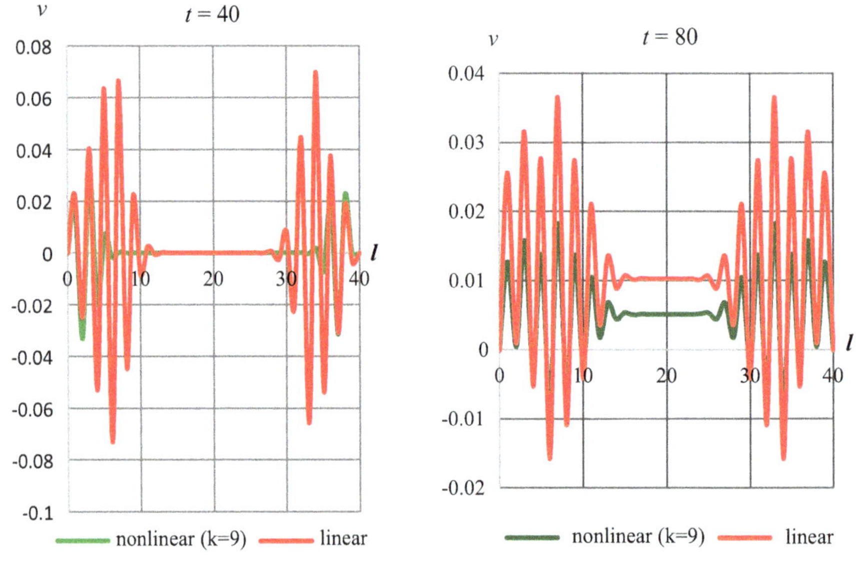

Fig. 9 Propagation of transverse vibrations $v \cdot 10^4$ along the length of the rod at $t = 40$ s and 80 s

excitations for transverse displacements $v(x, t)$ and $w(x, t)$ appear at the ends of the rod and gradually decrease in the middle of the rod.

Figure 11 shows graphs of vibration in the torsion angle $\theta \cdot 10^3$. This figure demonstrates the localization of large amplitudes closer to the ends of the rod. Here, the number of iterations is $k = 9$.

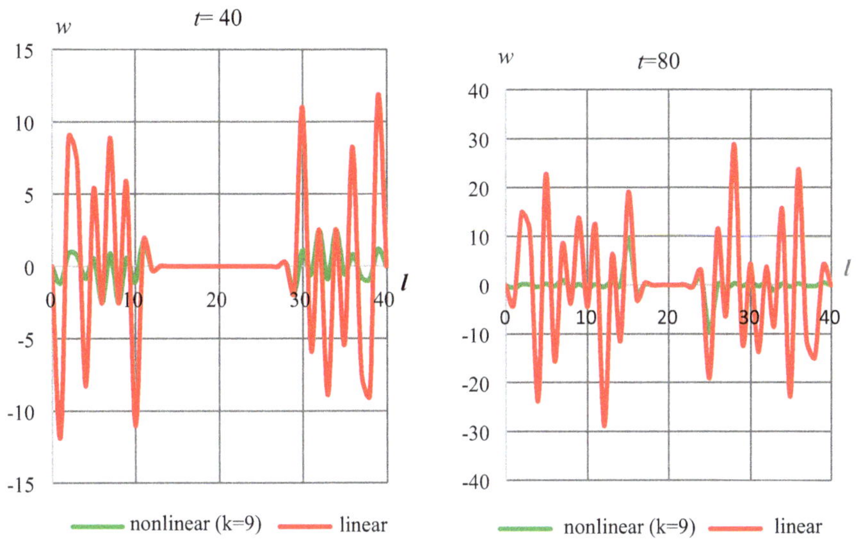

Fig. 10 Propagation of transverse vibrations $w \cdot 10^2$ along the length of the rod at $t = 40$ s and 80 s

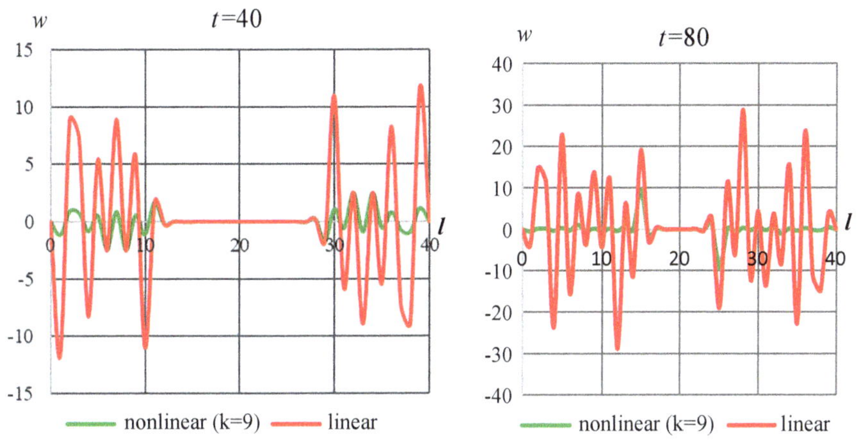

Fig. 11 Propagation of transverse vibrations $w \cdot 10^2$ along the length of the rod at $t = 40$ s and 80 s

Below are the graphs of results when one end of the rod is rigidly fixed and the other end is free.

Similar results were obtained for vibrations in the inclination angles α_1 and α_2 with corresponding intervals of change. Graphs comparing the results of inclination angles along the length of the rod in linear and nonlinear formulations are given in Fig. 13. As can be seen, the difference in results is 0.001. Here, the number of iterations is $k = 9$.

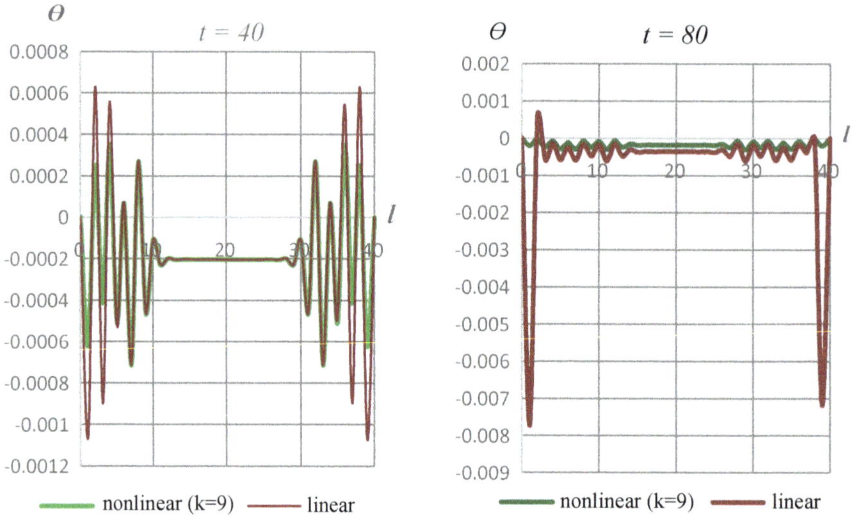

Fig. 12 Vibration propagation of torsion angle $\theta \cdot 10^3$ along the length of the rod at different time points t

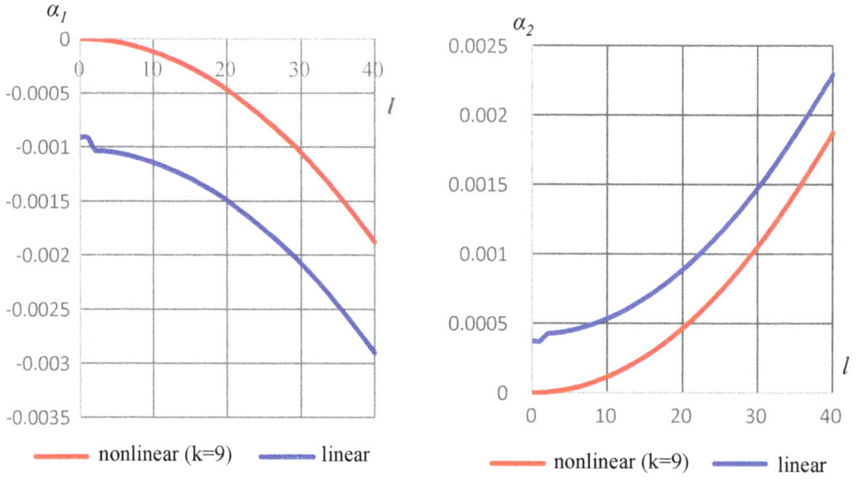

Fig. 13 Comparison of the results of the angle of inclination α_1 and α_2 along the length of the rod in linear and nonlinear formulations

A comparison of results for the torsion angle θ showed a similar pattern. The difference on average was 0.0001.

At longitudinal $u(x, t)$ and vertical $v(x, t)$ and $w(x, t)$ displacements, the difference is approximately 0.00003. All figures show that indicators $\alpha_2(x, t)$ decrease monotonically, and the difference interval is 0.0008. At the vertical displacement, the difference interval is 0.004. For these parameters, the indicators were monotonically increasing.

The comparison of results between linear and nonlinear formulations is illustrated in Figs. 8, 9, 10, 11, 12, and 13. It has been determined that taking nonlinear terms into account significantly improves the dynamic model of rods. This improvement results in reduced amplitudes of transverse vibrations and increased frequencies in comparison to the linear cases.

The results obtained are consistent with the results of studies on the nonlinear dynamics of rods, and when compared, the displacements of elastic elements in the linear formulation differ from the nonlinear formulation of the problem. This suggests that the design of the rods with the geometric and physical parameters considered here is significantly nonlinear. Therefore, linear modeling of this problem will lead to a first (overestimated) approximation of the solutions. In addition, it will reflect the variety of complex oscillatory processes that occur at multiple frequencies, which will remain outside the scope of linear dynamic models.

3 Conclusion

The following main aspects can be noted in conclusion.

A detailed numerical analysis of rods in linear and geometrically nonlinear formulations is presented. Vibrations of spatially loaded rods in linear formulations were studied. The statics and dynamics of spatially loaded rods in linear and geometrically nonlinear formulations were considered. Transverse vibrations of a rod rigidly fixed at two ends were shown in linear and nonlinear formulations for boundary conditions, at different time points.

In the figures, displacements and stresses are given in dimensionless form. The 30KhSA material was considered as an example [28].

Linear and geometrically nonlinear problems of structural materials such as rods solved in the practice of design and survey work were formulated, and numerical experiments were conducted. Systems of nine linear differential equations related to nine parameters and systems of six nonlinear second-order differential equations were solved. Analysis of the numerical results shows that the solution to a system of second-order differential equations, considering all the parameters of the displacement vector along the coordinate axes, provides the possibility of a detailed description of the stress–strain state and physical and mechanical properties of the object under consideration. This, in turn, serves as a fundamental basis for forming and presenting appropriate application proposals and recommendations to design engineers.

References

1. Rvachev, V.L.: The Theory of the R-Function and Some of Its Applications. Naukova Dumka, Kyiv (1987)
2. Rvachev, V.L., Goncharyuk, I.V.: Torsion of Rods of Complex Profile. Publishing House of Kharkov Polytechnic Institute, Kharkov (1973)
3. Obraztsov, I.F., et al.: Finite Element Method in Problems of Structural Mechanics of Aircraft. Higher School, Moscow (1985)
4. Nazirov, Sh.A., Anarova, Sh.A.: Mathematical model of vibration of spatially loaded thin-walled rods. In: Modern Problems of Mechanics and Mathematics: Proceedings of the International Scientific Conference Dedicated to the 85th Anniversary of Academician of the National Academy of Sciences of Ukraine Yaroslav Stepanovich Podstrigach and the 40th Anniversary of the Scientific Institution Created in the Field of Mechanics and Mathematics, pp. 81–83 (2013)
5. Nazirov, Sh.A., Yuldashev, T., Anarova, Sh.A.: Nonlinear mathematical model of vibration of thin-walled rods under spatial loading. Issues Comput. Appl. Math. Collect. Sci. Works Tashkent (131), pp. 5–26 (2014)
6. Anarova, Sh.A., Ismoilov, Sh.M.: Nonlinear mathematical model of stress-strain state of spatially loaded rods with account for temperature. AIP Conf. Proc. **2365**, 070019 (2021)
7. Anarova, Sh.A., Ismoilov, Sh.M., Shokirov, D.: Nonlinear mathematical model of vibration processes of spatially loaded rods with account for temperature. In: International Conference on Information Science and Communications Technologies (ICISCT) (2021). https://doi.org/10.1109/ICISCT52966.2021.9670072
8. Anarova, Sh.A.: Algorithm of solution of geometrically nonlinear problem of rods with arbitrary mechanical geometrical characteristics. Int. J. Adv. Res. Sci. Eng. Technol. **4**(11), 4796–4815 (2017). ISSN: 2350-0328
9. Anarova, Sh.A., Nuraliev, F.M., Dadenova, G.: Mathematical model of spatially loaded bars with account of torsion function and transverse shear. Int. J. Techn. Res. Appl. **4**(1), 22–32 (2016). e-ISSN: 2320-8163
10. Kabulov, V.K.: Algorithmization in the Theory of Elasticity and Deformation Theory of Plasticity. Fan, Tashkent (1966)
11. Anarova, Sh.A., Yuldashev, T.: Derivation of differential equations for vibration of rods in a geometrically nonlinear formulation. Probl. Comput. Appl. Math. Tashkent (2), 72–105 (2018)
12. Nazirov, Sh.A.: Calculation of the problem of torsion of rods of complex profile. Issues Comput. Appl. Math. Collect. Sci. Works Tashkent (89), 65–82 (1992)
13. Nazirov, Sh.A.: Computational algorithms that implement three-dimensional nonlinear mathematical models of the theory of elasticity and plasticity. Issues Comput. Appl. Math. Collect. Sci. Works Tashkent (129), 9–21 (2013)
14. Nazirov, Sh.A., Yuldashev, T., Anarova, Sh.A.: Derivation of variational equations for vibration of thin-walled rods under spatial loading. Issues Comput. Appl. Math. Collect. Sci. Works Tashkent (129), 155–166 (2013)
15. Anarova, Sh.A., Ismoilov, Sh.M.: Mathematical simulation of stress-strain state of loaded rods with account of transverse bending. Mech. Sci. Technol. Update IOP Publ. IOP Conf. Ser. J. Phys. Conf. Ser. **1260**, 7 (2019)
16. Anarova, Sh.A., Ismoilov, Sh.M., Abdirozikov, O.S.: Software of linear and geometrically nonlinear problems solution under spatial loading of rods of complex configuration. In: Singh, M., et al. (eds.) Intelligent Human Computer Interaction 12th International Conference, IHCI 2020 Proceedings, Part I 2021. IHCI 2020, LNCS 12615, pp. 380–389. Springer Nature Switzerland AG (2020). https://doi.org/10.1007/978-3-030-68449-5_39
17. Anarova, Sh.A., Nazirov, A.Sh.: Structure of a program complex for studying the stress-strain state of elastic prismatic bodies of arbitrary cross-section. TATU Khabarlari **1**(37), 51–59 (2016)
18. Anarova, Sh.A., Safarov, Sh.Sh.: Mathematical support for the stress-strain state of rods under spatial loading. Probl. Comput. Appl. Math. Tashkent (4), 20–34 (2016)

19. Nazirov, Sh.A., Nuraliev, F.M., Amanov, O.T.: Structure of a software package for solving boundary value problems for systems of partial differential equations. Uzb. J. Probl. Inf. Energy Tashkent (6), 40–43 (1998)
20. Nazirov, Sh.A., Anarova, Sh.A.: Modeling of the stress-strain state of prismatic bodies of arbitrary cross-section in problems of constrained torsion. Uzb. J. Probl. Inf. Energy Tashkent (3), 22–25 (2003)
21. Nazirov, Sh.A.: Three-dimensional nonlinear mathematical models of mechanics of a deformable rigid body. Issues Comput. Appl. Math. Collect. Sci. Works Tashkent (128), 14–46 (2012)
22. Washizu, K.: Variational Methods in the Theory of Elasticity and Deformation Plasticity: Transl. from English. Mir, Moscow (1987)
23. Mikhlin, S.G.: Variational Methods in Mathematical Physics. Nauka, Moscow (1970)
24. Lurie, A.I.: Nonlinear Theory of Elasticity. Nauka, Moscow (1980)
25. Anarova, Sh.A., Yuldashev, T.: Mathematical model of nonlinear equations of vibrations of rods under dynamic loading. Uzb. J. Probl. Inf. Energy (6), 36–42 (2014)
26. Anarova, Sh.A., Yuldashev, T.: Mathematical models of spatially loaded rods considering the function of torsion and transverse shear. TATU Khabarlari **4**(32), 76–86 (2014)
27. Anarova, Sh.A., Yuldashev, T.: Derivation of a mathematical model of spatially loaded rods considering the function of torsion and transverse shear. Probl. Comput. Appl. Math. Tashkent (1), 28–40 (2015)
28. Belyaev, N.M.: Strength of Materials. Nauka, Moscow (1965)

Stability Analysis of the Fractional Block Method for Solving FODEs

Nursyazwani Mohamad Noor and Siti Ainor Mohd Yatim

Abstract This paper analyzes the 2-point Fractional Block Method based on the Backward Differentiation Formula for solving stiff fractional-order differential equations. An analysis of the stability properties of the method is conducted to verify its A-stability. Since the method fulfills the characteristic, it can be employed to approximate solutions for stiff problems. Consequently, two test problems are solved to assess the method's performance. As a result, the derived method exhibits advantages over existing methods in terms of accuracy.

Keywords Backward Differentiation Formula · Block method · Fractional-order differential equations · Stability

1 Introduction

Classical differentiation and integration methods are generalized to arbitrary orders in fractional calculus. It is utilized in numerous disciplines of science and engineering, including biology, fluid mechanics, and medicine, which piques the interest of numerous academics. Similar to Ordinary Differential Equations (ODEs), it is typically challenging to obtain an analytical solution for fractional-order differential equations (FODEs) of the form

$$ {}_{t_0}^{C}D^{\alpha} y(t) = f(t, y(t)), \qquad t \in [a, b], \tag{1}$$

N. M. Noor · S. A. M. Yatim (✉)
School of Distance Education, Universiti Sains Malaysia, USM, Gelugor, Penang, Malaysia
e-mail: ainor@usm.my

N. M. Noor
School of Quantitative Sciences, College of Arts and Sciences, Universiti Utara Malaysia, UUM, Sintok, Kedah, Malaysia

A. Shuib et al. (eds.), *Mathematical Modeling and Numerical Analysis*,
Springer Proceedings in Mathematics & Statistics 510,
https://doi.org/10.1007/978-981-96-9350-4_2

where a is the initial point, b is the end point, $\alpha \in (0, 1]$ is the order of the FODEs and ${}_{t_0}^{C}D^{\alpha}$ is the fractional Caputo's derivative operator defined as follows [1].

$$ {}_{t_0}^{C}D^{\alpha}y(t) = \frac{1}{\Gamma(1-\alpha)}\int_{t_0}^{t}\frac{y'(\tau)d\tau}{(t-\tau)^{\alpha}}, \qquad 0 < \alpha < 1, \tag{2} $$

where $\Gamma(\cdot)$ is the gamma function. According to [2]:

$$ {}_{t_0}^{C}D^{\alpha} = {}_{t_0}^{\mathrm{RL}}D^{\alpha}\left(y(t) - y(t_0)\right), \tag{3} $$

where ${}_{t_0}^{\mathrm{RL}}D^{\alpha}y(t)$ is defined as the Riemann–Liouville differential operator as:

$$ {}_{t_0}^{\mathrm{RL}}D^{\alpha} = \frac{1}{\Gamma(1-\alpha)}\left(\frac{d}{dt}\right)\int_{t_0}^{t}\frac{y(\tau)\,d\tau}{(t-\tau)^{\alpha}}, \qquad 0 < \alpha < 1. \tag{4} $$

Reference [3] stated that the most challenging aspect of the derivation process is the determination of the convolution weights examined in [4–6]. Thus, they develop a method called a k-step Continuous Fractional Adams Moulton's Methods (CFAMM) that is easier to comprehend than the previous existing researches. This method was derived based on the idea from [7]. In addition, the previous research papers are more focused to non-block method until at one time the Backward Differentiation Formula (BDF) is modified to Block Backward Differentiation Formula (BBDF) method which is first introduced by [8]. The advantage of implement the block method is able to compute the solutions at r point simultaneously (r = 2, 3, 4, …).

In this paper, we will investigate the BBDF-related method derived from the concept of [3]. BBDF is one of the most dependable block numerical algorithms for solving complex initial value problems. References [9, 10] introduced the BBDF method for solving ODEs containing a 2-point block with certain points of back values to simultaneously derive y_{n+1} and y_{n+2}. The general formula for the 2-point BBDF method is given by

$$ \sum_{j=0}^{3}\gamma_{j,i}y_{n+j-1} = h\beta_i f\left(t_{n+i}, y_{n+i}\right), \qquad i = 1, 2. \tag{5} $$

Then, the BBDF method grew in popularity among researchers for solving stiff first-order and higher-order ODEs [11, 12]. Numerous numerical algorithms exist for solving classical differential equations (DEs), but very few exist for solving FODEs. The fractional derivatives are a generalization of the classical derivative, but not in a mathematical sense. Consequently, some numerical techniques for classical DEs can be extended to FODEs to some degree. To improve the procedure, we shall implement the fractional Linear Multistep Method (FLMM) for deriving the fractional BBDF, which can be used to solve Eq. (1). Throughout this paper, the Third Order 2-point

Fractional BBDF (2FBBDF(3)) studied by [13] is considered to solve fractional-order stiff ODEs. Hence, the numerical method of 2FBBDF(3) is given as follows (Refer to [13] for the particulars derivation):

$$\begin{aligned} y_{n+1} &= \frac{\alpha(-2+\alpha)}{3\left(\alpha^2-6\alpha+6\right)} y_{n-1} + \frac{\alpha^2-9\alpha+12}{2\left(\alpha^2-6\alpha+6\right)} y_n + \frac{\alpha(\alpha-5)}{6\left(\alpha^2-6\alpha+6\right)} y_{n+2} \\ &\quad + \frac{\Gamma(4-\alpha)}{\alpha^2-6\alpha+6} h^\alpha f_{n+1}, \\ y_{n+2} &= -\frac{2\alpha(\alpha+1)}{\alpha^2-5\alpha-18} y_{n-1} - \frac{3\left(\alpha^2-13\alpha+6\right)}{\alpha^2-5\alpha-18} y_n + \frac{6\alpha(\alpha-7)}{\alpha^2-5\alpha-18} y_{n+1} \\ &\quad - \frac{6\left(2^{\alpha-1}\right)\Gamma(4-\alpha)}{\alpha^2-5\alpha-18} h^\alpha f_{n+2}. \end{aligned} \tag{6}$$

This article is structured as follows: In Sect. 2, the stability properties of the method are analyzed. The numerical simulations are presented and discussed in Sect. 3 to validate the theoretical outcomes of the proposed method. Lastly, Sect. 4 summarizes this paper's findings and suggestions for future research.

2 Analysis of the Method

This section will explore into various definitions, fundamental concepts, and theories of FLMM from [2, 14]. The focus will be on determining the method's order and error constant, as well as conducting a stability analysis.

Definition 1 [15, 16] The FLMM in Eq. (6) is said to be of order p if, $C_0 = C_1 = \cdots = C_p = 0$ and $C_{p+1} \neq 0$.Thus, $C_{p+1} \neq 0$ is the error constant of the method.

Equation (6) will be evaluated using specific values of α within the range (0, 1.0]. This paper investigates the values of α, particular 0.7, 0.8, and 0.9, as α approaches 1.0 to determine how closely the approximate value aligns with the exact solution. Next, Definition 1 is utilized to determine the order and error constant of the method in the following calculation:

Table 1 The values of e_1 and e_2 for the 2FBBDF(3) method with $\alpha \in (0, 1]$

Error	$\alpha = 0.70$	$\alpha = 0.80$	$\alpha = 0.90$	$\alpha = 1.00$
e_1	$\frac{1841}{30,228}$	$\frac{181}{2208}$	$\frac{661}{5828}$	$\frac{1}{6}$
e_2	$-\frac{3829}{46,222}$	$-\frac{53}{534}$	$-\frac{1749}{14,942}$	$-\frac{3}{22}$

$$
\begin{aligned}
C_0 &:= \sum_{j=0}^{3} \gamma_j = \begin{bmatrix} 0 \\ 0 \end{bmatrix}, \\
C_1 &:= \sum_{j=0}^{3} \left(\frac{(j-1)}{1!}\gamma_j - \frac{(j-1)^{1-\alpha}}{\Gamma(2-\alpha)}\beta_j \right) = \begin{bmatrix} 0 \\ 0 \end{bmatrix}, \\
C_2 &:= \sum_{j=0}^{3} \left(\frac{(j-1)^2}{2!}\gamma_j - \frac{(j-1)^{2-\alpha}}{\Gamma(3-\alpha)}\beta_j \right) = \begin{bmatrix} 0 \\ 0 \end{bmatrix}, \\
C_3 &:= \sum_{j=0}^{3} \left(\frac{(j-1)^3}{3!}\gamma_j - \frac{(j-1)^{3-\alpha}}{\Gamma(4-\alpha)}\beta_j \right) = \begin{bmatrix} 0 \\ 0 \end{bmatrix}, \\
C_4 &:= \sum_{j=0}^{3} \left(\frac{(j-1)^4}{4!}\gamma_j - \frac{(j-1)^{4-\alpha}}{\Gamma(5-\alpha)}\beta_j \right) = \begin{bmatrix} e_1 \\ e_2 \end{bmatrix} \neq \begin{bmatrix} 0 \\ 0 \end{bmatrix},
\end{aligned}
\tag{7}
$$

where e_1 and e_2 are the error constants for y_{n+1} and y_{n+2}, respectively. The values of the error constants are tabulated in Table 1.

Therefore, it can be concluded that the 2FBBDF(3) method is of order 3 when $C_4 \neq 0$.

The parameter space of absolute stability is one of the practical requirements for a method to be practical [17]. In order to evaluate the stability of the procedure, it is necessary first to establish the following definitions.

Definition 2 [14] When all the roots, r_s of the stability polynomial, $\pi(r; \bar{h}) = 0$ satisfy $|r_s| \leq 1$, where $s = 1, 2, \ldots, k$, it is said that the method in Eq. (6) is absolutely stable for a given $\bar{h}$ in a region, R.

Definition 3 [14] If the entirety of the left-hand half-plane $R(h^\alpha \lambda) < 0$ encompasses the absolute stability region, then the method in Eq. (6) is considered A-stable.

The stability analysis in matrix form for the 2FBBDF(3) method is conducted using the following test Eq. [1]:

$$
\begin{aligned}
D^\alpha y(t) &= \lambda y(t), \quad \lambda \in \mathbb{C}, \quad \mathrm{Re}(\lambda) < 0, \\
y(t_0) &= y_0, \quad \alpha \in (0, 1],
\end{aligned}
\tag{8}
$$

where Re is the real part and the true solution can be expressed using the Mittag-Leffler function

$$E_{\alpha}(t)=\sum_{k=0}^{\infty}\left(\frac{t^{k}}{\Gamma\left(\alpha k+1\right)}\right), \tag{9}$$

as $y(t)=E_{\alpha}\left(\lambda(t-t_{0})^{\alpha}\right)y_{0}$. Substituting test Eq. (8) into Eq. (6) which resulted

$$\begin{aligned} y_{n+1}&=\frac{\alpha\left(-2+\alpha\right)}{3\left(\alpha^{2}-6\alpha+6\right)}y_{n-1}+\frac{\alpha^{2}-9\alpha+12}{2\left(\alpha^{2}-6\alpha+6\right)}y_{n}+\frac{\alpha\left(\alpha-5\right)}{6\left(\alpha^{2}-6\alpha+6\right)}y_{n+2}\\ &\quad+\frac{\Gamma\left(4-\alpha\right)}{\alpha^{2}-6\alpha+6}h^{\alpha}\lambda y_{n+1},\\ y_{n+2}&=-\frac{2\alpha\left(\alpha+1\right)}{\alpha^{2}-5\alpha-18}y_{n-1}-\frac{3\left(\alpha^{2}-13\alpha+6\right)}{\alpha^{2}-5\alpha-18}y_{n}+\frac{6\alpha\left(\alpha-7\right)}{\alpha^{2}-5\alpha-18}y_{n+1}\\ &\quad-\frac{6\left(2^{\alpha-1}\right)\Gamma\left(4-\alpha\right)}{\alpha^{2}-5\alpha-18}h^{\alpha}\lambda y_{n+2}, \end{aligned} \tag{10}$$

and let $h^{\alpha}\lambda=\bar{h}$, it yields the following equation.

$$\begin{aligned} y_{n+1}&=\frac{\alpha\left(-2+\alpha\right)}{3\left(\alpha^{2}-6\alpha+6\right)}y_{n-1}+\frac{\alpha^{2}-9\alpha+12}{2\left(\alpha^{2}-6\alpha+6\right)}y_{n}+\frac{\alpha\left(\alpha-5\right)}{6\left(\alpha^{2}-6\alpha+6\right)}y_{n+2}\\ &\quad+\frac{\Gamma\left(4-\alpha\right)}{\alpha^{2}-6\alpha+6}\bar{h}y_{n+1},\\ y_{n+2}&=-\frac{2\alpha\left(\alpha+1\right)}{\alpha^{2}-5\alpha-18}y_{n-1}-\frac{3\left(\alpha^{2}-13\alpha+6\right)}{\alpha^{2}-5\alpha-18}y_{n}+\frac{6\alpha\left(\alpha-7\right)}{\alpha^{2}-5\alpha-18}y_{n+1}\\ &\quad-\frac{6\left(2^{\alpha-1}\right)\Gamma\left(4-\alpha\right)}{\alpha^{2}-5\alpha-18}\bar{h}y_{n+2}. \end{aligned} \tag{11}$$

For simplicity, Eq. (11) can be represented in matrix form as

$$\begin{aligned} &\begin{bmatrix} 1-\frac{\Gamma(4-\alpha)}{\alpha^{2}-6\alpha+6}\bar{h} & -\frac{\alpha(\alpha-5)}{6\left(\alpha^{2}-6\alpha+6\right)}\\ -\frac{6\alpha(\alpha-7)}{\alpha^{2}-5\alpha-18} & 1+\frac{6\left(2^{\alpha-1}\right)\Gamma(4-\alpha)}{\alpha^{2}-5\alpha-18}\bar{h} \end{bmatrix}\begin{bmatrix} y_{n+1}\\ y_{n+2}\end{bmatrix}\\ &\quad=\begin{bmatrix} \frac{\alpha(-2+\alpha)}{3\left(\alpha^{2}-6\alpha+6\right)} & \frac{\alpha^{2}-9\alpha+12}{2\left(\alpha^{2}-6\alpha+6\right)}\\ -\frac{2\alpha(\alpha+1)}{\alpha^{2}-5\alpha-18} & -\frac{3\left(\alpha^{2}-13\alpha+6\right)}{\alpha^{2}-5\alpha-18}\end{bmatrix}\begin{bmatrix} y_{n-1}\\ y_{n}\end{bmatrix}, \end{aligned} \tag{12}$$

which is equivalent to

$$A_{0}Y_{j}=A_{1}Y_{j-1}. \tag{13}$$

The stability polynomial for the 2FBBDF(3) method can be derived by employing the following mathematical expression:

$$\pi\left(r;\bar{h}\right)=\det\left|\sum_{j=0}^{k}A_j r^{k-j}\right|=0. \tag{14}$$

Hence,

$$\begin{aligned}\pi\left(r;\bar{h}\right)=\det&\left|\begin{bmatrix}1-\frac{\Gamma(4-\alpha)}{\alpha^2-6\alpha+6}\bar{h} & -\frac{\alpha(\alpha-5)}{6(\alpha^2-6\alpha+6)}\\ -\frac{6\alpha(\alpha-7)}{\alpha^2-5\alpha-18} & 1+\frac{6(2^{\alpha-1})\Gamma(4-\alpha)}{\alpha^2-5\alpha-18}\bar{h}\end{bmatrix}r\right.\\ &\left.-\begin{bmatrix}\frac{\alpha(-2+\alpha)}{3(\alpha^2-6\alpha+6)} & \frac{\alpha^2-9\alpha+12}{2(\alpha^2-6\alpha+6)}\\ -\frac{2\alpha(\alpha+1)}{\alpha^2-5\alpha-18} & -\frac{3(\alpha^2-13\alpha+6)}{\alpha^2-5\alpha-18}\end{bmatrix}\right|.\end{aligned} \tag{15}$$

1. If $\alpha=0.7$:

$$\begin{aligned}\pi\left(r;\bar{h}\right)=&-\frac{8970\pi\csc(0.3\pi)\left((6.87r+0.91)2^{0.7}+21.01r+7.83\right)\bar{h}r}{481{,}129\Gamma(0.7)}\\ &+\frac{2{,}413{,}827(2^{0.7})\pi^2\csc(0.3\pi)^2\bar{h}^2r^2}{48{,}112{,}900\Gamma(0.7)^2}+\frac{613{,}870}{481{,}129}r^2-\frac{563{,}960}{481{,}129}r\\ &-\frac{49{,}910}{481{,}129}.\end{aligned} \tag{16}$$

2. If $\alpha=0.8$:

$$\begin{aligned}\pi\left(r;\bar{h}\right)=&-\frac{55\left(\left((5.52r+0.96)2^{0.8}\right)+21.36r+11.28\right)\pi\csc(0.2\pi)\bar{h}r}{4094\Gamma(0.8)}\\ &+\frac{1089(2^{0.8})\pi^2\csc(0.2\pi)^2)\bar{h}^2r^2}{51{,}175\Gamma(0.8)^2}+\frac{2915}{2047}r^2-\frac{2695}{2047}r-\frac{220}{2047}.\end{aligned} \tag{17}$$

3. If $\alpha=0.9$:

$$\begin{aligned}\pi\left(r;\bar{h}\right)=&-\frac{770\left(\left((4.23r+0.99)\,2^{0.9}\right)+21.69r+14.67\right)\pi\csc(0.1\pi)\bar{h}r}{101{,}943\Gamma(0.9)}\\ &+\frac{5929(2^{0.9})\pi^2\csc(0.1\pi)^2\bar{h}^2r^2}{1{,}132{,}700\Gamma(0.9)^2}+\frac{18{,}830}{11{,}327}r^2-\frac{17{,}640}{11{,}327}r\\ &-\frac{1190}{11{,}327}.\end{aligned} \tag{18}$$

4. If $\alpha=1.0$:

$$\pi\left(r;\bar{h}\right)=\frac{12}{11}\bar{h}^2r^2-\frac{28}{11}\bar{h}r^2-\frac{20}{11}\bar{h}r+\frac{23}{11}r^2-2r-\frac{1}{11}. \tag{19}$$

In approximating the problem (1), zero stability is a crucial property that ensures the solution remains well-behaved as we take smaller and smaller steps. Thus, the following definition is considered.

Definition 4 [16] The FLMM in Eq. (6) is considered to be zero stable if no root of the characteristic polynomial has modulus greater than one, and if every root with modulus one is simple.

From Eqs. (16)–(19), zero stability of the methods are determined by referencing Definition 4. By solving $\pi\ (r; 0) = 0$ when $\bar{h} = 0$, we obtained the following roots, r:

1. For $\alpha = 0.7$, $r_s = 1, -0.08130$.
2. For $\alpha = 0.8$, $r_s = 1, -0.07547$.
3. For $\alpha = 0.9$, $r_s = 1, -0.06320$.
4. For $\alpha = 1.0$, $r_s = 1, -0.04348$.

Therefore, the 2FBBDF(3) method (6) is zero-stable when the value of roots, $r_s \leq |1|$ and it can be concluded that the method possesses absolute stability within region R for a given $\bar{h}$, as defined in Definition 2.

Subsequently, we visualized the regions of stability for the method by using Eqs. (16)–(19) via Maple software and the graphs are presented in Fig. 1. Referring to the Definitions 2–3, we investigated the stability regions of the method.

Analysis on Fig. 1, the region inside the circle is unstable, whereas the region outside the circle is stable. As the absolutely stable region extends beyond the circles and encompasses the entire left half-plane, this method demonstrates A-stability. Furthermore, it can be inferred that the absolutely stable region of the 2FBBDF(3) method are larger when $\alpha = 0.7$ compared to the other examined values of α, and these regions decrease in size as the values of α approach 1.0.

3 Results and Discussion

This section presents the method's efficiency through considering two numerical illustrations. We will compare the approximate solutions yielded by this method with existing numerical solutions. The numerical results were computed using C++ programming, and the absolute error, ABERR, was evaluated using the subsequent formula.

$$\text{ABERR}_i = \left| y_i^{\text{exact}} - y_i^{2\Gamma\text{BBDF}(3)} \right|, \tag{20}$$

This paper considers the ABERR due to the availability of numerical analysis in the literature that provides the ABERR value at each specific point [18, 19].

Hence, the subsequent examples are considered to confirm the efficacy of the suggested numerical technique.

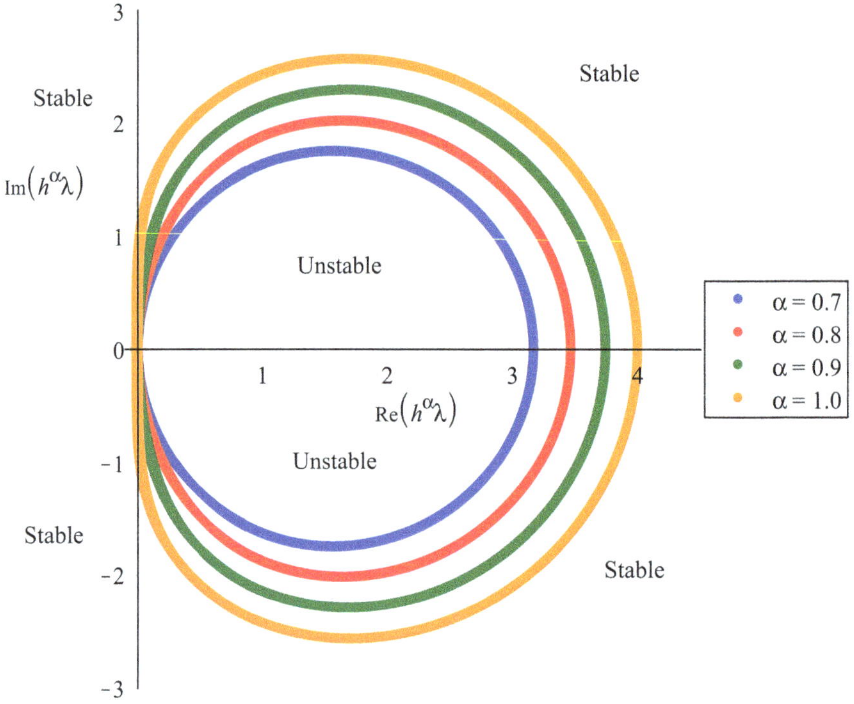

Fig. 1 Stability region of 2FBBDF(3) method with $\alpha = 0.7, 0.8, 0.9$, and 1.0

Example 1 Consider the following nonlinear fractional Abel ODEs [18].

$$D^{\alpha}u(t) = 4u - u^3, \qquad u(0) = \frac{1}{2}, \qquad t \in [0, 0.04],$$

with the provided mathematical solution when $\alpha = 1.0$ is

$$u(t) = 2\left(\frac{e^{8t}}{e^{8t} + 15}\right)^{\frac{1}{2}}.$$

Example 2 Consider the linear stiff system of FODE [19].

$$\begin{aligned} D^{\alpha_1}y_1(t) &= -y_1 + 95y_2, \qquad y_1(0) = 1, \\ D^{\alpha_2}y_2(t) &= -y_2 - 97y_2, \qquad y_2(0) = 1, \qquad t \in [0, 5], \end{aligned}$$

with the provided mathematical solution of the system when $\alpha_1 = \alpha_2 = 1.0$ as

$$y_1(t) = \frac{1}{47}\left(95e^{-2t} - 48e^{-96t}\right), \qquad y_1(t) = \frac{1}{47}\left(48e^{-96t} - e^{-2t}\right).$$

Table 2 Numerical results for Example 1 when $\alpha = 0.9$

t	ADM	FDTM	2FBBDF(3)	Exact
0.00	0.500000	0.500000	0.500000	0.500000
0.05	0.648654	0.648650	0.594835	0.601526
0.10	0.805543	0.805406	0.702843	0.718388
0.15	0.975880	0.974818	0.825500	0.850860
0.20	1.153150	1.149257	0.958479	0.995864
0.25	1.325740	1.317904	1.100727	1.148350
0.30	1.477020	1.473868	1.243914	1.301680
0.35	1.585070	1.628842	1.384203	1.446340
0.40	1.622310	1.839587	1.512225	1.575520

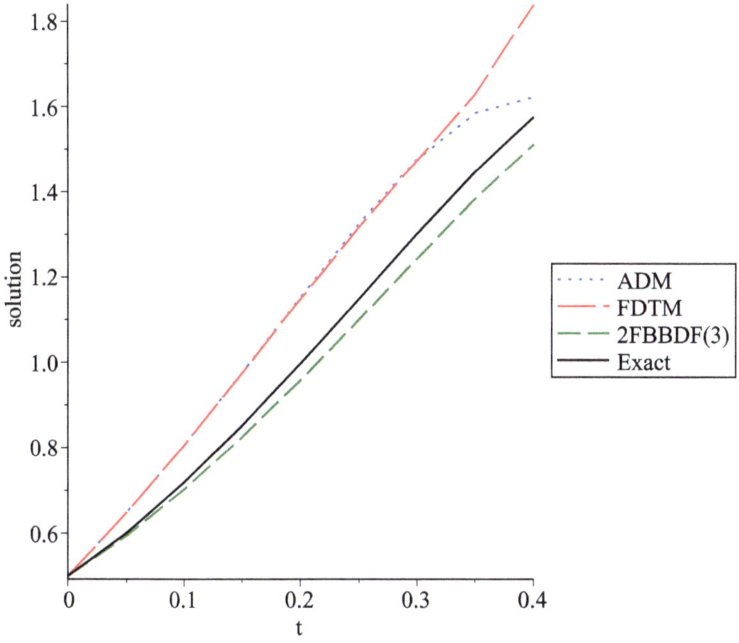

Fig. 2 Efficiency curves of Example 1 for $\alpha = 0.9$

Example 1 is approximated using the 2FBBDF(3) method with fractional order $\alpha = 0.9$, and the results are compared to the Adomian decomposition method (ADM) and fractional differential transform method (FDTM) from [18] (refer Table 2). The data presented in Table 2 are utilized to plot Fig. 2. In comparison to ADM and FDTM, the 2FBBDF(3) method approximates the exact solution more closely, as shown by the graphs in Fig. 2. In addition, the approximate solution to Example 1 is illustrated in Fig. 3 for different values of α.

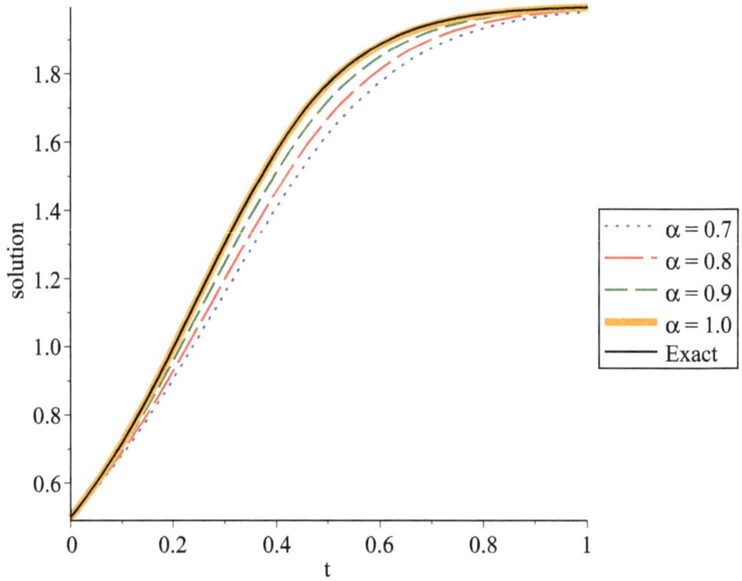

Fig. 3 The numerical solution of Example 1 when $\alpha = 0.7, 0.8, 0.9$, and 1.0

Table 3 Numerical results for Example 2 when $\alpha = 0.9$ and $h = 0.2$

t	y_i	Method					
		B-polynomials		MB-polynomials		2FBBDF(3)	
		Solution	ABERR	Solution	ABERR	Solution	ABERR
0.0	y_1	1.00E+00	0.00E+00	1.00E+00	0.00E+00	1.00E+00	0.00E+00
	y_2	1.00E+00	0.00E+00	1.00E+00	0.00E+00	1.00E+00	0.00E+00
1.0	y_1	3.20E−01	4.68E−02	3.30E−01	5.61E−02	− 5.13E+02	5.13E+02
	y_2	− 6.73E−03	3.85E−03	− 2.62E−03	2.59E−04	5.13E+02	5.13E+02
2.0	y_1	8.94E+03	8.94E+03	1.15E−01	5.95E−02	2.84E+01	2.83E+01
	y_2	− 8.94E+03	8.94E+03	− 8.20E−04	2.39E−04	− 2.83E+01	2.83E+01
3.0	y_1	7.88E+05	7.88E+05	4.03E−02	3.53E−02	9.30E−02	8.80E−02
	y_2	− 7.88E+05	7.88E+05	− 2.30E−04	1.77E−04	− 8.37E−02	8.37E−02
4.0	y_1	1.39E+07	1.39E+07	2.55E−02	2.45E−02	7.59E−04	2.53E−04
	y_2	− 1.39E+07	1.39E+07	− 1.30E−04	1.19E−04	− 5.31E−06	5.34E−06
5.0	y_1	1.16E+08	1.16E+08	1.31E−02	1.30E−02	1.46E−04	5.40E−05
	y_2	− 1.16E+08	1.16E+08	− 8.00E−05	7.90E−05	− 1.50E−06	5.30E−07

Table 3 compares the absolute error of approximate solutions for Example 2 obtained using the 2FBBDF(3) method, the B-polynomials method, and the MB-polynomials method by [19]. The solution is approximated by considering $\alpha = 0.9$ and the step length $h = 0.2$. Figure 4 depicts graphs of log ABERR versus t, which are

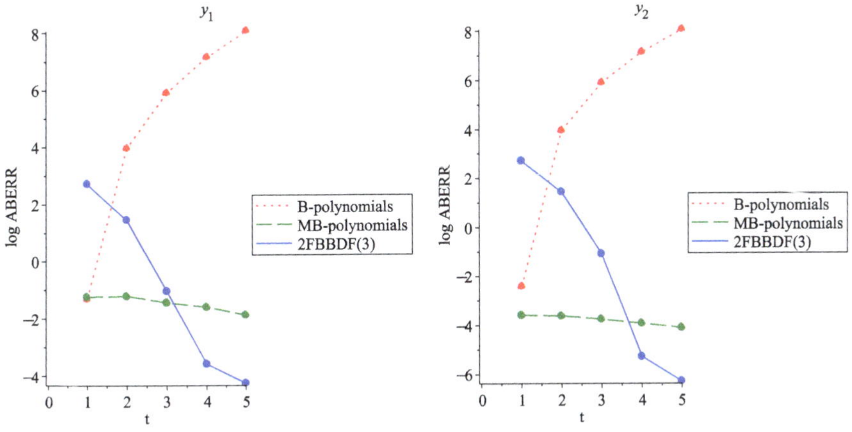

(a) ABERR for $y_1(t)$ (b) ABERR for $y_2(t)$

Fig. 4 The efficiency curves of Example 2 between the methods of comparison

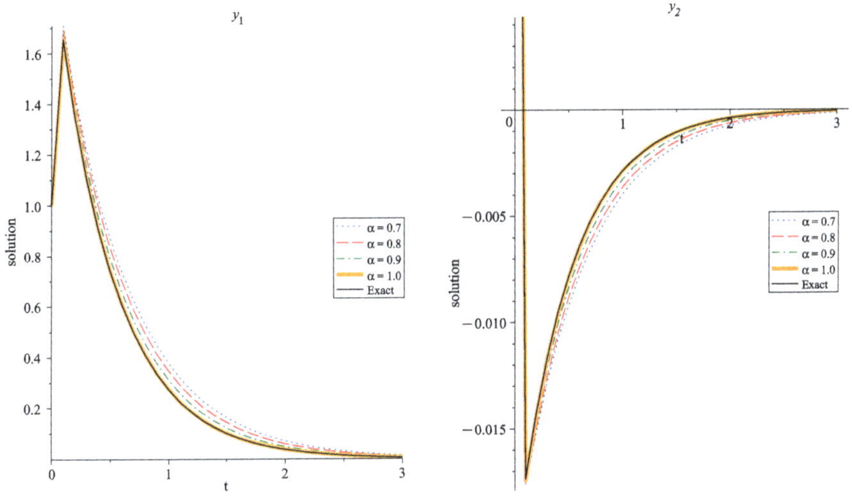

(a) Simulations for $y_1(t)$ (b) Simulations for $y_2(t)$

Fig. 5 The numerical solution of Example 2 when $\alpha_1 = \alpha_2 = 0.7, 0.8, 0.9$, and 1.0

used to evaluate the performance of the methods. As can be seen, the absolute error for 2FBBDF(3) method decreases for both y_1 and y_2 as t increases. Similar to the MB-polynomials method, but distinct from the B-polynomials method, the pattern diverges. The numerical simulations of Example 2 for the fractional orders $\alpha = 0.7$, 0.8, 0.9, and 1.0 are investigated by plotting the graphs in Fig. 5. As a result, the graphs' patterns converge to the exact solution.

4 Conclusions

In this paper, the stability analysis of the 2FBBDF(3) proposed by [13] revealed that the derived method in Eq. (6) is A-stable for fractional orders $\alpha = 0.7, 0.8, 0.9$, and the integer order 1.0. Using the proposed method in Eq. (6), numerical examples were solved. It was found that the 2FBBDF(3) method produces similar results to existing methods for solving both linear and nonlinear FODEs. Consequently, the 2FBBDF(3) method can serve as an alternative computational tool for solving stiff FODEs of linear and nonlinear systems. Moreover, the 2FBBDF(3) method is an appropriate choice for solving fractional-order differential equations (FODEs), especially when working with models that require taking into consideration memory effects, such as a dynamical system.

Acknowledgements This work was supported by the Universiti Sains Malaysia (USM), and the Universiti Utara Malaysia (UUM) through the Ministry of Higher Education Malaysia.

Disclosure of Interests The authors have no conflicts of interest to declare.

References

1. Zabidi, N.-A., Majid, Z.-A., Kilicman, A., Ibrahim, Z.-B.: Numerical solution of fractional differential equations with Caputo derivative by using numerical fractional predict–correct technique. Adv. Contin. Discrete Models **2022**(26), 1–23 (2022)
2. Galeone, L., Garrappa, R.: On multistep methods for differential equations of fractional order. Mediterr. J. Math. **3**, 565–580 (2006)
3. Biala, T.-A., Jator, S.-N.: Block implicit Adams methods for fractional differential equations. Chaos Solitons Fract. **81**, 365–377 (2015)
4. Zhao, J., Long, T., Xu, Y.: Generalised backward differentiation formulae for fractional differential equation. East Asian J. Appl. Math. **9**(3), 506–521 (2019)
5. Bonab, Z.-F., Javidi, M.: Higher order methods for fractional differential equation based on fractional backward differentiation formula of order three. Math. Comput. Simul. **172**, 71–89 (2020)
6. Irandoust-Pakchin, S., Abdi-Mazraeh, S.: Fractional second linear multistep methods: the explicit forms for solving fractional differential equations and stability analysis. Int. J. Comput. Math. **100**(1), 20–46 (2022)
7. Onumanyi, P., Sirisena, U.-W., Jator, S.-N.: Continuous finite difference approximations for solving differential equations. Int. J. Comput. Math. **72**, 15–27 (1999)
8. Ibrahim, Z.-B., Othman, K.-I., Suleiman, M.: Implicit r-point block backward differentiation formula for solving first-order stiff odes. Appl. Math. Comput. **186**, 558–565 (2007)
9. Ijam, H.-M., Ibrahim, Z.-B.: Diagonally implicit block backward differentiation formula with optimal stability properties for stiff ordinary differential equations. Symmetry **11**, 1342 (2019)
10. Zawawi, I.-S.-M., Ibrahim, Z.-B.: BBDF-α for solving stiff ordinary differential equations with oscillating solutions. Tamkang J. Math. **51**(2), 123–136 (2020)
11. Buhari, A., Hamisu, M.: Diagonally implicit extended 2-point super class of block backward differentiation formula with two off-step points for solving first order stiff initial value problems. Appl. Math. Comput. Intell. **12**(1), 101–124 (2023)

12. Asnor, A.-I., Yatim, S.-A.-M., Ibrahim, Z.-B.: Numerical solution of fourth-order stiff ordinary differential equations by implicit high order block backward differentiation formula. AIP Conf. Proc. **2423**, 020005-1–020005-8 (2021)
13. Noor, N.-M., Yatim, S.-A.-M.: A block backward differentiation formula for solving fractional pharmacokinetics model. AIP Conf. Proc. **2880**, 020002-1–020002-9 (2023)
14. Lambert, J.-D.: Computational Methods in Ordinary Differential Equations. Wiley, Great Britain (1973)
15. Galeone, L., Garrappa, R.: Explicit methods for fractional differential equations and their stability properties. J. Comput. Appl. Math. **228**, 54–560 (2009)
16. Soomro, H., Zainuddin, N., Daud, H., Sunday, J., Jamaludin, N., Abdullah, A., Mulono, A., Kadir, E.-A.: 3-point block backward differentiation formula with an off-step point for the solutions of stiff chemical reaction problems. J. Math. Chem. **61**, 75–97 (2023)
17. Aksah, S.-J., Ibrahim, Z.-B., Zawawi, I.-S.-M.: Stability analysis of singly diagonally implicit block backward differentiation formulas for stiff ordinary differential equations. Mathematics **7**(211), 1–16 (2019)
18. Alkresheh, H.-A.-H.: Approximate analytical methods for ordinary differential equations of fractional order. Ph.D. thesis, Universiti Sains Malysia (2017)
19. Alshbool, M.-H.-T., Hashim, I.: Multistage Bernstein polynomials for the solutions of the fractional order stiff systems. J. King Saud Univ.-Sci. **28**, 280–285 (2016)

Piecewise Constant Argument Method for Nonlinear Systems of Differential Equations

Mukhiddin I. Muminov and Zafar Z. Jumaev

Abstract This paper presents an efficient approach for determining an approximate solution of system of first order differential equations with variable coefficients. It is introduced as a differential equation with piecewise constant arguments corresponding to the considered initial value problem, which depend on a positive integer number n. It is proved that this equation has a unique piecewise-smooth solution, which will be an approximate solution for the considered initial value problem for large n. Numerical results are given on examples showing the efficiency and high accuracy of the suggested method.

Keywords Linear ordinary differential equation · Piecewise constant argument · Approximate solution · Absolute error · Residual error

1 Introduction

Differential equations form the basis of many physical systems, prominently appearing in fields such as engineering, population dynamics, applied mathematics, physics, economics, and astronomy [1–3]. Zill [4] illustrates their application in modeling phenomena like bacterial growth, radioactive decay, fossil aging, Newton's law of cooling, and salt solution mixing.

Systems of ODEs are similarly vital in various scientific and engineering contexts, such as tank mixing problems, electrical networks, and mass-spring systems [5]. Numerous numerical techniques have been developed to solve these systems, includ-

M. I. Muminov
Faculty of Mathematics, Samarkand State University, Samarkand, Uzbekistan
e-mail: mmuminov@mail.ru

V.I. Romanovskiy Institute of Mathematics, Uzbekistan Academy of Sciences, Tashkent, Uzbekistan

Z. Z. Jumaev (✉)
Faculty of Intelligent Systems and Computer Technologies, Samarkand State University, Samarkand, Uzbekistan
e-mail: zafarlangar8708@gmail.com
URL: http://www.springer.com/gp/computer-science/lncs

A. Shuib et al. (eds.), *Mathematical Modeling and Numerical Analysis*,
Springer Proceedings in Mathematics & Statistics 510,
https://doi.org/10.1007/978-981-96-9350-4_3

ing the differential transform approximation [6] and Haar wavelets collocation methods [7, 8]. Aziz et al. [9] introduced a new numerical technique using Haar wavelets for nonlinear integral-type equations.

Alternative numerical methods like the finite element approach have been widely applied to nonlinear equations [10]. Both mesh-based and meshless methods have advanced the solving of linear and nonlinear ODEs, PDEs, and fractional differential equations [11–13].

Nonnegativity of solutions and global unique existence of two systems of nonlinear first-order ODEs modeling Ca^{2+} is studied [14]. A finite-time stability for hybrid dynamical systems with deviating argument is explored using a hybrid control scheme in [15]. The reliability of the Non-Fixed Step-Size Algorithm offering efficient integration over varying intervals for stiff differential systems is confirmed through validation of its properties [16–18]. Finding exact periodic solutions of first-order nonhomogeneous differential equations with piecewise constant arguments, providing conditions and explicit solutions are given in [19, 20].

In this article is presented an efficient approach for determining an approximate solution of linear and nonlinear systems of ordinary differential equations of the form

$$\begin{cases} y_1'(t) = f_1(t, y_1(t), y_2(t), y_3(t)), \qquad t \in [0, 1], \\ y_2'(t) = f_2(t, y_1(t), y_2(t), y_3(t)), \\ y_3'(t) = f_3(t, y_1(t), y_2(t), y_3(t)), \\ y_1(0) = y_{10}, \quad y_2(0) = y_{20}, \quad y_3(0) = y_{30} \end{cases} \tag{1}$$

with initial conditions where $f_i(t, y_1(t), y_2(t), y_3(t))$, $i = 1, 2, 3$ is third order polynomial of functions $(y_1(t), y_2(t), y_3(t))$ of the form

$$f_i(t, y_1(t), y_2(t), y_3(t)) = \sum_{0 \le p+q+r \le 3} y_1^p y_2^q y_3^r A_{pqr}^{(i)}(t), \; i = 1, 2, 3, \tag{2}$$

here the coefficients $A_{pqr}^{(i)}(t)$ are continuous real valued functions on $[0, 1]$ and y_{10}, y_{20} and y_{30} are real constants.

We introduced a system of ordinary differential equations with piecewise constant arguments corresponding to the considered initial value problem, which depend on a positive integer number n. It is proved that this system of equations has a unique piecewise-smooth solution, which will be an approximate solution for the considered initial value problem for large n. Numerical results are given on examples showing the efficiency and high accuracy of the suggested method. The efficiency of the methods is illustrated with certain numerical examples, for which is more accurate with faster convergence.

2 Differential Equations with Piecewise Constant Arguments

Let $(y_1(t), y_2(t), y_3(t))$ be solution of the initial value problem (1) and $|y_i(t)| \leq d,\ i = 1, 2, 3$. Further, we suppose that the functions $f_i(t, y_1, y_2, y_3)$ has the compact support $D = [0, 1] \times [-2d, 2d]^3$, i.e. we assume

$$f_i(t, y_1, y_2, y_3) = 0 \quad \text{for} \quad (t, y_1, y_2, y_3) \notin D, \quad i = 1, 2, 3. \tag{3}$$

We introduce a system of differential equations with a piecewise constant argument corresponding to the nonlinear differential equation (1). Let n be a positive integer number and $t_k = \frac{k}{n},\ k = 0, 1, 2, \ldots, n$. For $t \in [t_{k-1}, t_k),\ k = 1, 2, \ldots, n$, consider the following differential equation with piecewise constant arguments of the form

$$\begin{aligned} y_1'(t) &= f_1\left(t, y_1\left(\tfrac{k-1}{n}\right), y_2\left(\tfrac{k-1}{n}\right), y_3\left(\tfrac{k-1}{n}\right)\right), \\ y_2'(t) &= f_2\left(t, y_1\left(\tfrac{k-1}{n}\right), y_2\left(\tfrac{k-1}{n}\right), y_3\left(\tfrac{k-1}{n}\right)\right), \\ y_3'(t) &= f_3\left(t, y_1\left(\tfrac{k-1}{n}\right), y_2\left(\tfrac{k-1}{n}\right), y_3\left(\tfrac{k-1}{n}\right)\right), \\ y_1(0) &= y_{10}, \quad y_2(0) = y_{20}, \quad y_3(0) = y_{30}. \end{aligned} \tag{4}$$

A solution of (4) is defined as follows [20].

Definition 1 A vector function $(y_{1n}(t), y_{2n}(t), y_{3n}(t))$ is called a solution of the initial value problem (4) if the following conditions are satisfied:

(i) $y_1(t)$, $y_2(t)$ and $y_3(t)$ are continuous on $[0, 1]$;
(ii) $y_1'(t)$, $y_2'(t)$ and $y_3'(t)$ exist and continuous on $[0, 1]$ with possible exception at points $t = \frac{k}{n},\ k = 0, 1, \ldots, n$, where one-sided derivatives exist;
(iii) $y_1(t)$, $y_2(t)$ and $y_3(t)$ satisfies the initial value problem (4) in $(0, 1)$, with the possible exception at the points $t = \frac{k}{n},\ k = 0, 1, \ldots, n$.

Theorem 1 *For any positive integer number n, the initial value problem (4) has a unique solution* $(y_{1n}(t), y_{2n}(t), y_{3n}(t))$ *defined in (12).*

We remark that under assumption (3), the functions $f_i(t, y_1, y_2, y_3),\ i = 1, 2, 3$ are bounded. Therefore, the functions $(y_{1n}(t),\ y_{2n}(t)$ and $y_{3n}(t))$ are uniformly bounded, i.e., there exists a positive number K such that

$$|y_{in}(t)| \leq K, \quad \text{for all} \quad n, t \quad \text{and} \quad i = 1, 2, 3. \tag{5}$$

The following theorem claims that the function (y_{1n}, y_{2n}, y_{3n}) can be approximated solution for the initial value problem (1).

Theorem 2 *For any* $\varepsilon > 0$, *there exists a positive number* $n_0 = n(\varepsilon)$ *such that for any positive integer number n with* $n > n_0$ *the inequalities*

$$\begin{aligned}
&\sup_{t\in[0,1]} |y'_{1n}(t) - f_1(t, y_{1n}(t), y_{2n}(t), y_{3n}(t))| < \varepsilon,\\
&\sup_{t\in[0,1]} |y'_{2n}(t) - f_2(t, y_{1n}(t), y_{2n}(t), y_{3n}(t))| < \varepsilon,\\
&\sup_{t\in[0,1]} |y'_{3n}(t) - f_3(t, y_{1n}(t), y_{2n}(t), y_{3n}(t))| < \varepsilon
\end{aligned} \tag{6}$$

hold, where (y_{1n}, y_{2n}, y_{3n}) *is solution of the initial value problem (4).*

3 Proof of the Main Results

Proof of the Theorem 1. Let n be a positive integer number. For $t \in [0, \frac{1}{n})$ Eq. (4) is

$$y'_i(t) = f_i(t, y_{1n}(0), y_{2n}(0), y_{3n}(0)). \tag{7}$$

Integrating these equations, we have

$$y_i(t) = \int_0^t f_i(s, y_{1n}(0), y_{2n}(0), y_{3n}(0))\mathrm{d}s + y_i(0), \qquad i = 1, 2, 3, \qquad t \in \left[0, \frac{1}{n}\right). \tag{8}$$

Since $y_i(t)$ is continuous $[0, \frac{1}{n})$, there exist the following limits

$$y_1\left(\frac{1}{n}\right) = \lim_{t\to\frac{1}{n}-0} y_1(t), \qquad y_2\left(\frac{1}{n}\right) = \lim_{t\to\frac{1}{n}-0} y_2(t), \qquad y_3\left(\frac{1}{n}\right) = \lim_{t\to\frac{1}{n}-0} y_3(t). \tag{9}$$

Let $y_{k-2}(t), i = 1, 2, 3$ be solution of equation (4) for $t \in \left[\frac{k-2}{n}, \frac{k-1}{n}\right)$ with

$$\begin{aligned}
y_1\left(\frac{k-1}{n}\right) &= \lim_{t\to\frac{k-1}{n}-0} y_1(t),\\
y_2\left(\frac{k-1}{n}\right) &= \lim_{t\to\frac{k-1}{n}-0} y_2(t),\\
y_3\left(\frac{k-1}{n}\right) &= \lim_{t\to\frac{k-1}{n}-0} y_3(t).
\end{aligned} \tag{10}$$

Further, integrating

$$y'_i(t) = f_i(t, y_{1n}(t), y_{2n}(t), y_{3n}(t)), \quad i = 1, 2, 3 \tag{11}$$

and using initial conditions (10) for $t \in \left[\frac{k-1}{n}, \frac{k}{n}\right)$ we obtain a solution of (4) as

$$y_i(t) = \int_{\frac{k-1}{n}}^{t} f_i\left(s, y_{1n}\left(\frac{k-1}{n}\right), y_{2n}\left(\frac{k-1}{n}\right), y_{3n}\left(\frac{k-1}{n}\right)\right) \mathrm{d}s + y_i\left(\frac{k-1}{n}\right), \quad i = 1, 2, 3, \tag{12}$$

for $t \in \left[\frac{k-1}{n}, \frac{k}{n}\right)$, $k = 2, \ldots, n$.

By construction, the functions y_1, y_2 and y_3 defined by (12) are differentiable on $(0, \frac{1}{n}) \cup (\frac{1}{n}, \frac{2}{n}) \cup \cdots \cup (\frac{k-1}{n}, \frac{k}{n})$ and continuous on $[0, 1]$. It is clear that this vector function (y_1, y_2, y_3) is unique solution for Eq. (4).

Proof of the Theorem 2. Let (y_{1n}, y_{2n}, y_{3n}) be solution of (4). Then for $t \in [t_k, t_{k+1})$ the equality

$$y'_{in}(t) - f_i(t, y_{1n}(t), y_{2n}(t), y_{3n}(t)) = f_i(t_k, y_{1n}(t_k), y_{2n}(t_k), y_{3n}(t_k)) - f_i(t, y_{1n}(t), y_{2n}(t), y_{3n}(t)),$$

holds. Since the function f_i is differentiable in D, there exists $F > 0$ such that

$$|f_i(t, y_{1n}(t), y_{2n}(t), y_{3n}(t)) - f_i(t_k, y_{1n}(t_k), y_{2n}(t_k), y_{3n}(t_k))| \le F\left(|y_{1n}(t) - y_{1n}(t_k)| + |y_{2n}(t) - y_{2n}(t_k)| + |y_{3n}(t) - y_{3n}(t_k)|\right). \tag{13}$$

By (12) for $t \in \left[\frac{k-1}{n}, \frac{k}{n}\right)$,

$$y_i(t) - y_i\left(\frac{k-1}{n}\right) = \int_{\frac{k-1}{n}}^{t} f_i\left(s, y_{1n}\left(\frac{k-1}{n}\right), y_{2n}\left(\frac{k-1}{n}\right), y_{3n}\left(\frac{k-1}{n}\right)\right) \mathrm{d}s, \tag{14}$$

holds, where $i = 1, 2, 3$ and $k = 2, \ldots, n$. According to (13) yields

$$\left|y_i(t) - y_i\left(\frac{k-1}{n}\right)\right| \le F\frac{1}{n}$$

for any $t \in \left[\frac{k-1}{n}, \frac{k}{n}\right)$, $i = 1, 2, 3$ and $k = 2, \ldots, n$. This proves the assertion of the theorem.

In order to verify the effectiveness of the proposed methods, we consider several examples and give the absolute errors of approximate and exact solutions.

4 Summary of Algorithms

Algorithms for solving the initial conditional problem are presented as follows:

Input: Initial conditions.
Step 1: Find $(y_1(t), y_2(t), y_3(t))$ in the system of equations (16) for $t \in [0, \frac{1}{n})$.
Step 2: Find limits $y_1(\frac{1}{n}) = \lim_{t\to\frac{1}{n}-0} y_1(t)$, $y_2(\frac{1}{n}) = \lim_{t\to\frac{1}{n}-0} y_2(t)$, $y_3(\frac{1}{n}) = \lim_{t\to\frac{1}{n}-0} y_3(t)$.
Step 3: According to (12) compute $(y_1(t), y_2(t), y_3(t))$ for $t \in [\frac{k-1}{n}, \frac{k}{n})$, $k = 2, \ldots, n$ and find the limits $y_1(\frac{k-1}{n}) = \lim_{t\to\frac{k-1}{n}-0} y_1(t)$, $y_2(\frac{k-1}{n}) = \lim_{t\to\frac{k-1}{n}-0} y_2(t)$, $y_3(\frac{k-1}{n}) = \lim_{t\to\frac{k-1}{n}-0} y_3(t)$.
Step 4: We construct an approximate solution from (12).
Step 5: We draw the graph of the approximate solution and compare it with the graph of the exact solution by placing it on one coordinate.
Step 6: Determine the absolute errors $L_\infty(Y_p) = \max_{0\le t\le 1} |Y_p(t) - Y_{pn}(t)|$ between exact and approximated solutions. If the exact solution missing determine the residual errors using the error formula $L_{e_i} = \max_{0\le t\le 1} |Y'_{in}(t) - f_i(t, y_{1n}(t), y_{2n}(t), y_{3n}(t))|$.
Output: If the errors at the specified point are acceptable, then the for loop ends.

5 Numerical Results

In order to obtain an approximate solution of equation (1) from Theorem 2, we need to calculate (y_{1n}, y_{2n}, y_{3n}) which has the explicit form (12). In this case, when the functions $f(t)$ have exact antiderivatives, the numerical calculation of (y_{1n}, y_{2n}, y_{3n}) is simple. We use a numerical method of calculation (y_{1n}, y_{2n}, y_{3n}) for the case, when the variable coefficients $f(t)$ have no exact antiderivatives. For this case, in the explicit form (12) of (y_{1n}, y_{2n}, y_{3n}), we can change $f(t)$ by the step functions $f(\frac{k-1}{n})$ for $t \in [\frac{k-1}{n}, \frac{k}{n})$. This method is based on the following well known assertion [21].

Lemma 1 *If the function f is continuous on the closed interval $[a, b]$, then the sequence of piecewise continuous functions f_n defined on $[a, b]$ as $f_n(b) = f(b)$ and $f_n(t) = f\left(\frac{k(b-a)}{n}\right), t \in \left[a + \frac{(k-1)(b-a)}{n}, a + \frac{k(b-a)}{n}\right)$, $k = 1, 2, \ldots, n$ uniformly converges to f on $[a, b]$.*

Let

$$Y_{in}(t) = \int_{\frac{k-1}{n}}^{t} f_i\left(\frac{k-1}{n}, Y_{1n}\left(\frac{k-1}{n}\right), Y_{2n}\left(\frac{k-1}{n}\right), Y_{3n}\left(\frac{k-1}{n}\right)\right) \mathrm{d}s + Y_{in}\left(\frac{k-1}{n}\right), \tag{15}$$

for $t \in \left[\frac{k-1}{n}, \frac{k}{n}\right)$, where

$$Y_{in}(0) = y_{i0}, \quad Y_{in}\left(\frac{k-1}{n}\right) = \lim_{t \to \frac{k-1}{n} - 0} Y_{in}(t), \qquad i = 1, 2, 3.$$

According to Lemma 1 and Theorem 2, we have the following theorem:

Theorem 3 *For any $\varepsilon > 0$, there exists a positive integer number $n_0 = n(\varepsilon)$ such that for any $n > n_0$ the estimation*

$$\begin{aligned}
&\sup_{t, t \neq \frac{k}{n}} |Y'_{1n}(t) - f_1(t, Y_{1n}(t), Y_{2n}(t), Y_{3n}(t))| < \varepsilon \\
&\sup_{t, t \neq \frac{k}{n}} |Y'_{2n}(t) - f_2(t, Y_{1n}(t), Y_{2n}(t), Y_{3n}(t))| < \varepsilon \\
&\sup_{t, t \neq \frac{k}{n}} |Y'_{3n}(t) - f_3(t, Y_{1n}(t), Y_{2n}(t), Y_{3n}(t))| < \varepsilon
\end{aligned} \tag{16}$$

holds.

In order to verify the effectiveness of the suggested methods, we consider several examples and give the absolute errors of the deference between approximate and exact solutions.

In the Test Problem 1 and Test Problem 2 investigated, the maximum absolute error $L_\infty(Y_p - Y_{pn})$ between the exact solutions $Y_p(t)$ and the approximate solutions $Y_{pn}(t)$ over a given interval [0, 1], where

$$L_\infty(Y) = \max_{0 \leq t \leq 1} |Y(t)|.$$

Test Problem 1. *Taking the linear stiff system*

$$\begin{cases}
y_1'(t) = t^3 + \cos(t) - 2\sin^2(t) - \sin^4(t) - 2 + y_3(t) + y_2^2(t) \\
y_2'(t) = t^3 + \sin(2t) - 2\sin(t) - \sin^2(t) - 2 + y_1^2(t) + y_3(t) \\
y_3'(t) = -3t^2 - \sin(t) - \sin^2(t) - 2 + y_1(t) + y_2(t)
\end{cases} \tag{17}$$

with given initial conditions

$$y_1(0) = 1, \qquad y_2(0) = 1, \qquad y_3(0) = 1.$$

The exact solution is

$$y_1(t) = 1 + \sin(t), \qquad y_2(t) = 1 + \sin^2(t), \qquad y_3(t) = 1 - t^3.$$

The absolute errors at the points t_k between the approximate solution calculated by the proposed method and the exact solution of the problem are presented in Table 1. The graphs of the derivatives of the exact (black color) and approximate solutions for the Test Problem 1 are presented in Fig. 1. The graphs of the derivatives of the

Table 1 Comparison of the piecewise constant argument method solution with those solutions available in by different methods for Test Problem 1

n	$L_\infty(Y_1)$	$L_\infty(Y_2)$	$L_\infty(Y_3)$
10	4.43288E−02	6.28134E−02	9.6331E−02
100	7.03431E−03	8.91406E−03	8.36823E−03
1000	7.47418E−04	9.36396E−04	8.15049E−04
10,000	7.48754E−05	9.37866E−05	8.14369E−05
100,000	9.39250E−06	1.12475E−05	7.07753E−06

exact (black color) and approximate solutions for the Test Problem 1 are presented in Fig. 2.

Test Problem 2. *Consider the system of highly nonlinear and nonhomogeneous differential equations*

$$\begin{cases} y_1'(t) = -\sin(t) - \cos(t) + 2\cos^2(t) + \cos^4(t) - e^t - t + y_1(t) - y_2^2(t) + y_3(t) \\ y_2'(t) = -e^{2t} - 2te^t - t^2 - \sin(2t) - \cos(t) - 1 + y_1(t) + y_3^2(t) \\ y_3'(t) = -t - \cos^3(t) - 4\cos^2(t) - 3\cos(t) - 1 + y_2(t) + y_3(t) + y_1^3(t) \end{cases} \tag{18}$$

with initial conditions

$$y_1(0) = 2, \qquad y_2(0) = 2, \qquad y_3(0) = 1.$$

The exact solution is

$$y_1(t) = \cos(t) + 1, \qquad y_2(t) = 1 + \cos^2(t), \qquad y_3(t) = e^t + t.$$

For this problem, absolute errors at the points t_k between the approximate solution and the exact solution are presented in Table 2. In this case, the graphs of the approximate solution y_{1n}, y_{2n}, y_{3n}, and the exact solution $y_1(t)$, $y_2(t)$, $y_3(t)$ of the system are compared in Fig. 3.

Test Problem 3. *Consider the nonlinear system of differential equations with variable coefficients*

$$\begin{cases} y_1'(t) = \frac{3+\sin(t)}{1+t} y_1(t) y_2(t) y_3(t) \\ y_2'(t) = \frac{\cos(t)}{4+\sin(t)} y_2^2(t) + y_3^2(t) \\ y_3'(t) = \frac{t}{1+t} y_2^2(t) + \frac{3+t^2}{(2+t)^2} y_3(t) y_1(t) \end{cases} \tag{19}$$

with given initial conditions

$$y_1(0) = 1.5, \qquad y_2(0) = 0.5, \qquad y_3(0) = 0.25.$$

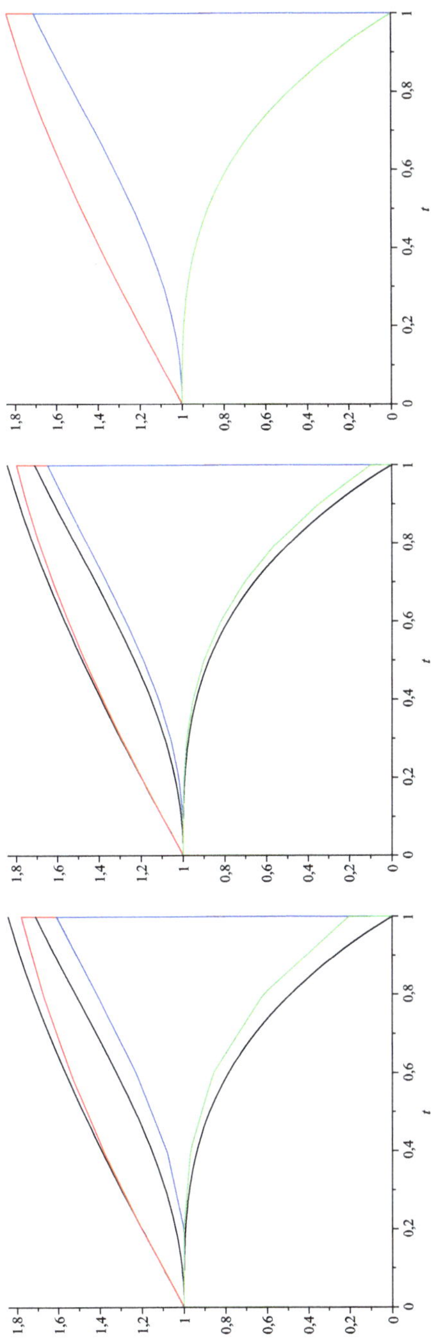

Fig. 1 Graphs of original (black color) and approximate (Y_1 red color, Y_2 blue color and Y_3 green color) solutions for Test Problem 1, respectively as $n = 5$, $n = 10$ and $n = 100$

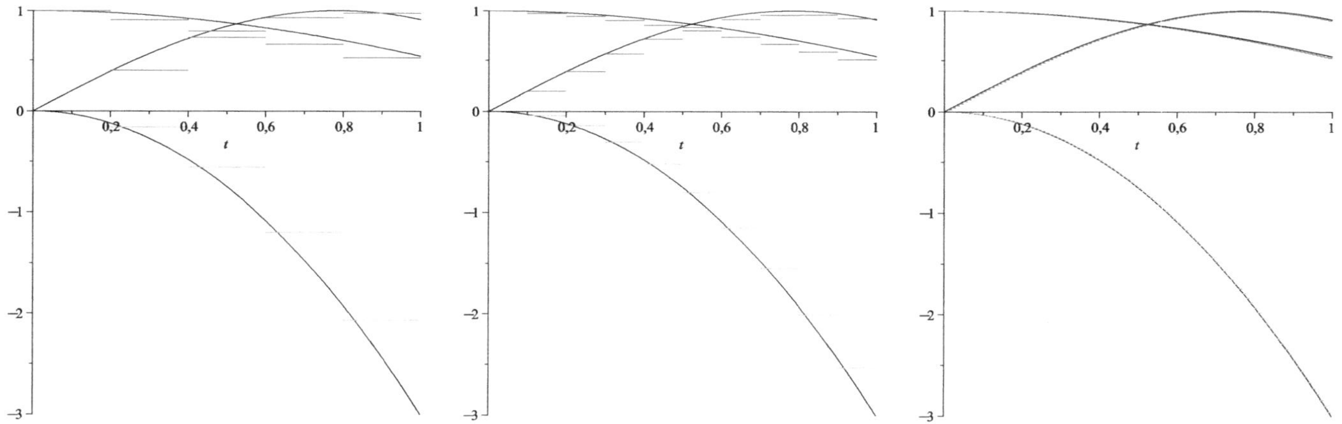

Fig. 2 Graphs of the derivatives of exact (black color) and approximate (Y_1' red color, Y_2' blue color and Y_3' green color) solutions for Test Problem 1, respectively as $n = 5$, $n = 10$ and $n = 100$

Table 2 Comparison of the piecewise constant argument method solution with those solutions available in by different methods for Test Problem 2

n	$L_\infty(Y_1)$	$L_\infty(Y_2)$	$L_\infty(Y_3)$
10	6.30813E−02	1.95015E−01	2.54873E−01
100	9.30123E−03	4.50419E−02	2.75886E−02
1000	1.1802E−03	4.5197E−03	2.50369E−03
10,000	1.20025E−04	4.51441E−04	2.47876E−04
100,000	1.37357E−05	4.78776E−05	2.51616E−05

Note that in this example an exact solution of equation is missed. However, we can evaluate the residual error L_{e_i} for the approximated solutions $y_{1n}(t)$, $y_{2n}(t)$ and $y_{3n}(t)$), where

$$L_{e_i} = \max_{0 \le t \le 1} |Y'_{in}(t) - f_i(t, y_{1n}(t), y_{2n}(t), y_{3n}(t))|.$$

The residual errors for the approximated solutions $y_{1n}(t)$, $y_{2n}(t)$ and $y_{3n}(t)$) in the Test Problem 3 are presented in Table 3 as $n = 10, \ldots, n = 10^5$.

The graphs of the approximate solutions for Test Problem 3 are presented in Fig. 4.

Test Problem 4. *Consider the system of highly nonlinear and nonhomogeneous differential equations*

$$\begin{cases} y_1'(t) = \cos(t) + y_1^2(t) + y_2(t) - (1 + t^2 + \sin^2(t)), t \in [0, 2], \\ y_2'(t) = 2t - (1 + t^2)\sin(t) + y_1(t)y_2(t) \end{cases} \tag{20}$$

with given initial conditions

$$y_1(0) = 0 \quad and \quad y_2(0) = 1.$$

The exact solution is

$$y_1(t) = \sin(t), \quad y_2(t) = 1 + t^2.$$

The Test Problem 4 was investigated by the Haar wavelet collocation method (HWCM) and the higher-order Haar wavelet collocation method (H-HWCM) in [22]. The comparison of the maximum absolute error $L_\infty(Y_p - Y_{pn})$ between the exact Y_p and approximated solutions Y_{pn} is presented in Table 4, where

$$L_\infty(Y) = \max_{0 \le t \le 2} |Y(t)|.$$

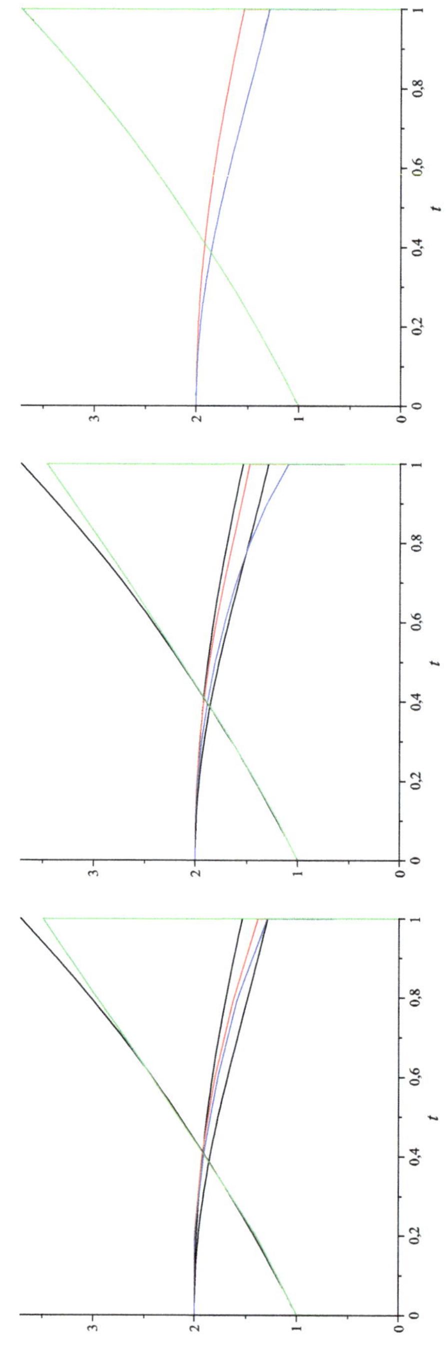

Fig. 3 Graphs of original (black color) and approximate (Y_1 red color, Y_2 blue color and Y_3 green color) solutions for Test Problem 2, respectively, as $n = 5$, $n = 10$ and $n = 100$

Table 3 Residual error for the Test Problem 3

n	$Le_i(Y_1)$	$Le_i(Y_2)$	$Le_i(Y_3)$
10	9.33635E−01	1.6406E−01	2.76384E−01
100	2.03977E−01	3.05677E−02	5.19257E−02
1000	2.27101−02	3.32197E−03	5.66681E−03
10,000	2.29657E−03	3.35081E−04	5.71873E−04
100,000	2.29918E−04	335377E−05	5.72399E−05

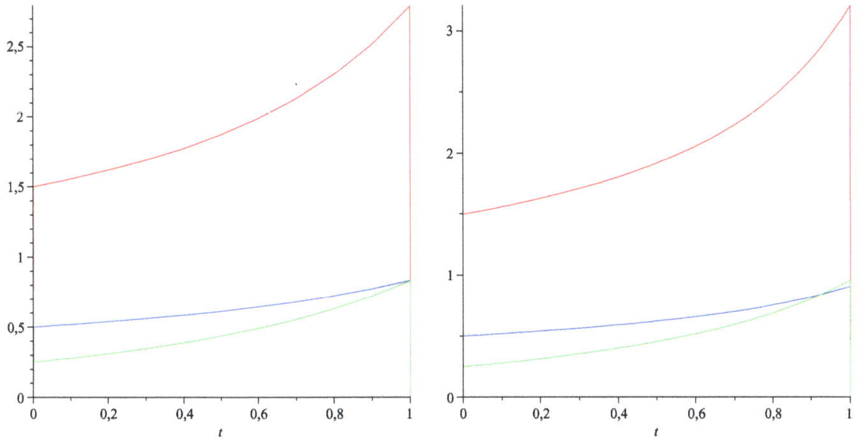

Fig. 4 Graphs of approximate (y_{1n} red color, y_{2n} blue color and y_{3n} green color) solutions for Test Problem 3, respectively as $n = 10$ and $n = 100$

Table 4 Comparison of numerical results obtained by different methods for Test Problem 4

HWCM		H-HWCM		Present method	
$L_\infty(y_1)$	$L_\infty(y_2)$	$L_\infty(y_1)$	$L_\infty(y_2)$	$L_\infty(y_1)$	$L_\infty(y_2)$
2.9324E−05	4.1138E−05	1.9399E−05	1.5708E−07	5.0040E−11	1.0001E−10

6 Conclusion

A new method is presented for finding approximate solution of the system of equations. First, the differential equation with piecewise constant arguments corresponding to the considering equation has been constructed. The definition of solution for ordinary differential equations with piecewise constant argument has been introduced as a piecewise-smooth function with respect to the positive integer number n. It has been proved that this piecewise-smooth solution is an approximate solution of the considering initial value problem for large n. Numerical results are provided to show the effectiveness of the proposed methods and the errors between the approximate and the exact solutions are evaluated.

Acknowledgements We would like to thank the anonymous reviewers for their constructive comments that have greatly contributed to improving the manuscript.

Disclosure of Interests We declare that we have no known competing financial interests or personal relationships that could have appeared to influence the work reported in this paper.

References

1. Guo, C., et al.: Fixed-time safe tracking control of uncertain high-order nonlinear pure-feedback systems via unified transformation functions. Eur. Digit. Math. Libr. **59**(3), 342–364 (2023). arXiv:2305.00505
2. Meng, Q., et al.: Adaptive fixed-time stabilization for a class of uncertain nonlinear systems. IEEE Trans. Autom. Control 1–8 (2023). https://doi.org/10.1109/TAC.2023.3244151
3. Ricardo, H.J.: A Modern Introduction to Differential Equations, 3rd edn. Academic Press (2020)
4. Zill, D.G.: Introduction to the Finite Element Method, 9th edn. (2018)
5. Kreyszig, E.: Advanced Engineering Mathematics, 10th edn. Wiley, New York (2020)
6. Kurnaz, A., Oturanç, G.: The differential transform approximation for the system of ordinary differential equations. Int. J. Comput. Math. **82**(6), 709–719 (2005). https://doi.org/10.1080/00207160512331329050
7. Hsiao, C.: Haar wavelet approach to linear stiff systems. Math. Comput. Simul. **64**(5), 561–567 (2004). https://doi.org/10.1016/j.matcom.2003.11.011
8. Ahsan, M., et al.: A high-order multi-resolution wavelet method for nonlinear systems of differential equations. Math. Comput. Simul. **215**, 543–559 (2024). https://doi.org/10.1016/j.matcom.2023.08.032
9. Aziz, I., et al.: A new method based on Haar wavelet for the numerical solution of two-dimensional nonlinear integral equations. J. Comput. Appl. Math. **272**, 70–80 (2014). https://doi.org/10.1016/j.cam.2014.04.027
10. Reddy, J.: Introduction to the Finite Element Method, 4th edn. McGraw-Hill Education, New York (2019)
11. Zaheer-Ud-Din, et al.: Meshless analysis of nonlocal boundary value problems in anisotropic and inhomogeneous media. Mathematics **8**(11), 2045 (2020)
12. Fu, Z., et al.: A robust kernel-based solver for variable-order time fractional PDEs under 2D/3D irregular domains. Appl. Math. Lett. **94**, 105–111 (2019). https://doi.org/10.1016/j.aml.2019.02.025
13. Fu, Z., et al.: Localized collocation schemes and their applications. Acta Mech. Sin. **38**(7), 422167 (2022). https://doi.org/10.1007/s10409-022-22167-x
14. Wacker, B., Schluter, J.C.: Qualitative analysis of two systems of nonlinear first-order ordinary differential equations for biological systems. Math. Models Appl. Sci. **45**, 4597–4624 (2022). https://doi.org/10.1002/mma.8056
15. Xi, Q., Liu, X.: Finite-time stability and controller design for a class of hybrid dynamical systems with deviating argument. Nonlinear Anal. Hybrid Syst. **39**, 100952 (2021). https://doi.org/10.1016/j.na.2020.100952
16. Sunday, J., et al.: A pair of three-step hybrid block methods for the solutions of linear and nonlinear first-order systems. Util. Math. **118**, 1–15 (2021)
17. Sunday, J., et al.: Numerical integration of stiff differential systems using non-fixed step-size strategy. Symmetry **14**(8), 1575 (2022)
18. Sunday, J., et al.: Variable step hybrid block method for the approximation of Kepler problem. Fractal Fract. **6**(6), 343 (2022)
19. Muminov, M.I., Jumaev, Z.Z.: Exact periodic solutions of second-order differential equations with piecewise constant arguments. Adv. Math. **10**, 3113–3128 (2021)

20. Muminov, M.I., Radjabov, T.A.: Existence conditions for 2-periodic solutions to non-homogeneous differential equations with piecewise constant argument. Exam. Counterexam. **5**, 100–145 (2024)
21. Kolmogorov, A., Fomin, S.: Elements of the Theory of Functions and Functional Analysis (1957)
22. Ahsan, M., et al.: A high-order multi-resolution wavelet method for nonlinear systems of differential equations. Math. Comput. Simul. **215**, 543–559 (2024). https://doi.org/10.1016/j.matcom.2023.08.032

Numerical Solution of the SIR Model of COVID-19 Transmission with Vaccination by Using Adams–Bashforth Method

Siti Rahimah Batcha, Nur Zafirah Mohd Sidek, Siti Nur Aleeza Mohd Kamal, Nurdini Sofea Sharudin, and Harliza Mohd Hanif

Abstract The global COVID-19 pandemic has underscored the importance of understanding and managing infectious disease dynamics. This paper focuses on the numerical solution of the Susceptible-Infectious-Recovered (SIR) model, a widely used mathematical framework for studying disease transmission. The numerical solution is implemented through the Adams–Bashforth method, known for its efficiency in solving ordinary differential equations. The project incorporates the element of vaccination into the classical SIR model to explore the impact of vaccination campaigns on disease dynamics. This project contributes to the ongoing efforts to understand and mitigate the effects of COVID-19 by offering a computational perspective on disease dynamics and vaccination strategies. The combined utilization of the Adams–Bashforth and RK4 methods enhances the accuracy and efficiency of the numerical solution, making it a valuable tool for researchers and policymakers alike in addressing the challenges posed by infectious diseases. The results analysis shows that the vaccinated model outperformed the unvaccinated model, with fewer estimated infections in general, indicating that the Adams–Bashforth method is efficient in approximating the further behavior of vaccine and number of infected people interaction.

S. R. Batcha (✉) · N. Z. M. Sidek · S. N. A. M. Kamal · N. S. Sharudin · H. M. Hanif
Faculty of Computer and Mathematical Sciences, Universiti Teknologi MARA, Darul Ehsan, Selangor, Malaysia
e-mail: rahimahbatcha@uitm.edu.my

N. Z. M. Sidek
e-mail: nurzafirah@uitm.edu.my

S. N. A. M. Kamal
e-mail: 32022908017@student.uitm.edu.my

N. S. Sharudin
e-mail: 42022931181@student.uitm.edu.my

H. M. Hanif
e-mail: harliza50@uitm.edu.my

A. Shuib et al. (eds.), *Mathematical Modeling and Numerical Analysis*,
Springer Proceedings in Mathematics & Statistics 510,
https://doi.org/10.1007/978-981-96-9350-4_4

Keywords COVID-19 · Susceptible-Infectious-Recovered (SIR) · Adams–Bashforth · Vaccination · Ordinary differential equations

1 Introduction

The current outbreak of the SARS-CoV-2 influenza virus, coronavirus (coronavirus disease 2019; previously known as 2019-nCoV), epi-centered in Hubei Province of the People's Republic of China, has spread to many other countries [1]. COVID-19 was declared a pandemic by WHO on March 11 [15]. Numerous numbers of models have been developed. Compartmental models are frequently used in studies on epidemiology that split a population into groups using the SIR model.

1.1 SIR Model

The SIR model, the most widely used model, was invented by Kermack and McKendrick in 1927. It is arguably the most basic yet useful disease model for studying infectious illnesses in mathematical epidemiology [16]. As stated by [7], the SIR model is mainly utilized to evaluate and forecast the course of diseases. The population is divided into three segments as the model's basis. Each compartment is determined by the population's infectious condition [13]. As a result, the population is separated into three susceptible classes (S), infected (I), and recovered (R) [6]. The SIR model has received significant interest in many studies, particularly in the SIR model of COVID-19. Previous researchers have also developed an SIR model that consists of vaccination. Due to that, the first prevalent concern in this research is to analyze the difference in the result of the mathematical model in the total spread of COVID-19 when vaccination is added as a variable and compare it with the SIR model without vaccination.

1.2 Unvaccinated Model

Model of the SIR by [5] indicates model without the influence of vaccination

$$\frac{\mathrm{d}S(t)}{\mathrm{d}t} = \frac{-\beta I(t)S(t)}{N} \tag{1}$$

$$\frac{\mathrm{d}I(t)}{\mathrm{d}t} = \frac{\beta I(t)S(t)}{N} - \gamma I(t) \tag{2}$$

$$\frac{dR(t)}{dt} = \gamma I(t) \tag{3}$$

$$N = S(t) + I(t) + R(t), \quad \forall T. \tag{4}$$

The model, $S(t)$ denotes the count of individuals in each area at time t who are susceptible to contracting the disease and can be infected by those who are currently infectious $I(t)$. $I(t)$ represents the count of individuals in the same area at time t who are currently infected and capable of spreading the virus. $R(t)$ indicates the count of individuals in the area at time t who have recovered from the infection and are no longer in the infected state. N represents the total population in the area at time t.

Additionally, β is a constant representing the infectivity rate, indicating the expected number of people that an infectious individual may infect. γ is another constant representing the removal rate, signifying the expected number of people recovering from the infected state. The ratio of β to γ is referred to as the reproduction number, denoted as $R_0 = \beta/\gamma$. This reproduction number reflects the average number of secondary infections expected from a single infected person, assuming everyone else is in a susceptible state.

1.3 Vaccinated Model

The vaccinated model refers to the SIR model with vaccination effect given by the following expressions as by [17].

$$\frac{dS(t)}{dt} = -d \in -\frac{\beta I(t)S(t)}{N}, \quad \text{s.t. } S(t) > 0 \tag{5}$$

$$\frac{dI(t)}{dt} = \frac{\beta I(t)S(t)}{N} - \hat{\gamma} I(t) \tag{6}$$

$$\frac{dR(t)}{dt} = d(1 - \in) + \hat{\gamma} I(t). \tag{7}$$

1.4 Adams–Bashforth Moulton Method

Adams–Bashforth method was developed specifically for solving differential equations [10]. The Adams–Bashforth methods allow us to compute explicitly the approximate solution at an instant time from the solutions in previous instants [12]. As for this study, the Adams–Bashforth method is chosen to solve the SIR model. The first method of solving the differential equations numerically, provided by Bashforth and Adams, consists of an arbitrary second-order differential equation, with y and t as the variables. The derivation of the method consists of using the initial equation and differentiating it repeatedly to get formulae for higher derivatives of the same function. By substituting the initial value, we can evaluate the initial value and expand the solution using the Taylor series.

In the entire family of linear multistep formulas for the numerical solution of nonstiff problems of ordinary differential equations, Adams formulas are the most widely used techniques. While Adams–Moulton formulae (implicit) need an interactive process before implementation to generate an explicit expression for the term Z_{n+1}, Adams–Bashforth formulas (explicit) have a straightforward implementation [11].

The formula shown below is the fourth-order Adams–Bashforth. It is used to predict rough approximation at new points [2].

$$S_{i+1} = S_i + \frac{h}{24}(55f_i - 59f_{i-1} + 37f_{i-2} - 9f_{i-3}) \tag{8}$$

$$I_{i+1} = I_i + \frac{h}{24}(55f_i - 59f_{i-1} + 37f_{i-2} - 9f_{i-3}) \tag{9}$$

$$R_{i+1} = R_i + \frac{h}{24}(55f_i - 59f_{i-1} + 37f_{i-2} - 9f_{i-3}). \tag{10}$$

The fourth-order Adams–Moulton below is used as a corrector and it will improve the initial value [2].

$$S_{i+1}^P = S_i + \frac{h}{24}\left(9f_{i+1}^{(j)} + 19f_i - 5f_{i-1} + f_{i-2}\right) \tag{11}$$

$$I_{i+1}^P = I_i + \frac{h}{24}\left(9f_{i+1}^{(j)} + 19f_i - 5f_{i-1} + f_{i-2}\right) \tag{12}$$

$$R_{i+1}^P = R_i + \frac{h}{24}\left(9f_{i+1}^{(j)} + 19f_i - 5f_{i-1} + f_{i-2}\right), \tag{13}$$

where $i = 0, 1, \ldots, n - 1$.

2 Methodology

To ensure the accuracy and effectiveness of numerical solutions, the results were validated against the previous method. The paper by [3] used the fourth-order Runge–Kutta method (RK4) for solving the DDEs. The equations used:

$$y_1' = y_1(t - 1) + y_2(t),$$

Table 1 Parameters used by [3]

Variable	Normal case	Diabetes case
r_c	12	21.6
r_1	0.39	0.39
K_G	100	125
K_1	8	7
h	6	5.8

$$t \geq 0,$$

$$y_2' = y_1(t) + y_1(t-1).$$

With the initial functions given as,

$$y_1 = e^t,$$

$$y_2 = 1 - e^{-1}.$$

And the exact solutions,

$$y_1 = e^t,$$

$$y_2 = 1 - e^{(t-1)}.$$

With the value of t in the interval [0, 3].

The parameters used by [3] are shown in Table 1.

To validate the results from the paper by [3] that stated at the literature review of DDEs, the Adams–Bashforth method was applied to the same dataset used in the paper. The original results, obtained using the RK4, are shown in Fig. 1. In comparison, the results obtained from the Adams–Bashforth method are presented in Fig. 2.

Figure 2 shows the numerical solution along with the exact solutions of the DDEs model from paper by [3]. The x-axis represents time (t), and the y-axis represents the function $y(t)$. The graph includes y_1 (blue solid line) and y_2 (green solid line) as the numerical solutions for the first and second components of the DDEs, respectively, and yExact1 (red dashed line) and yExact2 (blue dashed line) as their exact solutions for validation. The exact solutions initially closely match the numerical solutions, but differences appear over time, showing where the RK4 method may not fully coincide with the numerical solutions, suggesting the possibility of using the Adams–Bashforth method for better accuracy and continuity.

Figure 2 shows the numerical solution along with the exact solutions of the DDEs modeling using the Adams–Bashforth method. The x-axis represents time (t), and

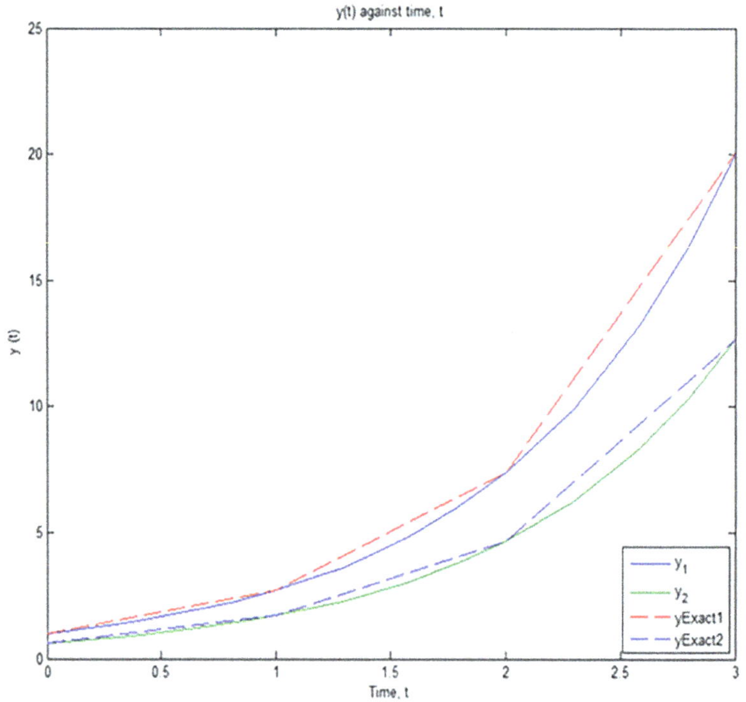

Fig. 1 Numerical solutions of equation in [3] using RK4

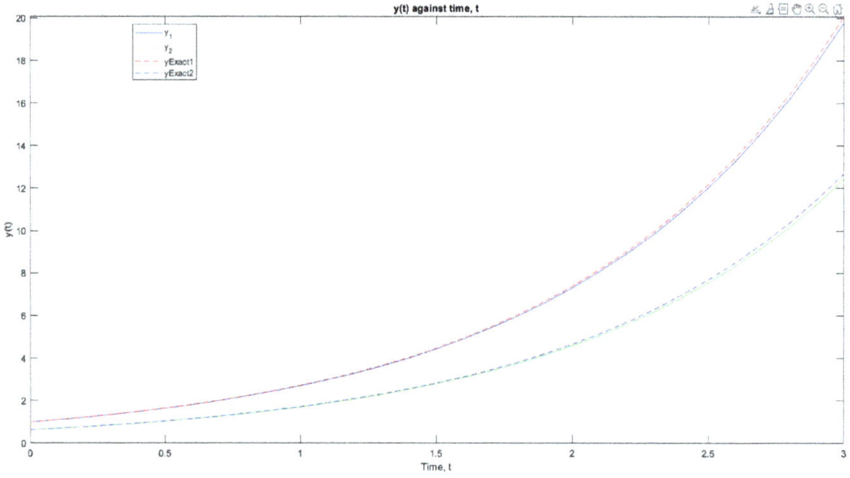

Fig. 2 Numerical solutions of equation in [3] using Adams–Bashforth

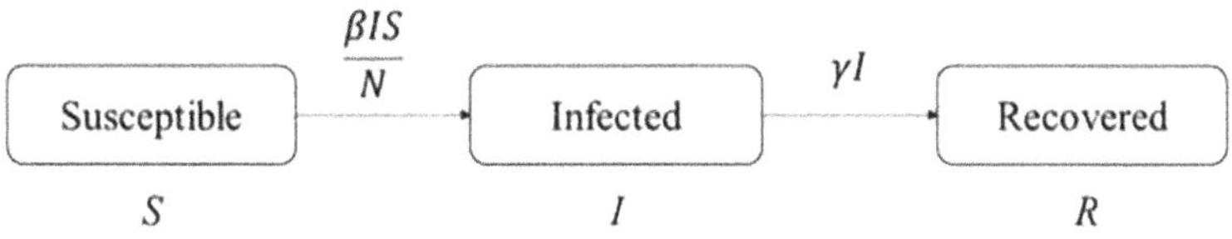

Fig. 3 Illustration of demographics-based SIR model

the y-axis represents the function $y(t)$. The graph includes y_1 (blue solid line) and y_2 (green solid line) as the numerical solutions for the first and second components of the DDEs, respectively, and yExact1 (red dashed line) and yExact2 (blue dashed line) as their exact solutions for validation. The exact solutions closely follow the numerical solutions throughout the time interval, showing enhanced smoothness and reduced variations compared to the RK4 results. This indicates that the Adams–Bashforth method provides a more accurate and reliable calculation for modeling the delay differential equations.

Adams–Bashforth method produced better results compared to those of the RK4 method. As for the Adams–Bashforth methods, it presented even more exact representation of the glucose and insulin levels over time, which very close to the origin solution. Thus, it can be assumed that the Adams–Bashforth methods are more appropriate while modeling the vaccine and number of infected people interaction.

The following expression offers the basic SIR model without the isolation and vaccination effects [8, 13, 14]; and it can be written as shown in Fig. 3. The number of people who are susceptible changes over time.

The number of people who are suspected changes over time in the form of Eq. (1).

$$\frac{dS(t)}{dt} = \frac{-\beta I(t)S(t)}{N}.$$

The number of people who are infected changes over time in the form of Eq. (2).

$$\frac{dI(t)}{dt} = \frac{\beta I(t)S(t)}{N} - \gamma I(t).$$

The number of people who recovered changes over time in the form of Eq. (3).

$$\frac{dR(t)}{dt} = \gamma I(t).$$

The total number of populations in the form of Eq. (4).

$$N = S(t) + I(t) + R(t), \quad \forall T,$$

where

N Total number of populations.

Table 2 Parameter's value for without and with vaccination

Parameters	Without vaccination	With vaccination
N	5,625,454	5,625,464
S	4,260,461	3,981,344
I	754,235	475,118
R	610,758	331,641
β	0.000000164	0.000000176
γ	0.08	0.08
R_0	0.94	0.81

$S(t)$ The number of people in a particular area at time t who are susceptible of getting the disease from infected people.
$I(t)$ The number of people in a particular area at time t who are infected and contagious as due to the virus spread.
$R(t)$ The number of people that are recovered in a particular area at time t.
β Constant of infectivity/transmission rate.
γ Constant recovery rate.

with initial guess $S(0) = S_0$, $I(0) = I_0$, $R(0) = R_0$.

The model then is changed where it will be the SIR model with vaccination, and it is based on the article by [17] in the form of Eqs. (5), (6), and (7).

$$\frac{\mathrm{d}S(t)}{\mathrm{d}t} = -d \in -\frac{\beta I(t)S(t)}{N}, \quad \text{s.t. } S(t) > 0$$

$$\frac{\mathrm{d}I(t)}{\mathrm{d}t} = \frac{\beta I(t)S(t)}{N} - \hat{\gamma} I(t)$$

$$\frac{\mathrm{d}R(t)}{\mathrm{d}t} = d(1- \in) + \hat{\gamma} I(t).$$

$\in$ is defined as the ratio of positive cases in the vaccinated samples. The vaccine efficacy is given by $(1- \in) \cdot \hat{\gamma} = \gamma - \hat{p}$ and $\hat{p}$ is the isolation rate of positive infected individuals in the population. Note that the reproduction number is now $R_0 = \beta/\hat{\gamma}$.

The parameters from research carried out by [9] focusing on Selangor are shown in Table 2.

3 Results and Discussion

This section uses the Adams–Bashforth method to solve the SIR model without and with vaccination by using the Selangor parameter for 2021, as indicated in Table 2. The computation's results will be presented graphically as this kind of presentation

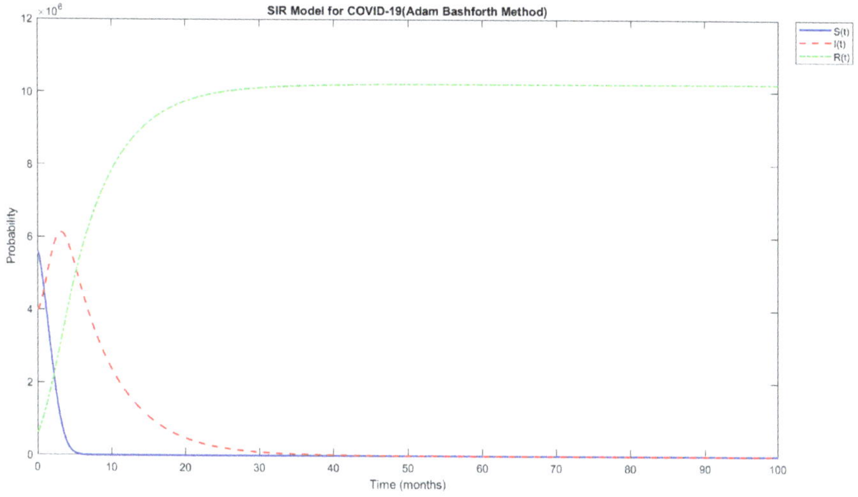

Fig. 4 SIR model without vaccination

makes the dynamics and implications of the SIR model's results easier for readers to comprehend and more visually appealing.

The graph in Fig. 4 displays the outcomes of the SIR model for COVID-19 that solved the differential equations using the Adams–Bashforth approach. The chances of a person falling into each category which are susceptible (S), infected (I), and recovered (R) are represented by the y-axis, while the x-axis shows the amount of time in months. The solid blue line on the graph indicates those vulnerable to infection, the dashed red line indicates those who are infected, and the dotted green line indicates those who have recovered. The dynamics of the COVID-19 pandemic are depicted in the graph across time, accounting for immunity and recovery.

Figure 5 shows the Adams–Bashforth technique outcome graph for the COVID-19 vaccine. The y-axis represents the likelihood related to the population of Selangor, while the x-axis represents time in months. The population is represented by the y-axis, with a range of -12×10^9 to 6×10^9. The graph displays four lines representing various populations: Vaccinated (V), recovered (R), infected (I), and susceptible (S). Over time, fewer people are vulnerable to the sickness, as seen by the 'vulnerable' line, which drops dramatically before leveling off. The 'infected' line grows at first and subsequently drops, following a typical infection curve pattern. The 'recovered' line increases over time, indicating that more individuals recover from the infection. The 'vaccinated' line remains constant throughout, indicating that the number of vaccinated individuals does not change over time.

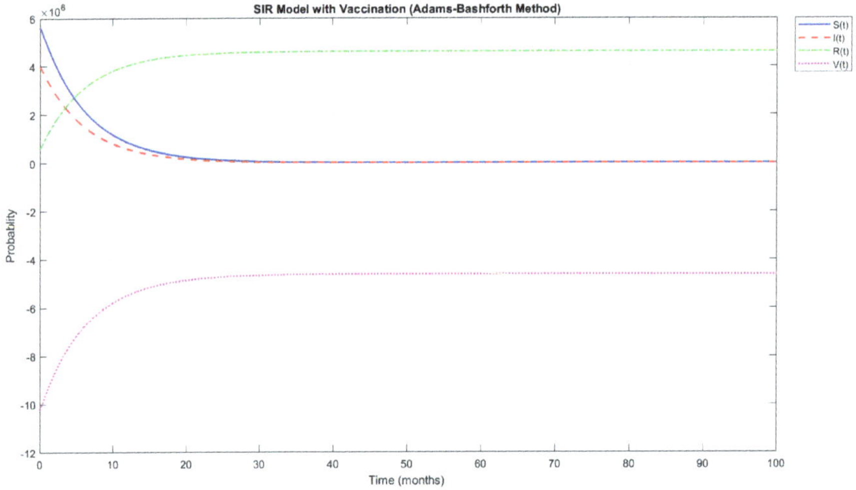

Fig. 5 SIR model with vaccination

4 Conclusion and Recommendation

The SIR model for COVID-19 in this study serves as a basic model to illustrate the state of the disease. This research uses data from the Ministry of Health Malaysia in 2021 to simulate the one-year transmission process of COVID-19. This study found that the vaccinated model outperformed the unvaccinated model, with fewer estimated infections in general. Furthermore, vaccination will significantly reduce the reproduction rate to less than one.

In addition, there are some recommendations to enhance the study such as by applying the method to solve an advanced model of COVID-19 for example Susceptible-Exposed-Infected-Susceptible (SEIS) model that describes the situation in which a person has been infected but is yet to have started to release infectious viruses or bacteria [4] or applying the method to other infected disease such as influenza.

Acknowledgements This research has been supported by the Universiti Teknologi MARA.

Funding The author(s) received no specific funding for this work.

References

1. Ahmad, S., Ullah, A., Al-Mdallal, Q.M., Khan, H., Shah, K., Khan, A.: Fractional order mathematical modeling of COVID-19 transmission. Chaos Solitons Fract. **139**, 110256 (2020). https://doi.org/10.1016/j.chaos.2020.110256

2. Ascher, U.M., Petzold, L.R.: Computer Methods for Ordinary Differential Equations and Differential-Algebraic Equations (1997)
3. Batcha, S.R., Sidek, N.Z.B.M.: Application approach for diabetes delay differential equations. Math. Lett. **2**(2), 287–295 (2023)
4. Britton, T., House, T., Lloyd, A.L., Mollison, D., Riley, S., Trapman, P.: Five challenges for stochastic epidemic models involving global transmission. Epidemics **10**, 54–57 (2015). https://doi.org/10.1016/j.epidem.2014.05.002
5. Brauer, F.: The Kermack–McKendrick epidemic model revisited. Math. Biosci. **198**(2), 119–131 (2005). https://doi.org/10.1016/j.mbs.2005.07.006
6. Ebraheem, H.K., Alkhateeb, N., Badran, H., Sultan, E.: Delayed dynamics of SIR model for COVID-19. Open J. Model. Simul. **09**(02), 146–158 (2021). https://doi.org/10.4236/ojmsi.2021.92010
7. Ghanbari, B., Kumar, S., Kumar, R.: A study of behaviour for immune and tumor cells in immunogenetic tumour model with non-singular fractional derivative. Chaos Solitons Fract. **133** (2020). https://doi.org/10.1016/j.chaos.2020.109619
8. Kousar, N., Mahmood, R., Ghalib, M.: A numerical study of SIR epidemic model. Int. J. Sci. Basic Appl. Res. **25**(2), 354–363 (2016). http://gssrr.org/index.php?journal=JournalOfBasicAndApplied
9. Ministry of Health: COVID-19 · Malaysia I KKMNOW. KKMNow (n.d.). https://data.moh.gov.my/dashboard/covid-19
10. Odunayo, M.A., Ogunwobi, Z.O.: Comparison of Adams–Bashforth–Moulton method and Milne–Simpson method on second order ordinary differential equation. Turk. J. Anal. Number Theory **9**(1), 1–8 (2021). https://doi.org/10.12691/tjant-9-1-1
11. Orvay, F.C., Palanca, M.T.M., Alcaide, A.S.: Adams methods: implementation as predictor-corrector methods. Far East J. Appl. Math. **66**(2), 93–105 (2012)
12. Peinado, J., Ibáñez, J., Arias, E., Hernández, V.: Adams–Bashforth and Adams–Moulton methods for solving differential Riccati equations. Comput. Math. Appl. **60**(11), 3032–3045 (2010). https://doi.org/10.1016/j.camwa.2010.10.002
13. Petard, H.: A contribution to the mathematical theory of big game hunting. Am. Math. Mon. **45**(7), 446 (1938). https://doi.org/10.2307/2304150
14. Smith, D., Moore, L., Chuang, J.C., Michel, J.: The SIR model for spread of disease. J. Online Math. Appl. **1**(3), 1–5 (2001)
15. Tang, B., Wang, X., Li, Q., Bragazzi, N.L., Tang, S., Xiao, Y., Wu, J.: Estimation of the transmission risk of the 2019-nCoV and its implication for public health interventions. J. Clin. Med. **9**(2), 1–13 (2020). https://www.ncbi.nlm.nih.gov/pmc/articles/PMC7074281/pdf/jcm-09-00462.pdf
16. Teoh, Y.K., Hamdan, N.F., Abu Hasan, S., Ariffin, A.F., Mahat, A.: Application of susceptible-infected-removed model with vital dynamics for COVID-19 outbreak in Malaysia. J. Comput. Res. Innov. **7**(2), 82–87 (2022). https://doi.org/10.24191/jcrinn.v7i2.292
17. Wong, W.K., Juwono, F.H., Chua, T.H.: SIR simulation of COVID-19 pandemic in Malaysia: will the vaccination program be effective? 1–19 (2021). http://arxiv.org/abs/2101.07494

Validating Advanced Numerical Techniques Based on BBDF Structure for Precision in CLL Therapy Studies

Hazizah Mohd Ijam, Norshakila Abd Rasid, Noorazlina Mohamid Salih, Nur Azila Yahya, and Rusliza Ahmad

Abstract This study evaluates the performance of advanced numerical techniques based on the block backward differentiation formula (BBDF) in solving therapy models for Chronic Lymphocytic Leukemia (CLL). The CLL treatment model uses deferential equations to illustrate the interactions between cancer cells and treatment. The approximation uses BBDF-based model known as diagonal implicit BBDF with intermediate points to capture the dynamic interactions between cancer cells, immune cells, and therapeutic agents. The effectiveness of the new solver is assessed by comparing its results with both analytical solutions and existing solver. The findings demonstrate that the new method is effective for solving CLL therapy models.

Keywords ODE · CLL · BBDF

1 Introduction

Ordinary Differential Equations (ODEs) play a crucial role in cancer research by providing an alternative approach to understanding the dynamics of cancer and establishing its fundamental principles. Researchers can create ODE-based models of tumor growth patterns that focus on cell proliferation, apoptosis, and interactions with the surrounding microenvironment. These models help in understanding tumor growth and its response to therapeutic interventions. Moreover, ODE models can aid

H. M. Ijam
Faculty of Computer and Mathematical Sciences, Universiti Teknologi MARA, Shah Alam, Selangor, Malaysia

N. A. Rasid (✉) · N. A. Yahya · R. Ahmad
Faculty of Computer and Mathematical Sciences, Universiti Teknologi MARA, Perak Branch, Tapah Campus, Tapah Road, Perak, Malaysia
e-mail: norshakila@uitm.edu.my

N. M. Salih
Electrical Engineering Technology Section, Universiti Kuala Lumpur Malaysian Institute of Marine Engineering Technology, Lumut, Perak, Malaysia

A. Shuib et al. (eds.), *Mathematical Modeling and Numerical Analysis*, Springer Proceedings in Mathematics & Statistics 510,
https://doi.org/10.1007/978-981-96-9350-4_5

in predicting tumor progression and treatment response, thus assisting clinicians in making informed decisions about drug usage [1, 2]. For instance, [3] employed a tumor-immune ODE-based model to optimize chemotherapy dosing in breast cancer, while [4] developed a multi-tumor cell model to evaluate radiotherapy approaches for lung cancer. Similarly, [5] explored the use of curative vaccines for prostate cancer treatment through ODE modeling. Validating these models with patient data allows for refinement and adjustment as new information emerges, making them essential tools in understanding cancer cell and immune system interactions [6–8].

Chronic Lymphocytic Leukemia (CLL) is a prevalent form of leukemia that primarily affects individuals in the Western world [9]. It is characterized by the abnormal proliferation and accumulation of B lymphocytes in the blood and bone marrow, leading to enlarged spleen and lymph nodes. This slowly progress cancer can be diagnosed at an early stage. Despite its gradual progression, CLL can result in mortality if left untreated. Understanding it characteristics and treatment options for CLL is crucial due to its mortality rate. In recent years, advances in CLL therapies have introduced various treatment strategies that can be effectively modeled with ODEs [10]. One such treatment is using Bruton's tyrosine kinase (BTK) inhibitors like Ibrutinib, which targets the BTK pathway crucial for the survival and proliferation of malignant B cells. By inhibiting this pathway, Ibrutinib prevents the growth of CLL cells, and its effects are well-captured by the ODE-based model employed in this study [11]. Another important treatment for CLL is chemoimmunotherapy, which combines chemotherapy and immunotherapy. Chemotherapy kills cancer cells directly, while immunotherapy boosts the immune system's ability to target malignant cells. The combination reflects the biological complexity of CLL treatment, addressing both direct cell destruction and immune system modulation. The ODE models used in this research capture these interactions, providing a robust framework to evaluate the efficacy of chemoimmunotherapy in CLL [12, 13].

Modeling cancer treatment with ODEs often involves handling stiff phenomena, which necessitates specialized numerical solvers to achieve accurate approximations. To address the challenges of stiffness and prevent numerical instability, solvers like the block backward differentiation formula (BBDF) have been employed in stiff ODE problems [14]. These methods have demonstrated significant efficacy across various stiff applications [15, 16], particularly in oncology and cancer area which stiff ODEs are commonly encountered [17–19]. Recent developments have introduced more time-efficient and accurate solvers known as hybrid block methods. These methods play a crucial role in combining multiple techniques to offer flexibility in capturing complex dynamics and eventually improve stability and accuracy [20–22]. The methods incorporate off-step points as additional evaluation points between standard time steps. This strategy provides improve precision by capturing intermediate behaviors in stiff ODE systems [23, 24].

Building on these advancements, we propose a new block BDF-based solver called Diagonally Implicit BBDF with Off-step Points (DOBBDF) to approximate the dynamics of cancer ODE-based models. Our solver incorporates intermediate off-step points to enhance accuracy, leveraging the benefits of hybrid methods to address stiff ODEs more effectively. The solver captures complex behaviors within the step size with off-step points for allowing better approximation in stiff cancer models like those used in the study of Chronic Lymphocytic Leukemia (CLL) treatment. By developing this new solver, the study aims to address the challenges posed by stiff ODEs in cancer models, offering a robust numerical approach to accurately capture the dynamics of targeted and combination therapies in CLL treatment.

This paper is structured as follows: In Sect. 2, we provide a detailed explanation of the ODE-based models for the treatment of Chronic Lymphocytic Leukemia (CLL). Section 3 demonstrates the formulation of the new solver method using the Lagrange interpolation polynomial and presents the numerical findings. Finally, in Sect. 4, we conclude with a summary of the research findings.

2 Material and Methods

This section demonstrated the models and method for analyzing Chronic Lymphocytic Leukemia (CLL) therapy. We first lay out two different therapeutic ODE-based models representing BTK inhibitor Ibrutinib therapy and chemoimmunotherapy. We analyze these models through specific parameter sets and numerical methods. Following the model descriptions, we present the derivation and implementation of the new solver name as a Diagonally Block Backward Differentiation Formula (DOBBDF) with intermediate points to solve the CLL-based model of ODEs. The parameters and their biological validation are discussed to ensure the relevance of the models to clinical scenarios.

2.1 Treatment Models for Chronic Lymphocytic Leukemia (CLL)

Bruton Tyrosine Kinase (BTK) Inhibitor Ibrutinib Therapy

First ODE-based model we considered was developed by [12]. This model demonstrates the interaction within a two-compartment system resulting from the effects of the drug known as the BTK inhibitor Ibrutinib. The first compartment of the system indicates the volume of the mortality rate per second of CLL in tissue, whereas the second compartment shows the volume of the mortality rate per second of CLL in blood. The model is described as follows:

$$
\begin{aligned}
x' &= -A_1x + A_1B - Cx, \\
y' &= -A_2x + Cx, \\
(t_0) &= x_0, \quad y(t_0) = y_0.
\end{aligned}
\tag{1}
$$

The parameters A_1, A_2, B, and C are explained in detail in Table 1.

Chemoimmunotherapy

We consider a combination of chemotherapy and immunotherapy, known as chemoimmunotherapy, for our second model. Over the past decade, this model has been the subject of study as an effective alternative treatment for CLL [13]. Based on the nonlinear ODE model developed by [14], the dynamics of the system are mathematically demonstrated in Eq. (2):

$$
\begin{aligned}
H' &= -K_1HI - \frac{FH^2}{J} - \frac{DHG}{E+G} + FH, \\
I' &= z_0 + z(t) - K_2HI - sI + \frac{LHI}{M+H} - \frac{NIG}{O+G} \\
G' &= -PG + s(t).
\end{aligned}
\tag{2}
$$

The ODE system in Eq. (2) demonstrates the interplay between pharmacodynamics and pharmacokinetics. In the first compartment, $H(t)$ denotes the number of B-CLL lymphocytes, and in the second compartment, $I(t)$ represents the number of T-lymphocytes (immune cells). These two compartments represent the pharmacodynamics process, showing the biochemical effects of drugs interacting with cellular receptors. Conversely, the third term in Eq. (2), $G(t)$, describes the pharmacokinetics,

Table 1 Parameters for the bruton tyrosine kinase (BTK) inhibitor ibrutinib therapy model

Parameters	Definition	Values/unit
A_1	The rate at which CLL cells are depleted within the tissue	2.7×10^{-2}
A_2	The rate at which CLL cells are diminished in the bloodstream	2×10^{-3}
B	This parameter is included to determine the degree to which the number of CLL cells converges and stabilizes over the long term during treatment	4×10^{-1}
C	The speed at which tissue cells are redistributed into the bloodstream	9.6×10^{-3}
x_0	The quantity of CLL cells measured in volume within the tissue	3034 ml
y_0	The volume-based count of CLL cells present in the bloodstream	153 ml

Table 2 Parameters for the chemoimmunotherapy model

Parameters	Definition	Values/unit
F	The rate at which cancer advances	1×10^{-2}
J	The maximum capacity of cancer cells	1×10^{12}
K_1	The parameter indicating the relationship between cancer and immune cells that significantly affects cancer populations	0.5×10^{-10}
K_2	The parameter that characterizes the connection between cancer and immune cells that influences immune populations	0.1×10^{-12}
z_0	The natural influx of immune cells to the site of interaction	0.3×10^{6}
D	The mortality rates resulting from the exposure of chemotherapeutic drugs to cancer cells	0.08×10^{2}
N	The mortality rates resulting from the exposure of chemotherapeutic drugs to immune cells	1000
E and O	The drug volume that impacts half of its maximum effectiveness in each cell population	$E = 2000$ $O = 0.05 \times 10^{8}$
s	The rate at which immune cells are produced in response to stimulation by cancer	0.001
L	The inherent mortality rate of immune cells	0.01×10^{-10}
M	The volume of cancer cells triggering an immune system response is at half of its maximum capacity	100
P	The rate of ineffectiveness of a non-specific chemotherapeutic drug within a given cycle	416×10^{-2}

specifically the concentration of the chemotherapeutic drug used during treatment. Based on the study by [14], the parameters and their values are detailed in Table 2.

2.2 *Biological Validation of Parameters*

The parameters for the CLL models in Sect. 2.1 were based on existing biological and experimental data. The values and definitions for each parameter are provided in Tables 1 and 2, respectively. These parameters are derived from empirical studies and literature to ensure their relevance to CLL treatment. The BTK inhibitor model parameters are from the experimental data in [12], which measures cell depletion rates. Similarly, parameters in the chemoimmunotherapy model are collected in the studies by [13, 14].

2.3 Diagonally BBDF with Intermediate Points

We employ DOBBDF within the PECE mode, with P, C, and E denoting predictor, corrector, and function evaluation of $f(x, y)$, respectively. The calculation process of PECE mode is: $P{:}\,y_{n+i}^{k} \rightarrow E{:}\,f_{n+i}^{k} \rightarrow C{:}\,y_{n+i}^{k+1} \rightarrow E{:}\,f_{n+i}^{k+i}$ where k takes values of $\frac{1}{2}, 1, \frac{3}{2}, 2$. To begin, the predictor formula is derived using a Lagrange interpolation polynomial as described in Eqs. (3) and (4).

$$P_k(x) = P(x_n + sh) = \sum_{j=0}^{k} L_{k,j}(x) f\left(x_{n+1-j}\right), \tag{3}$$

where

$$L_{k,j}(x) = \prod_{i=0, i\neq j}^{k} \frac{(x - x_{n+1-i})}{\left(x_{n+1-j} - x_{n+1-i}\right)} \quad j = 0, \frac{1}{2}, \ldots, k. \tag{4}$$

The back-values $y_{n-1}, y_{n-\frac{1}{2}}$ and y_n are use in interpolate points to computes $y_{n+\frac{1}{2}}^{p}, y_{n+1}^{p}, y_{n+\frac{3}{2}}^{p}$, and y_{n+2}^{p}.

$$\begin{aligned} P_k(x) = P\left(x_{n+\frac{1}{2}} + sh\right) &= \frac{2(y_{n-1})\left(hs + \frac{3h}{2}\right)(hs + h)}{h^2} \\ &- \frac{4\left(y_{n-\frac{1}{2}}\right)(hs + 2h)(hs + h)}{h^2} + \frac{2(y_n)\left(hs + \frac{3h}{2}\right)(hs + 2h)}{h^2}. \end{aligned} \tag{5}$$

Substituting $s = 0, \frac{1}{2}, \ldots, k$ in (5) to obtain the following predictor formula.

$$\begin{aligned} y_{n+\frac{1}{2}}^{p} &= P\left(x_{n+\frac{1}{2}}\right) = y_{n-1} - 3y_{n-\frac{1}{2}} + 3y_n \\ y_{n+1}^{p} &= P(x_{n+1}) = \frac{27}{9} y_{n-1} - \frac{16}{2} y_{n-\frac{1}{2}} + \frac{18}{3} y_n \\ y_{n+\frac{3}{2}}^{p} &= P\left(x_{n+\frac{3}{2}}\right) = \frac{18}{3} y_{n-1} - 15 y_{n-\frac{1}{2}} + \frac{20}{2} y_n \\ y_{n+1}^{p} &= P(x_{n+1}) = \frac{20}{2} y_{n-1} - 24 y_{n-\frac{1}{2}} + 15 y_n. \end{aligned} \tag{6}$$

The corrector formula for the point y^C_{n+2} is derived using Eqs. (3) and (4), which involves interpolating points (x_{n-1}, y_{n-1}), (x_n, y_n), $\left(x_{n+\frac{1}{2}}, y_{n+\frac{1}{2}}\right)$, (x_{n+1}, y_{n+1}), $\left(x_{n+\frac{3}{2}}, y_{n+\frac{3}{2}}\right)$ and (x_{n+2}, y_{n+2}). This process results in a Lagrange Polynomial, which is then expressed as follows:

$$\begin{aligned} P_k(x) = P\left(x_{n+\frac{1}{2}} + sh\right) &= \frac{-\left(\frac{2}{45}\right)\left[(y_{n-1})\left(hs+\frac{h}{2}\right)\left(hs-\frac{h}{2}\right)(hs-h)\left(hs-\frac{3h}{2}\right)\right]}{h^4} \\ &- \frac{\left(\frac{16}{9}\right)\left[\left(y_{n+\frac{1}{2}}\right)\left(hs+\frac{3h}{2}\right)\left(hs+\frac{h}{2}\right)\left(hs-\frac{h}{2}\right)(hs-h)\left(hs-\frac{3h}{2}\right)\right]}{h^4} \\ &+ \frac{(2)\left[(y_{n+1})\left(hs+\frac{3h}{2}\right)s\left(hs+\frac{h}{2}\right)(hs-h)\left(hs-\frac{3h}{2}\right)\right]}{h^4} \\ &- \frac{\left(\frac{16}{15}\right)\left[\left(y_{n+\frac{3}{2}}\right)\left(hs+\frac{3h}{2}\right)\left(hs+\frac{h}{2}\right)s\left(hs-\frac{h}{2}\right)\left(hs-\frac{3h}{2}\right)\right]}{h^4} \\ &+ \frac{\left(\frac{2}{9}\right)\left[(y_{n+2})\left(hs+\frac{3h}{2}\right)s\left(hs+\frac{h}{2}\right)\left(hs-\frac{h}{2}\right)(hs-h)\right]}{h^4}. \end{aligned} \tag{7}$$

By substituting $s = \frac{3}{2}$ into Eq. (7), we can formulate the corrector formula for the point y^C_{n+2} as follows:

$$y^C_{n+2} = \frac{2}{135}y_{n-1} - \frac{1}{3}y_n + \frac{32}{27}y_{n+\frac{1}{2}} - 2y_{n+1} + \frac{32}{15}y_{n+\frac{3}{2}} + \frac{2}{9}hf_{n+2}. \tag{8}$$

We achieve the lower triangular matrix form by formulating $y^C_{n+\frac{3}{2}}, y^C_{n+1}$ and $y^C_{n+\frac{1}{2}}$, wherein the derivation at each stage excludes one interpolated point from the previous stage using Eqs. (3) and (4). Consequently, the general corrector formula for DOBBDF is as follows:

$$\begin{aligned} y^C_{n+\frac{1}{2}} &= -\left(\frac{1}{8}\right)y_{n-1} + \left(\frac{9}{8}\right)y_n + \left(\frac{3}{8}\right)hf_{n+\frac{1}{2}} \\ y^C_{n+1} &= \left(\frac{1}{21}\right)y_{n-1} - \left(\frac{4}{7}\right)y_n + \left(\frac{32}{21}\right)y_{n+\frac{1}{2}} + \left(\frac{2}{7}\right)hf_{n+1} \\ y^C_{n+\frac{3}{2}} &= -\left(\frac{3}{122}\right)y_{n-1} + \left(\frac{25}{61}\right)y_n - \left(\frac{75}{61}\right)y_{n+\frac{1}{2}} + \left(\frac{225}{122}\right)y_{n+1} + \left(\frac{15}{16}\right)hf_{n+\frac{3}{2}} \\ y^C_{n+2} &= \left(\frac{2}{135}\right)y_{n-1} - \left(\frac{1}{3}\right)y_n + \left(\frac{32}{27}\right)y_{n+\frac{1}{2}} - 2y_{n+1} + \left(\frac{32}{15}\right)y_{n+\frac{3}{2}} + \left(\frac{2}{9}\right)hf_{n+2}. \end{aligned} \tag{9}$$

3 Result and Discussion

This section presents a quantitative analysis of the new solver's performance applied to the ODE dynamic models outlined in Eqs. (1) and (2). We created two CLL test models by applying the parameter values from Tables 1 and 2 into Eqs. (1) and (2) accordingly.

Test Model 1: BTK Inhibitor Ibrutinib Therapy

$$\begin{aligned}\frac{dx}{dt} &= -\left(2.7 \times 10^{-2}\right)\left(x - \left(4 \times 10^{-1}\right)\right) - \left(9.6 \times 10^{-3}\right)x, \\ \frac{dy}{dt} &= -\left(2 \times 10^{-3}\right)y + \left(9.6 \times 10^{-3}\right)x.\end{aligned} \tag{10}$$

We can derive the analytical solution for Eq. (10). By focusing on the first compartment of Eq. (10) and solving it then we obtain the following analytical solution:

$$x(t) = \left(\frac{9315}{61}\right)\exp\left(-\frac{183}{5000}t\right) + \frac{18}{61}. \tag{11}$$

Next, to derive the analytical solution for the second compartment in Eq. (10), we rearrange it as follows:

$$\frac{dy}{dt} + \frac{y}{500} = \frac{6}{625}x. \tag{12}$$

The integrating factor is $\exp\left(\frac{t}{500}\right)$. Using this factor, we derive the following solution for the blood compartment of y

$$y(t) = -\frac{447{,}120}{10{,}533}\exp\left(-\frac{183}{5000}t\right) + \frac{432}{305} + \frac{2{,}659{,}834}{865}\exp\left(-\frac{t}{500}\right) \tag{13}$$

Test Model 2: Chemoimmunotherapy

$$\begin{aligned}H' &= \left(0.5 \times 10^{-10}\right)HI + \left(1 \times 10^{-2}\right)H - \frac{H^2}{1 \times 10^{14}} - \frac{\left(0.08 \times 10^{2}\right)HG}{2 \times 10^{3} + G}, \\ I' &= \left(0.3 \times 10^{6}\right) + z(t) - \left(0.1 \times 10^{-12}\right)HI \\ &\quad - 0.001I + \frac{0.01 \times 10^{-10}HI}{100 + H} - \frac{1000IG}{\left(0.05 \times 10^{8}\right) + G} \\ G' &= -416 \times 10^{-2}G + s(t),\end{aligned} \tag{14}$$

The DOBBDF method is implemented in C++ to solve test models 1 and 2. The numerical results are presented in Tables 3, 4 and Figs. 1, 2, 3. The results are compared with existing solvers and the theoretical solutions derived in Eqs. (12) and (13). The following notations are used in the numerical tables and graph results:

t	Denotes days of treatment.
BBDF	Block Backward Differentiation Formula developed by [15].
ode15s	Variable-step, variable-order numerical differentiation formula ranging from order 1 to 5.
Maple	Maple software.
DOBBDF	Derived method.

The results from both test models highlight the accuracy and effectiveness of the DOBBDF method in solving complex CLL treatment scenarios. For **Test Model 1**, Table 3 presents the CLL cell counts in tissue per day of treatment, while Table 4 shows the CLL cell counts in blood, both covering a period from day 1 ($t = 0$) to day 100 ($t = 100$). The data assumes a high progression rate of the disease. In particular, Table 3 and Fig. 1 demonstrate the effects of Ibrutinib therapy on lymphocytosis, revealing a consistent reduction in lymphocyte count, indicating that

Table 3 Test Model 1: CLL cell counts in tissue per day of therapy

t	DOBBDF	BBDF	ode15s	Theoretical
0	153	153	153	153
5	125.055	125.055	152.372	127.462
10	102.223	102.223	150.019	106.196
15	83.5710	83.5710	147.079	88.4860
20	68.9610	68.9744	143.468	73.7379
25	57.3204	57.3791	138.559	61.4559
30	47.8235	45.5320	133.271	51.2279
35	39.8895	35.2009	126.371	42.7103
40	33.1951	26.2543	120.108	35.6171
45	27.6572	18.9671	111.233	29.7102
50	23.0814	13.3675	103.631	24.7910
55	19.2771	9.27487	93.3457	20.6945
60	16.0651	6.35642	83.5986	17.2831
65	13.4102	4.46165	75.2343	14.4422
70	11.2172	3.14495	59.9212	12.0763
75	9.39364	2.37247	46.3412	10.1061
80	7.85409	1.80828	32.5831	8.46540
85	6.58154	1.48016	16.0624	7.09906
90	5.53034	1.20814	5.87900	5.96122
95	4.65628	1.02684	1.58281	5.01366

Table 4 Test Model 1: CLL cells counts in the blood per day of therapy

t	DOBBDF	BBDF	ode15s	Theoretical
0	3034.00	3034.00	3034.00	3034.00
5	3011.00	3011.00	3011.20	3010.00
10	2986.90	2986.90	2975.12	2986.10
15	2961.93	2961.93	2950.30	2961.00
20	2910.09	2910.07	2910.89	2910.89
25	2910.09	2910.07	2910.89	2910.89
30	2883.63	2883.86	2863.52	2883.17
35	2857.02	2857.36	2838.20	2856.71
40	2830.36	2830.76	2810.77	2830.16
45	2803.65	2803.97	2782.03	2803.55
50	2776.96	2777.01	2760.01	2776.95
55	2750.34	2749.93	2732.40	2750.40
60	2723.83	2722.80	2711.18	2723.94
65	2697.43	2695.68	2684.30	2697.59
70	2671.18	2668.67	2664.27	2671.38
75	2645.09	2641.79	2637.20	2645.33
80	2619.18	2615.13	2619.22	2619.45
85	2593.47	2588.68	2592.34	2593.76
90	2567.95	2562.48	2576.00	2568.26
95	2542.64	2536.52	2534.57	2542.97

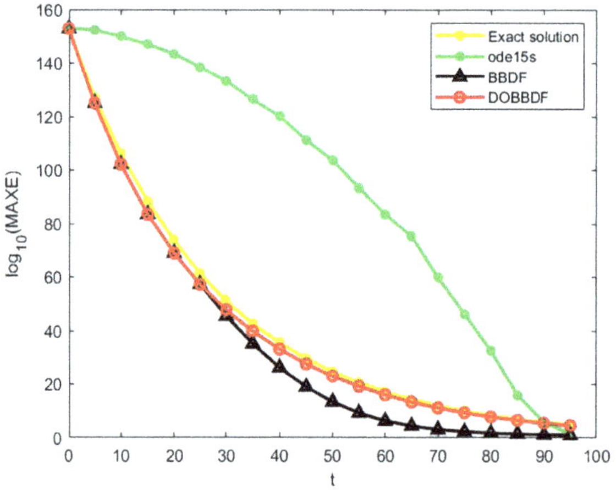

Fig. 1 Comparison between analytical and numerical solutions for the estimated number of tumor cells in tissue for the CLL model

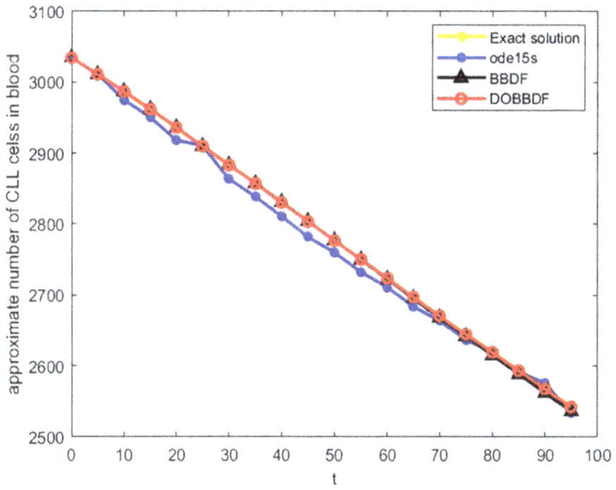

Fig. 2 Comparison of analytical and numerical solutions for the total CLL cell count in blood based on the CLL model

Ibrutinib significantly enhances the death rate of CLL cells in tissue. This decrease in CLL cells is further supported by the results in Table 4 and Fig. 2, where the drug's impact on increasing CLL cell mortality in blood is clearly observed. Overall, the DOBBDF method exhibits good accuracy, as its results closely align with the theoretical solution and those produced by other solvers like BBDF and ode15s. For **Test Model 2**, the focus shifts to the impact of chemoimmunotherapy over a longer treat treatment duration, spanning up to 1000 days. Numerical results are presented graphically using Maple, with the Maple solver included in Fig. 3a–c, which plot variables $H(t)$, $I(t)$, and $G(t)$ against time. These variables represent the growth of cancer cells, the immune response, and the concentration of chemotherapy doses, respectively. Figure 3a demonstrates a gradual decrease in cancer cell growth $H(t)$ over time due to the sustained effect of the treatment. In contrast, Fig. 3b shows a decline in immune cell levels $I(t)$, reflecting the proportional impact of chemotherapy on both immune and cancer cells. Finally, Fig. 3c shows that the concentration of chemotherapeutic doses remains stable throughout treatment. All solvers show consistency in these graphs, though slight divergence is observed with the ode15s solver in Fig. 3a, b, where it deviates slightly from the Maple solver, DOBBDF, and BBDF results. However, in Fig. 3c, all solvers maintain consistency. In both models, DOBBDF displays robust performance, matching well with theoretical solutions and other established solvers.

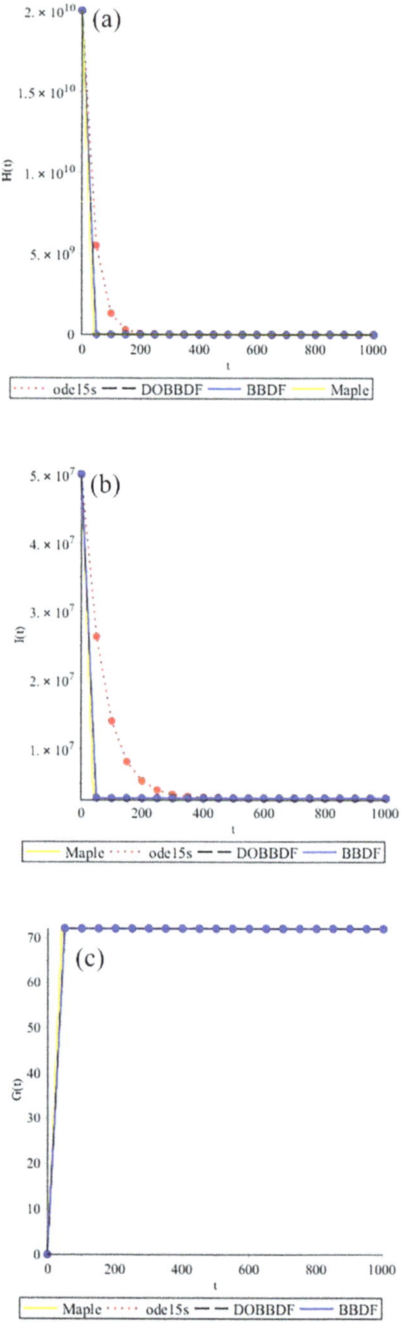

Fig. 3 Graphs of the variables $H(t)$, $I(t)$, and $G(t)$ for different methods compared over time t

4 Conclusion

This research presents the successful derivation and implementation of an altcr-native solver, DOBBDF, for approximating solutions to ODE-based models of Chronic Lymphocytic Leukemia (CLL). These models illustrate the dynamic interaction of CLL and the immune response under BTK Ibrutinib inhibitor therapy and chemoimmunotherapy. We compared DOBBDF with the existing classical block BDF, MATLAB's ode15s, and the Maple solver within the required interval. The results indicate that DOBBDF is an effective numerical solver for the CLL model.

However, certain limitations must be acknowledged, as DOBBDF uses a fixed step size, and its performance may be less optimal for ODE systems that exhibit rapid dynamic changes or discontinuities, where an adaptive step size approach might be more efficient. Additionally, while the solver performs well within the tested interval, further investigation is needed to assess its applicability to more complex or high-dimensional biomedical model.

Future work will improve the solver's adaptability by exploring ways to optimize its performance for different types of stiff ODEs and extending its application to more complex and high-dimensional models. Additionally, improving the computational efficiency of DOBBDF, possibly through parallel computing techniques, will be an area of focus to better support large-scale biomedical simulations.

Acknowledgements The authors gratefully acknowledge to Mathematical Sciences Studies, College of Computing Informatics and Mathematics, UiTM, for supporting this research.

Author Contribution All authors contributed equally to this work.

Funding This research was funded by MyRA Research Grant (MYRA-LPHD 2023) 600-RMC/GPM LPHD 5/3 (055/2023), Universiti Teknologi MARA (UiTM).

Conflict of Interest The authors have no conflicts of interest to declare.

References

1. Jarrett, A.M., Faghihi, D., Hormuth, D.A., Lima, E.A., Virostko, J., Biros, G., et al.: Optimal control theory for personalized therapeutic regimens in oncology: background, history, challenges, and opportunities. J. Clin. Med. **9**(5), 1314 (2020)
2. Bruno, R., Bottino, D., De Alwis, D.P., Fojo, A.T., Guedj, J., Liu, C., et al.: Progress and opportunities to advance clinical cancer therapeutics using tumor dynamic models. Clin. Cancer Res. **26**(8), 1787–1795 (2020)
3. Mehdizadeh, R., Shariatpanahi, S.P., Goliaei, B., Peyvandi, S., Rüegg, C.: Dormant tumor cell vaccination: a mathematical model of immunological dormancy in triple-negative breast cancer. Cancers **13**(2), 245 (2021)
4. Hong, W.S., Wang, S.G., Zhang, G.Q.: Lung cancer radiotherapy: simulation and analysis based on a multicomponent mathematical model. Comput. Math. Methods Med. **2021**(1), 6640051 (2021)
5. Salim, S.S., Mureithi, E., Shaban, N., Malinzi, J.: Mathematical modelling of the dynamics of prostate cancer with a curative vaccine. Sci. Afr. **11** (2021)

6. de Pillis, L., Renee Fister, K., Gu, W., Collins, C., Daub, M., Gross, D., Moore, J., Preskill, B.: Mathematical model creation for cancer chemo-immunotherapy. Comput. Math. Methods Med. **10**(3), 165–184 (2009)
7. Itik, M., Banks, S.P.: Chaos in a three-dimensional cancer model. Int. J. Bifurcation Chaos **20**(1), 71–79 (2010)
8. León-Triana, O., Pérez-Martínez, A., Ramírez-Orellana, M., Pérez-García, V.M.: Dual-target CAR-Ts with on- and off-tumour activity may override immune suppression in solid cancers: a mathematical proof of concept. Cancer **13**(4), 703 (2021)
9. Huh, J.: Epidemiologic overview of malignant lymphoma. Korean J. Hematol. **47**(2), 92 (2012)
10. Valle, P.A., Coria, L.N., Plata, C.: Personalized immunotherapy treatment strategies for a dynamical system of chronic myelogenous leukemia. Cancers **13**(9), 2030 (2021)
11. Messmer, B.T., Messmer, D., Allen, S.L., Kolitz, J.E., Kudalkar, P., Cesar, D., et al.: In vivo measurements document the dynamic cellular kinetics of chronic lymphocytic leukemia B cells. J. Clin. Investig. **115**(3), 755–764 (2005)
12. De Pillis, L.G., Radunskaya, A.: A mathematical tumor model with immune resistance and drug therapy: an optimal control approach. Comput. Math. Methods Med. **3**(2), 79–100 (2001)
13. Rodrigues, D.S., Mancera, P.F., Carvalho, T., Gonçalves, L.F.: A mathematical model for chemoimmunotherapy of chronic lymphocytic leukemia. Appl. Math. Comput. **349**, 118–133 (2019)
14. Ibrahim, Z.B., Othman, K.I., Suleiman, M.: Implicit r-point block backward differentiation formula for solving first-order stiff odes. Appl. Math. Comput. **186**(1), 558–565 (2007)
15. Rasid, N.A., Ibrahim, Z.B., Majid, Z.A., Ismail, F., Ismail, A.: An efficient direct diagonal hybrid block method for stiff second order differential equations. In: Ismail, A., Dahalan, W.M., Öchsner, A. (eds.) Advanced Maritime Technologies and Applications. Advanced Structured Materials, vol. 166. Springer, Cham (2022)
16. Zainuddin, N., Soomro, H., Jamaludin, N., Yatim, S.A.M., Ibrahim, Z.B.: Numerical solution of oil spill weathering with two points block backward differentiation formula. In: Abdul Karim, S.A., Shafie, A. (eds.) Towards Intelligent Systems Modeling and Simulation. Studies in Systems, Decision and Control, vol. 383. Springer, Cham (2022)
17. Jana Aksah, S., Ibrahim, Z.B.: Singly diagonally implicit block backward differentiation formulas for HIV infection of CD4+T cells. Symmetry **11**(5), 625 (2019)
18. Mohd Ijam, H., Ibrahim, Z.B., Abdul Majid, Z., Senu, N.: Stability analysis of a diagonally implicit scheme of block backward differentiation formula for stiff pharmacokinetics models. Adv. Differ. Equ. **2020**, 1–22 (2020)
19. Noor, N.M., Yatim, S.A.M., Zawawi, I.S.M.: The design of the new fractional block method in the solution of the three-compartment pharmacokinetics model. Eng. Lett. **32**(2) (2024)
20. Qureshi, S., Ramos, H., Soomro, A., Akinfenwa, O.A., Akanbi, M.A.: Numerical integration of stiff problems using a new time-efficient hybrid block solver based on collocation and interpolation techniques. Math. Comput. Simul **220**, 237–252 (2024)
21. Kaur, A., Kanwar, V., Ramos, H.: A conservative algorithm based on a hybrid block method and tension B-spline differential quadrature method for Rosenau–KdV–RLW equation. Math. Methods Appl. Sci. **47**(11), 8638–8650 (2024)
22. Abdulganiy, R.I., Ramos, H., Osilagun, J.A., Okunuga, S.A., Qureshi, S.: A functionally-fitted block hybrid Falkner method for Kepler equations and related problems. Comput. Appl. Math. **42**(8), 327 (2023)
23. Nasarudin, A.A., Ibrahim, Z.B., Rosali, H.: On the integration of stiff odes using block backward differentiation formulas of order six. Symmetry **12**(6), 952 (2020)
24. Abd Rasid, N., Ibrahim, Z.B., Majid, Z.A., Ismail, F.: Formulation of a new implicit method for group implicit BBDF in solving related stiff ordinary differential equations. Math. Stat. **9**(2), 144–150 (2021)

Modelling Tuberculosis Transmission with Vaccination in Malaysia: An SVIR Approach

Nur 'Izzati Hamdan, Noor Fatihah Latif, Nurshamira Shaaril, and Noor Afiqah Noor Asmadi

Abstract Tuberculosis (TB) is a highly infectious disease caused by Mycobacterium tuberculosis (MTB) that primarily affects the lungs. It spreads through the air when infected individuals cough, sneeze, or spit. TB remains a significant global health problem, including in Malaysia, where it poses substantial challenges. This study formulates a deterministic model of TB incorporating vaccination to study the dynamics of TB. The model is based on the susceptible-vaccinated-infectious-recovered (SVIR) framework. The disease-free equilibrium (DFE) point is identified, and the basic reproduction number R_0 is computed. The stability result shows that the DFE is stable when R_0 is less than 1 ($R_0 < 1$), indicating that TB will not persist in the population long term. Additionally, we analyse the effectiveness of the BCG vaccine by examining two models: one with and one without vaccination. The model without vaccination shows an increasing number of infected individuals, while the vaccinated model yields different results. This comparison underlines the essential role of immunisation in TB infection control. By understanding the stability of the DFE and assessing the vaccine's effectiveness, we provide valuable insights for TB control efforts and inform public health authorities of strategies to combat the virus.

Keywords Tuberculosis · BCG · Stability · Vaccination

1 Introduction

Tuberculosis (TB) is a chronic infectious disease primarily impacting the lungs but can also affect other organs. TB spreads through contaminated air when patients cough [1]. Respiratory droplets containing bacteria can be transmitted from one person to another through inhalation. During sneezing, coughing, or talking, close contact with a TB patient can lead to transmission. Symptoms of TB include constant fatigue, fever, night sweats, chest pain during breathing or coughing, back pain if it

N. 'I. Hamdan (✉) · N. F. Latif · N. Shaaril · N. A. N. Asmadi
Faculty of Computer and Mathematical Sciences, Universiti Teknologi MARA, Shah Alam, Selangor, Malaysia
e-mail: izzatihamdan@uitm.edu.my

A. Shuib et al. (eds.), *Mathematical Modeling and Numerical Analysis*,
Springer Proceedings in Mathematics & Statistics 510,
https://doi.org/10.1007/978-981-96-9350-4_6

affects the spine, and blood in urine if it affects the kidneys. Skin tests, blood tests, chest X-rays, and sputum analyses can diagnose TB [2].

Discovered by Robert Koch in 1882 in Germany, mycobacterium bacillus bacteria cause TB [3]. Since 2002, Malaysia has been fighting tuberculosis with the directly observed short-term treatment approach (DOTS). DOTS is a World Health Organisation-recommended method to manage and control TB [1]. The National Strategic Plan for Tuberculosis Control (2016–2020) states a 17.2% increase in cases in 2015 with 24,220 reported cases. According to the Ministry of Health Malaysia, in 2016, the notification rate was 79 cases per 100,000 people in 2015, making Malaysia a country of concern. In 2023, Malaysia recorded 26,781 cases of TB, a 5.47% increase from the 25391 TB cases reported in 2022. Figure 1 shows the latest update of the total number of TB cases in Malaysia from 2018 to 2023. The decline in TB cases in 2020 and 2021 was due to the COVID-19 pandemic [4].

Tuberculosis (TB) is the second infectious killer caused by a single infectious agent, affecting many people worldwide, especially in Asia and Africa. TB was responsible for 1.6 million deaths in 2021, making it a significant global health issue [1]. The World Health Organisation (WHO) states that the most critical TB cases occurred in Southeast Asia, Africa, and the Western Pacific. In addition, the World Health Organisation reported 87% of these cases in 30 high-tuberculosis countries. There are two types of TB: active and latent [5]. People with weak immune systems are more likely to develop TB. Before the COVID-19 pandemic, TB was causing more deaths than HIV and AIDS combined. HIV-positive individuals with weakened immune systems were at risk of developing active TB. The deadly combination of HIV and TB led to the death of about 187,000 people by 2021. The percentage of

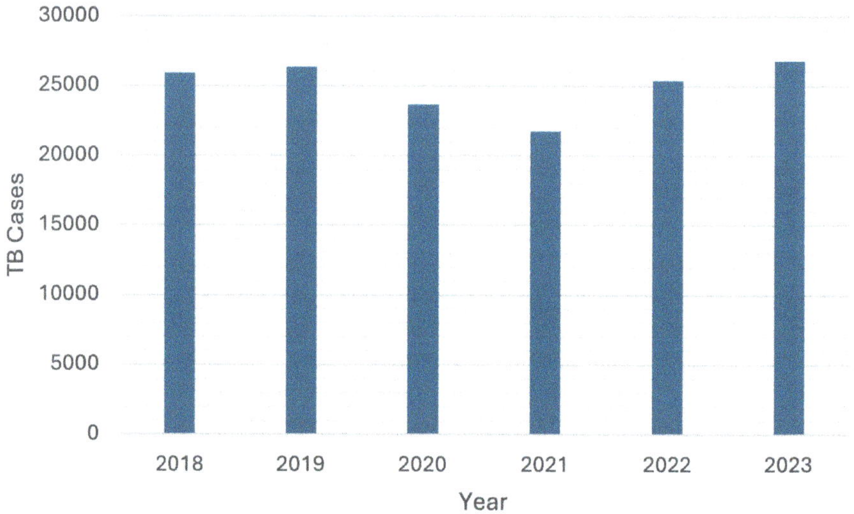

Fig. 1 Total number of cases and the notification rate of Tuberculosis (TB) in Malaysia from 2018 to 2023 [4]

notified TB patients who had an HIV test increased from 73% in 2020 to 76% in 2021 [6].

A mathematical model of infectious disease transmission is mainly constructed using theoretical and numerical methods to comprehend its pattern better. This model is classically known as the SIR model, and it is attributed to Kermack and McKendrick's fundamental research on epidemic theory in 1927 [7, 8]. Liu et al. developed an SVIR epidemic model with lethal diseases and a non-constant population. They considered the local stability analysis of an SVIR epidemic model with a deadly disease and a non-constant population based on the SVIR epidemic model [9]. The model was initially developed to utilise its ability to study and evaluate the appearance of disease outbreaks efficiently. This effort aimed to make it possible for the model to offer insightful information that would assist the implementation of adequate safety precautions.

Furthermore, since the twentieth century, interest has increased in investigating mathematical modelling of infectious diseases. The capacity for anticipating illnesses was intended to allow scientists to analyse inoculation or isolation strategies, which would substantially impact the death rate of a particular epidemic. The fundamental structure is commonly linked with disease transmission in a population. Following a probable first epidemic, the infected population declined until zero or achieved a steady endemic level. Many versions of the SIR model have been researched in recent years to more precisely reflect more complicated illnesses and infection pathways [10]. Leontine et al. in 2019 look at three variables in the model: susceptible (S), vaccinated (V), early latent (E), and late latent (L), which created the S-V-E-L-I model for TB transmission [11]. In this study, the vaccination did not confer total protection, although it lowered the chance of illness by a certain percentage. Widyaningsih et al. developed an SVIR model with relapse for TB transmission, measuring its accuracy using 2004–2014 data in Indonesia. The model showed a relative error of less than 0.1 for all variables, indicating high accuracy. It predicts a yearly decline in TB prevalence, incidence, and mortality in Indonesia, with average decreases of 1.90% for prevalence and mortality and 1.68% for incidence [12]. Tamhaji et al. in [13] conclude that immigration has a significant impact on the dynamics of TB transmission because it significantly affects the prevalence of the disease in the population despite the BCG-vaccine implementation.

Ucakan et al. developed three mathematical models, the SIR, SEIR, and BSEIR models, to represent the tuberculosis dynamics in Turkey [14]. The fundamental reproduction showed that any model with a basic reproduction number smaller than one would not lead to an endemic but could assist in characterising the evolution of TB and create a precautionary strategy. Following that, the SEIR model was developed with the inclusion of a vaccination and re-infected variables, yielding S-V-E-I-R-E. This research established how vaccination, exogenous reinfection, and transmission rate affect the model framework. The model bifurcated backwards, and the endemic equilibrium was more significant in the lowered scenario where the illness may persist in the population than in the higher case. To avoid the problem mentioned in this article, Sulayman [15] decreased the adequate reproduction number below and raised the vaccination rate over the threshold value for playing a substantial role in the model. His work aimed to estimate TB disease transmission patterns using the SIR model and the Runge-Kutta fourth-ordered approach. The report analysed the performance of the two typed and forecasted gearbox trends in Malaysia in 2021. The results demonstrate that integrating the SIR Model with demography enhances the prediction of disease spread. The study also discovered that more excellent transmission rates result in lowered incidence rates per 100,000 people, whereas higher incidence rates result in lowered recovery rates. The SIR demographic model could estimate disease transmission and analyse vaccine effectiveness. The immunisation strategy aimed to improve recovery rates and lower tuberculosis mortality rates. Quarantined sick individuals are recommended as a preventive measure, as BCG vaccination is ineffective in underdeveloped countries [16].

The classical SIR model has been widely implemented and modified for various infectious diseases such as measles [17], dengue [18], COVID-19 [19], and tuberculosis [14, 20] as an uncontrolled infectious disease which neither nation will ever get rid of. In the future, the best defence against tuberculosis is to bring the illness under control of infectious disease, globally. WHO recommends using the Bacillus Calmette-Guerin (BCG) vaccination as a preventive strategy against TB. Even though its efficacy is currently being discussed, the BCG vaccine remains one of the methods used to prevent tuberculosis. In all countries with a high risk of tuberculosis, the WHO recommends immunising infants with a single intradermal dose of BCG as soon as possible after birth. The characteristic features of TB disease in different age groups are not only seen in individuals who respond to the BCG vaccine. Children (1âŁ“15 years) are noted to have lower morbidity and mortality rates than adults (15âŁ“65 years). Hence, this study aims to propose a mathematical SVIR model for tuberculosis infection for surveillance that considers vaccine effectiveness, analyses the impact of the vaccine on tuberculosis transmission, and investigates the local stability analysis of equilibrium points.

2 Model Formulation

The proposed model in this study is motivated by the work of Widyaningsih et al. [12] and Tamhaji et al. [13]. However, in this study, we considered the Malaysia data cases and excluded the immigration parameter to focus on the vaccination measure. In the formulation of the model given by the schematic diagram in Fig. 2, the following assumptions are made:

- The population was heterogeneous, meaning that individuals could be grouped into different compartments or groups based on their epidemiological states: susceptible (S), vaccination (V), infectious (I), and recovered (R).
- The population size in each compartment varied with time (S(t), V(t), I(t), and R(t)), and the epidemic process was deterministic. It means changes in the population of a compartment can be calculated using historical data or data collection to develop the model.

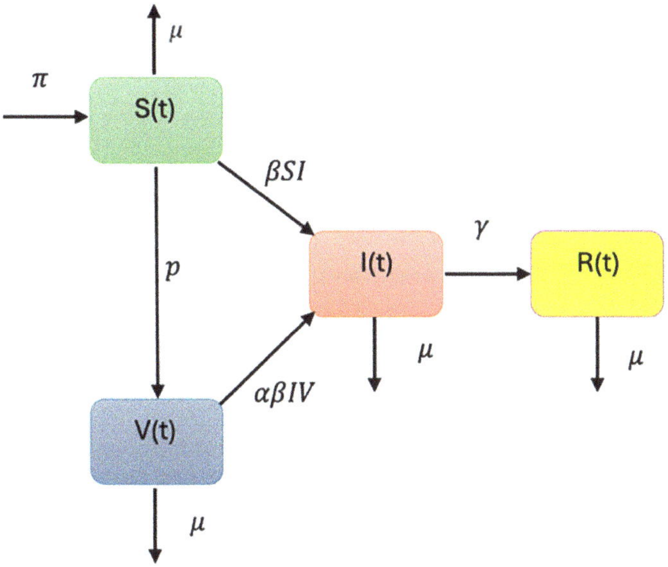

Fig. 2 Schematic diagram of SVIR model

- The BCG vaccine's efficacy is incomplete, so some vaccinated individuals will be infected by bacteria as indicated by the parameter α, $(0 < \alpha < 1)$.
- A proportion of the newborn population is immunised against TB infection through vaccination.
- People in each compartment had an equal natural death rate (μ).
- All newborns were previously unvaccinated and joined the susceptible compartment.
- There were no immigrants or emigrants. The only way to enter the population was through newborn babies, and the only way to exit was through death from natural causes.

First, a TB spread model was constructed by considering a vaccine compartment. The population was assumed to be closed and divided into compartments: susceptible (S), vaccinated (V), infected (I), and recovered (R). The SVIR epidemic model was used as a modification of the SIR model. This modification contributes to the imperfect vaccines, where vaccinated individuals can still get infected with a reduced transmission probability denoted by α where $(0 < \alpha < 1)$ [21]. It was assumed that the population remained constant throughout the analysis. The proposed model is as follows

Table 1 Description of variables of model (1)

Variable	Description
N	Total number of human population
S	Susceptible individuals
V	Vaccinated newborns
I	TB-infected individuals
R	Recovered individuals

Table 2 Description of parameters of model (1)

Parameter	Description
b	Birth rate
π	Recruitment rate
μ	Natural death rate
p	Fraction of the successfully vaccinated
α	The reduced probability of transmission due to immune protection from the imperfect vaccine
β	Transmission rate
γ	Recovery rate

$$
\begin{aligned}
\frac{\mathrm{d}S}{\mathrm{d}t} &= \pi + \mu SI - \beta SI - pS \\
\frac{\mathrm{d}V}{\mathrm{d}t} &= pS - \mu V - \alpha\beta IV \\
\frac{\mathrm{d}I}{\mathrm{d}t} &= \alpha\beta IV + \beta SI - \mu I - \gamma I \\
\frac{\mathrm{d}R}{\mathrm{d}t} &= \gamma I - \mu R.
\end{aligned}
\tag{1}
$$

with subject to $S(0) \geq 0$, $V(0) \geq 0$, $I(0) \geq 0$, $R(0) \geq 0$, and the parameters used in the model (1) were assumed to be constant and non-negative. The total population, $N = S(t) + V(t) + I(t) + R(t)$ where the birth and natural death rates are not assumed to be equal. Tables 1 and 2 describe the variables and parameters used in the model (1).

3 Model Analysis

3.1 Disease-Free Equilibrium Point (DFE)

The disease-free equilibrium (DFE) represents a state where the population has no infection. In this equilibrium, all infected classes will be zero, and only susceptible individuals will exist in the population. The disease-free equilibrium point of the system (1) is determined by setting the derivatives to zero,

$$\frac{dS}{dt} = \frac{dV}{dt} = \frac{dI}{dt} = \frac{dR}{dt} = 0. \tag{2}$$

Hence, the DFE point is given by:

$$x_0 = (S^*, V^*, I^*, R^*) = \left(\frac{\pi}{\mu + p}, \frac{\pi p}{\mu(\mu + p)}, 0, 0\right). \tag{3}$$

3.2 *Basic Reproduction Number*

In this study, the basic reproduction number, R_0 is computed using the next-generation method introduced in [22]. The computed R_0 for the system (1) is as follows:

$$R_0 = \frac{\pi p\alpha\beta + \pi\mu\beta}{\mu(\mu + p)(\mu + \gamma)}. \tag{4}$$

3.3 *Stability Analysis of the Disease-Free Equilibrium Point*

The following result is established for the local stability of the disease-free equilibrium point of system (1).

Theorem 1 *The disease-free equilibrium x_0 is locally asymptotically stable if $R_0 <$ 1 and is unstable if $R_0 > 1$.*

Proof The Jacobian matrix of system (1) at the DFE in (3), x_0 is presented by

$$J(x_0) = \begin{pmatrix} -\mu - p & 0 & -\beta\left(\frac{\pi}{\mu+p}\right) & 0 \\ p & -\mu & -\alpha\beta\left(\frac{\pi p}{\mu(\mu+p)}\right) & 0 \\ 0 & 0 & \alpha\beta\left(\frac{\pi p}{\mu(\mu+p)}\right) + \beta\left(\frac{\pi}{\mu+p}\right) - \mu - \gamma & 0 \\ 0 & 0 & \gamma & -\mu \end{pmatrix}. \tag{5}$$

The characteristic equation is obtained from the determinant of the Jacobian with the eigenvalues λ, $|J(x_0) - \lambda I| = 0$. Therefore, we have the following equation

$$-(\mu + p + \lambda)\left((\mu + \lambda)^2\left(\frac{\alpha\beta\pi p + \beta\pi\mu}{\mu(\mu + p)} - (\mu + \gamma + \lambda)\right)\left(\frac{1}{\mu + \gamma}\right)\right) = 0. \tag{6}$$

From (6), by performing basic algebraic manipulations to isolate variables by setting each factor to zero, we have the following relation:

$$-(\mu + p + \lambda) = 0 \quad (7)$$

$$(\mu + \lambda)^2 = 0 \quad (8)$$

$$R_0 - \frac{(\mu + \gamma + \lambda)}{\mu + \gamma} = 0. \quad (9)$$

The eigenvalues for the equation are $\lambda_1 = -(\mu + p)$, $\lambda_2, \lambda_3 = -\mu$, $\lambda_4 = -(\mu + \gamma)(1 - R_0)$. From λ_4, in regards to the Routh-Hurwitz criterion, $\lambda_4 < 0$, hence, the DFE point, x_0, is locally asymptotically stable if $R_0 < 1$ and unstable if $R_0 > 1$.

4 Results and Discussion

This section presented the results of the SVIR and SIR models, representing the dynamics of susceptible (S), vaccinated (V), infected (I) and recovered (R) individuals in the total population N. In the numerical analysis of the TB model (1), the statistics for each province in 2020 were derived from the Department of Statistics Malaysia's surveillance system [23]. The total population in 2020 was 32,447,385, with an initial number of BCG-vaccinated individuals $V(0)$ reported as 461,396 and an initial number of infected individuals $I(0)$ as 23,644, both considered in yearly intervals. Additionally, $R(0)$ represents the number of successfully treated tuberculosis patients in Malaysia in 2020. We assume that TB disease will not disappear even if the individual has BCG vaccination. Therefore, we assume that $R(0) = 0$. The SVIR formula for initial vulnerable populations is taken into consideration by $S(0) = N - V(0) - I(0) - R(0)$.

Here, we use the scaling method as it makes the graph easy to interpret, where the total population is scaled to 1, which means that y-axis values show the relative proportions of each compartment [24]. By dividing the initial values of S, V, I, and R by N, the sum of the scaled initial values always equals 1, representing the total population—the title y-axis "Proportion of Population" to convey this context. The graph illustrates the changes in relative population sizes over time, making it more straightforward for viewers to understand the dynamics without getting overwhelmed by the large numbers.

Figure 3 presents the SVIR graph of model (1). Figure 3b is the close-up view of the infected curve in Fig. 3a, as it cannot be seen clearly in Fig. 3a. The y-axis represents the population proportion in each compartment, and "In" denotes the infected compartment due to technical constraints in MAPLE. The x-axis represents time in yearly intervals.

Figure 4 for the susceptible curve shows a rapid decrease in individuals susceptible to TB within populations, indicating a significant reduction. The susceptible

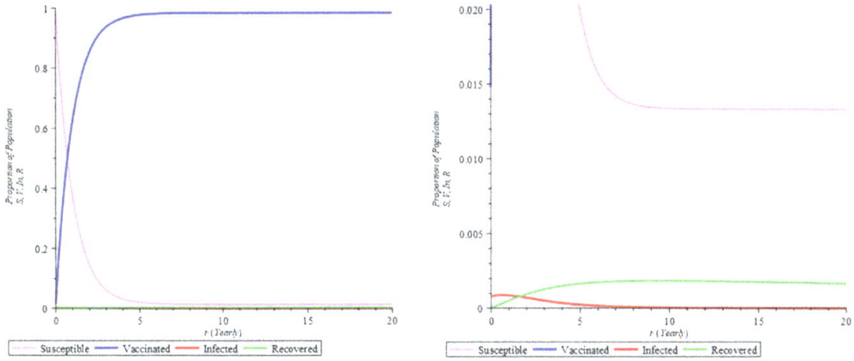

(a) The solution curves (b) The close-up view of the infected curve

Fig. 3 Numerical simulation of SVIR model (1) with BCG vaccination

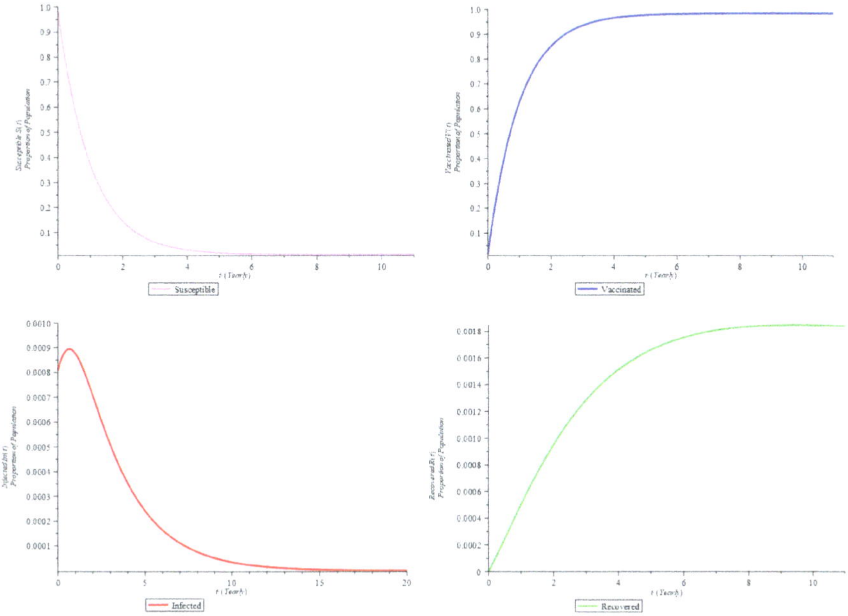

Fig. 4 Solution curve for every component of model (1) with BCG vaccination

graph also shows a decline from the first to the fifth year. Various factors, including vaccination impact, can change this. This vaccination is also included in the SVIR formula model (1) for the population we use. The rise in the y-axis implies that more individuals are becoming susceptible to TB disease, and the x-axis implies time. Vaccination lessens a person's vulnerability to the disease, making them less likely to become infected. As a result, the number of vulnerable people reduces with time.

Among the historical evidence is that BCG has been used as a TB vaccine for several decades and has demonstrated efficacy in preventing severe forms of TB. Malaysia is one of the countries that uses the BCG vaccination. It can also be claimed that there are fewer and fewer people who are vulnerable to tuberculosis, but the disease has not reached an empty or non-existent level. After vaccination, it is still present at a lesser incidence.

In contrast to the BCG-vaccinated compartment in Fig. 4, the vaccine graph is initialised with a zero value on the x and y-axis, indicating a year-to-year growth. It demonstrates Malaysia's growing population's acceptance of the vaccine. However, vaccination will reach saturation at some point. This is visible when the vaccination graph reaches a plateau and remains flat. This can happen when most Malaysians who want to get vaccinated have already received the vaccine, and the remaining susceptible population is minimal. We can also say that the link between the vaccinated and susceptible populations demonstrates the continual interplay between vaccination efforts and the vulnerable population. Moreover, the infected trajectory in Fig. 4 depicts a declining infected population. We can see a very significant decrease in the infected graph in a very short time. This decrease can also occur quickly, from the first to the fifth year. This can be attributed to the change in the graph due to efforts such as efficient treatment and successful vaccination to limit the spread of TB in Malaysia and the current number of cases. Also, the graph illustrates that as time passes, the curve begins to approach a stable state, which is the disease-free equilibrium point, after ten years. This behaviour verified the local stability of the disease-free equilibrium, which satisfied Theorem 1. When R_0 is less than one, the graph reaches a stable level, which indicates that this TB disease will likely die out or that the outbreak can be effectively controlled in Malaysia within the next 20 years. In the recovered graph in Fig. 4, the population begins at zero in the first year of disease ecology. As time goes on and more people recover from the sickness of TB, the graph will show an increasing trend or curve. We can see that after a specific amount of time, individuals in the SVIR model (1) transition from the infected (I) compartment to the recovered (R) compartment. As more infected people recover, resulting in a rising recovered graph. This may occur due to effective vaccination, which enhances the development of immunity in everyone who has received the vaccine. The increasing recovery graph illustrates that we can assume some of the population has obtained immunity to the disease due to past infections. When more people recover, they risk becoming infected with tuberculosis again if they do not take care of their health.

The TB transmission model without vaccination sheds light on disease propagation dynamics and the influence of treatments such as vaccination or social distancing, is represented in Fig. 5. The individual graph of every compartment of the model without vaccination is given in Fig. 6. Figure 6a represents a susceptible graph for the SIR model. These individuals have never been affected and may become infected if they meet contagious individuals. Because there is no vaccine, there is an increase in this vulnerable graph towards the 28th year. As a result, the number of persons susceptible to tuberculosis will rise. Figure 6b shows an epidemic period of 40 years, and there is a decrease in infected in a relatively short period before the infected graph

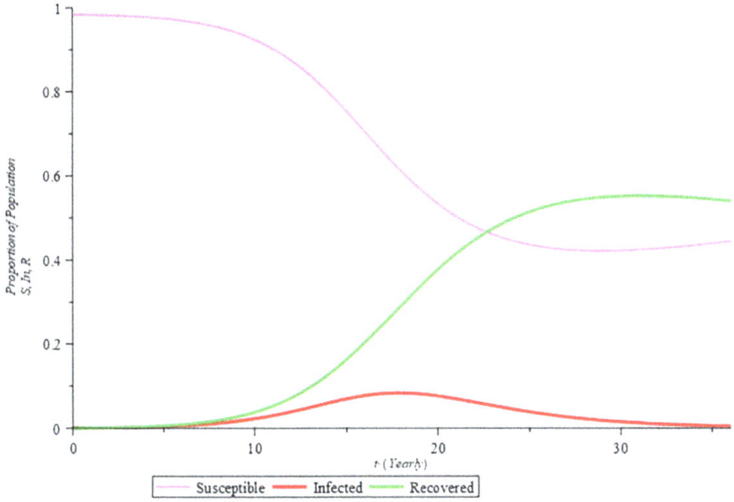

Fig. 5 Model without BCG vaccination; $p = 0$ and variable, V is not counted

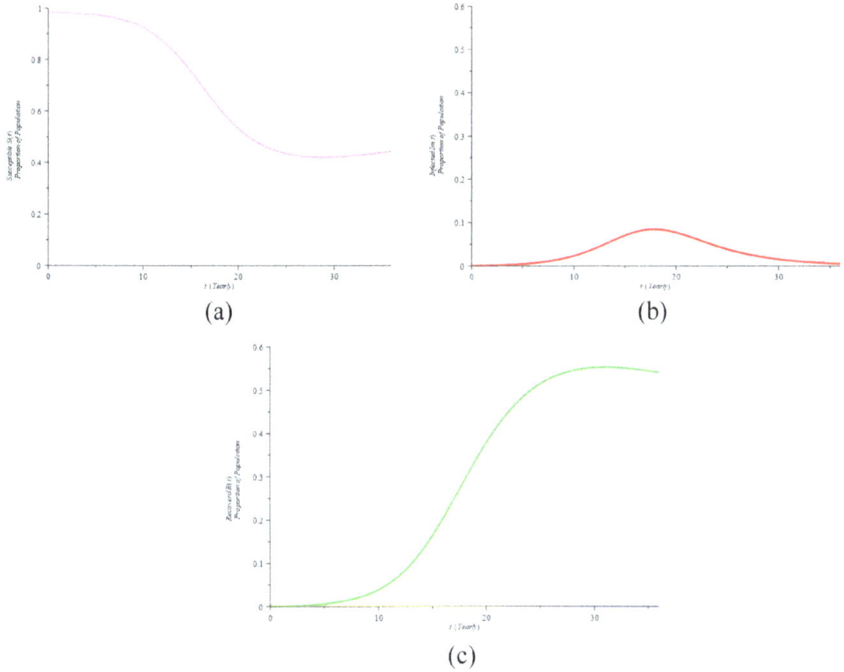

Fig. 6 Solution curve for every component of model (1) without BCG vaccination

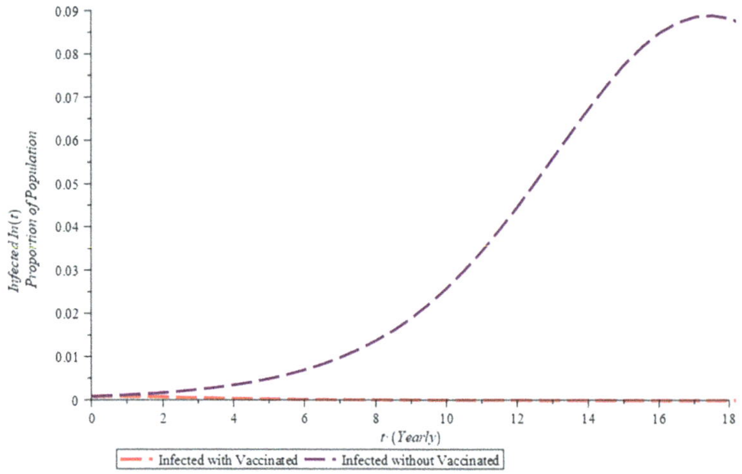

Fig. 7 Infected curves for with vaccination and without vaccination of model (1)

rises again. This tells us that TB disease is endemic in Malaysia without a vaccine; it will not go away and will always be there for a long time when there is no vaccine. Moreover, Fig. 6c depicts the number of people who recovered from tuberculosis in Malaysia in 2020, based on data from the Department of Statistics Malaysia, DOSM. These people have recovered from infections and developed immunity. Individuals who have recovered are no longer contagious. In this graph, the fall occurs after 50 years, and the number of people who recover decreases because we can assume they lack a strong immune system due to not being vaccinated.

Figure 7 presents a comparison of infected graphs from Figs. 4 and 5. The use of vaccination in one model but not in the other highlights significant differences. In the model without vaccination, the infected curve initially rises as more individuals become infected, reaches a peak, and then declines as people recover over a span of 15 years. In contrast, incorporating vaccination in the SVIR model alters the infection curve. As shown in Fig. 7, due to the protective effects of BCG vaccination, the infected curve has a slower rise, a lower peak, and a quicker decline.

5 Conclusion

This study investigated the influence of BCG vaccination on the transmission of tuberculosis (TB) caused by Mycobacterium tuberculosis in Malaysia, employing an SVIR model to account for the dynamics of vaccination within the population. Our findings reveal that including vaccine efficacy in the model highlights its critical role in TB infection surveillance and management. The results emphasise the significance of robust vaccination programmes in controlling TB transmission. This agrees with the report by the Disease Control Division [4] that BCG vaccination remains important for preventing severe forms of TB in children despite no fully effective vaccine being available for the prevention of TB at the moment. Future research should focus on continuously monitoring the effectiveness of these vaccination campaigns and exploring additional strategies to enhance public awareness and compliance. With concerted efforts in vaccination and public health education, Malaysia can make substantial progress in mitigating the impact of TB.

The main limitation of this study is that the parameters used were insufficiently precise to represent accurately the transmission dynamics of TB in Malaysia. Future research could focus on refining these parameters, particularly significant factors such as vaccination, through statistical methods or parameter estimation modelling approaches. Additionally, the model could be enhanced by incorporating parameters such as age, sex, isolation measures, and treatment strategies to improve its predictive accuracy ([25]).

Acknowledgements The author is very grateful for the financial support from the College of Computing, Informatics and Mathematics, Universiti Teknologi MARA Shah Alam, Selangor.

References

1. World Health Organization: Tuberculosis. https://www.who.int/news-room/fact-sheets/detail/tuberculosis (2023)
2. James, R.: Pulmonary tuberculosis: Types, symptoms, and treatments. Healthline. https://www.healthline.com/health/pulmonary-tuberculosis (2018)
3. Department of Health and Human Services, Historical Perspectives Centennial: Koch's Discovery of the Tubercle Bacillus. https://www.cdc.gov/mmwr/preview/mmwrhtml/00000222.htm (2019)
4. Division, D.C.: National Strategic Plan to End TB (2021–2030). Ministry of Health Malaysia (2021)
5. Mayo Clinic: Tuberculosis—Symptoms and Causes. https://www.mayoclinic.org/diseases-conditions/tuberculosis/symptoms-causes/syc-20351250 (2021)
6. World Health Organization: Tb Mortality. https://www.who.int/teams/global-tuberculosis-programme/tb-reports/global-tuberculosis-report-2021/disease-burden/mortality (2021)
7. Kermack, W.O., McKendrick, A.G.: A contribution to the mathematical theory of epidemics. Proc. R. Soc. A Math. Phys. Eng. Sci. **115**(772), 700–721 (1927)
8. Pfab, F., Nisbet, R.M., Briggs, C.J.: A time-since-infection model for populations with two pathogens. Theor. Popul. Biol. **144**, 1–21 (2022)
9. Liu, X., Takeuchi, Y., Iwami, S.: SVIR epidemic models with vaccination strategies. J. Theor. Biol. **253**(1), 1–11 (2008)
10. Chauhan, S., Misra, O., Dhar, J.: Stability analysis of sir model with vaccination. Am. J. Comput. Appl. Math. **4**(1), 17–23 (2014)
11. Leontine, N., Thomas, T.M., Franklin, A., Martin, L.M.: Mathematical model to assess vaccination and effective contact rate impact in the spread of tuberculosis. J. Biol. Dyn. **13**(1), 26–42 (2019)
12. Widyaningsih, P., Ngroho, A., Saputro, D., Sutanto: Tuberculosis transmission with relapse in Indonesia: susceptible vaccinated infected recovered model. J. Phys. Conf. Ser. (2019)
13. Tamhaji, N., Hamdan, N.: The dynamics of tuberculosis through BSEIR model with immigration in Malaysia. Malays. J. Fundam. Appl. Sci. **19**, 1176–1189 (2023)

14. Ucakan, Y., Gulen, S., Koklu, K.: Analysing of tuberculosis in turkey through SIR, SEIR and BSEIR mathematical models. Math. Comput. Model. Dyn. Syst. **27**(1), 179–202 (2021)
15. Sulayman, F., Abdullah, F.A.: Analysis of a tuberculosis infection model considering the influence of saturated recovery (treatment). Complexity (2021)
16. Azizan, F.L., Sathasivam, S., Khan, M., Ali, M., Hii, C., Yion, K.: Study of transmission of tuberculosis by sir model using runge-kutta method. J. Qual. Meas. Anal. **18**(2) (2022)
17. Witbooi, P.J.: An SEIR model with infected immigrants and recovered emigrants. Adv. Differ. Equ. **1** (2021)
18. Chanprasopchai, P., Tang, I.M., Pongsumpun, P.: Sir model for dengue disease with effect of dengue vaccination. Comput. Math. Methods Med. (2018)
19. Tong, Z., Lv, Y., Din, R.U., Mahariq, I., Rahmat, G.: Global transmission dynamic of sir model in the time of sars-cov-2. Results Phys. **25** (2021)
20. Nasution, H., Sitompul, P., Sinaga, L.P.: Effect of the vaccine on the dynamics of spread of tuberculosis sir models. J. Phys. Conf. Ser. **1819**(1) (2021)
21. Li, M.Y.: An introduction to mathematical modeling of infectious diseases. In: An Introduction to Mathematical Modeling of Infectious Diseases. Springer International Publishing (2018)
22. Van den Driessche, P., Watmough, J.: Reproduction numbers and sub-threshold endemic equilibria for compartmental models of disease transmission. Math. Biosci. **180**(1–2), 29–48 (2002)
23. Department of Statistics Malaysia: Vital Statistics. https://www.dosm.gov.my/portal-main/release-content/vital-statistics-malaysia-2021 (2021)
24. Mata, A.S., Dourado, S.M.P.: Mathematical modeling applied to epidemics: an overview. Sao Paulo J. Math. Sci. **15**, 1025–1044 (2021)
25. Ministry of Health Malaysia, Petunjuk Kesihatan 2021. https://www.moh.gov.my/moh/resources/Penerbitan/Penerbitan (2021)

A Computational Approach to Solving Second-Order Nonlinear Systems of Differential Equations by Using Piecewise Constant Argument Methods

Mukhiddin I. Muminov and Navruz M. Usmonov

Abstract This paper represents an efficient approach for solving approximated solution of the second-order nonlinear systems of differential equations using piecewise constant argument methods. A system of differential equations with piecewise constant arguments that depends on a positive integer n is introduced. It is proved that this system of equations has a unique piecewise solution, which is an approximate solution to the considered initial value problem for large n. Numerical results are presented on examples that demonstrate the effectiveness and high accuracy of the proposed method.

Keywords Approximate solution · Piecewise constant argument · Absolute error · Residual absolute error · System of nonlinear second-order differential equations

1 Introduction

In fact that second-order differential equations arise from many physical phenomena in a wide range of applications particularly in science and engineering. Such type differential equations are used to model problems in astrophysics, control theory, mechanics, circuit theory, biology, among others [1]. Over the years, several authors have proposed different interpolation and collocation methods for the direct solution of problems of the form (1) to (2). These include methods with constant coefficients [2], trigonometrically-fitted methods [3], variable step-size methods [4], exponentially fitted methods [5], and adapted block hybrid method [6, 7]. Allogmany and

M. I. Muminov (✉)
Faculty of Mathematics, Samarkand State University, Samarkand, Uzbekistan
e-mail: navruzusmonov417@gmail.com
URL: http://www.springer.com/gp/computer-science/lncs

V. I. Romanovskiy Institute of Mathematics, Uzbekistan Academy of Sciences, Tashkent, Uzbekistan

N. M. Usmonov
Department of Mathematics, Faculty of Information Technologies and Physics-Mathematics, Gulistan State University, Gulistan City, Uzbekistan

A. Shuib et al. (eds.), *Mathematical Modeling and Numerical Analysis*, Springer Proceedings in Mathematics & Statistics 510,
https://doi.org/10.1007/978-981-96-9350-4_7

Ismail [8] developed an implicit block method of order eight for the approximation of second-order differential equations of the form (1). The authors adopted the Hermite polynomial as basis function in constructing the method. The basic properties of the method were also analyzed and the method was implemented on some linear and nonlinear second order differential equations. In the mathematical literature, numerous direct numerical integrators are available to solve Problems (1)–(2) [9, 10], linear multi-step (LMS) and Runge-Kutta-Nystrom (RKN) are common methods that have been used significantly for this purpose. It is known that LMS methods are more effective compared to RKN methods for solving the IVP (1)–(2) because the latter requires more function evaluations to achieve the same order. However, some disadvantages of LMS methods include: (a) they suffer from instability [8, 11] and (b) majority of LMS methods use predictor-corrector mode in their implementation. An initial condition can be found inside the problem of a system of second-order differential equations in various fields of engineering and applied sciences, such as the many-body problem in celestial mechanics, thermal combustion of gases, Lennard-Jones potential in molecular dynamics, etc. [12]. Using a piecewise constant argument method the stability and bifurcation analysis of a discrete Leslie predator-prey system was investigated in [13].

This paper presents a new method for finding an approximated solution of the following system of differential equations:

$$\begin{aligned} x''(t) &= f_1(t, x(t), y(t), x'(t), y'(t)), \qquad t \in [0, 1] \\ y''(t) &= f_2(t, x(t), y(t), x'(t), y'(t)). \end{aligned} \tag{1}$$

with initial conditions

$$x(0) = \alpha, \quad x'(0) = \beta, \quad y(0) = \lambda, \quad y'(0) = \mu. \tag{2}$$

where α, β, λ, μ are real constants and $f_i(t, x(t), y(t), x'(t), y'(t)), \quad i = 1, 2$ is a second-order polynomial of x, y, x', y' of the form

$$f_i(t, x, y, x', y') = \sum_{0 \le p+m+l+r \le 2} x^p y^l x'^m y'^r A^{(i)}_{\mathrm{pmlr}}(t), \; i = 1, 2, \tag{3}$$

here the coefficients $A^{(i)}_{\mathrm{pmlr}}(t)$ are continuous real-valued functions. To construct approximated solution for the problem we first introduce the second-order nonlinear systems of differential equations with piecewise constant arguments that depends on a positive integer n. We prove that the constructed system of differential equations has a unique solution. Further, we show this solution is an approximate solution to the considered initial value problem for large n. We give numerical results in several examples that demonstrate the effectiveness and high accuracy of the proposed method.

2 A System of Differential Equations with Piecewise Constant Arguments

Let n be a positive integer number and $t_k = \frac{k}{n}, \quad k = 0, 1, 2, \ldots, n.$

For $t \in [t_{k-1}, t_k), \quad k = 1, 2, \ldots, n$ consider the following initial value problem with piecewise constant arguments of the form

$$\begin{aligned} x''(t) &= f_1(t, x(t_{k-1}), y(t_{k-1}), x'(t_{k-1}), y'(t_{k-1})), \\ y''(t) &= f_2(t, x(t_{k-1}), y(t_{k-1}), x'(t_{k-1}), y'(t_{k-1})), \\ x(0) &= \alpha, \quad x'(0) = \beta, \quad y(0) = \lambda, \quad y'(0) = \mu. \end{aligned} \tag{4}$$

A solutions of is defined as follows [14, 15].

Definition 1 *A vector function* $(x(t), y(t)) := (x_n(t), y_n(t))$ *is called a solution of the initial value problem (4) if the following conditions are satisfied:*

(I) $x(t), x'(t)$ *and* $y(t), y'(t)$ *are continuous on* $[0, 1]$;

(II) $x''(t)$ *and* $y''(t)$ *exist and continuous in* $[0, 1]$ *with possible exception at points* $t_k, \quad k = 0, 1, \ldots, n$ *where one-sided derivatives exist;*

(III) $x(t)$ *and* $y(t)$ *satisfy initial value problem (4) in (0, 1), with the possible exception at the points* $t_k, \quad k = 0, 1, \ldots, n.$

The following theorem claims uniqueness solution for the initial value problem (4) with piecewise constant arguments.

Theorem 1 *For any positive integer number n the initial value problem* (4) *has a unique solution* $(x_n(t), y_n(t))$ *defined by* (18).

Further we suppose that there exists a positive number C such that for any n and $k \in \{1, \ldots, n\}$ the inequalities

$$|x(t_k)| < C, \ |x'(t_k)| < C, \ |y(t_k)| < C, \ |y'(t_k)| < C \tag{5}$$

for all n and $k = 0, 1, \ldots, n$ hold.

We remark that for any continiuous function φ on $[0, 1]$, the sequence $\varphi_n(t) = \varphi(t_k), \ t \in [t_{k-1}, t_k), \quad k = 1, 2, \ldots, \ n$ with $\varphi_n(1) = \varphi(1)$ approaches uniformly to $\varphi(t)$ on $[0, 1]$ (See, for example the book [16]) . This statement shows that if $(x(t), y(t))$ is solution of the problem (1)–(2) then it will be approximated solution for the initial value problem (4). Therefore, we can prove the following theorem that claims x_n and y_n can be approximated solution of the initial value problem (1)–(2).

Theorem 2 *For any* $\varepsilon > 0$ *there exists a positive number* $n_0 = n_0(\varepsilon)$ *such that for any positive integer number n with* $n > n_0$ *the inequalities*

$$\begin{aligned} \sup_{t \in [0,1]} |x_n''(t) - f_1(t, x_n(t), y_n(t), x_n'(t), y_n'(t))| < \varepsilon \\ \sup_{t \in [0,1]} |y_n''(t) - f_2(t, x_n(t), y_n(t), x_n'(t), y_n'(t))| < \varepsilon \end{aligned} \tag{6}$$

hold, where (x_n, y_n) *is solution of the initial value problem* (4).

To numerical solving of the problem we introduce for $t \in [t_{k-1}, t_k)$, $k = 1, 2, \ldots, n$ the following system of differential equation with piecewise constant arguments of the form

$$\begin{aligned} &x''(t) = f_1(t_{k-1}, x(t_{k-1}), y(t_{k-1}), x'(t_{k-1}), y'(t_{k-1})), \\ &y''(t) = f_2(t_{k-1}, x(t_{k-1}), y(t_{k-1}), x'(t_{k-1}), y'(t_{k-1})), \\ &x(0) = \alpha, \quad x'(0) = \beta, \quad y(0) = \lambda, \quad y'(0) = \mu. \end{aligned} \tag{7}$$

By Theorem 1 follows that the system of Eq. (7) has a unique solution (X_n, Y_n). The proof of the following theorem is similar to the proof of Theorem 2.

Theorem 3 *For any* $\varepsilon > 0$ *there exists a positive number* $n_0 = n_0(\varepsilon)$ *such that for any positive integer number* n *with* $n > n_0$ *the inequalities*

$$\begin{aligned} &\sup_{t\in[0,1]} |X_n''(t) - f_1(t, X_n(t), Y_n(t), X_n'(t), Y_n'(t))| < \varepsilon \\ &\sup_{t\in[0,1]} |Y_n''(t) - f_2(t, X_n(t), Y_n(t), X_n'(t), Y_n'(t))| < \varepsilon \end{aligned} \tag{8}$$

hold, where (X_n, Y_n) is solution of the initial value problem (4).

3 Proof of the Theorem 1

Proof of the Theorem 1. Let n be a positive integer number. For $t \in [0, t_1)$ Eq. (4) is

$$\begin{aligned} &x''(t) = f_1(t, x(0), y(0), x'(0), y'(0)), \\ &y''(t) = f_2(t, x(0), y(0), x'(0), y'(0)), \\ &x(0) = \alpha, \quad x'(0) = \beta, \quad y(0) = \lambda, \quad y'(0) = \mu. \end{aligned} \tag{9}$$

To solve this system of differential equations, integrating both sides of these equations twice we obtain

$$\begin{aligned} &x'(t) = \int_0^t f_1(s, x(0), y(0), x'(0), y'(0))ds + x'(0), \\ &y'(t) = \int_0^t f_2(s, x(0), y(0), x'(0), y'(0))ds + y'(0) \end{aligned} \tag{10}$$

and

$$\begin{aligned} &x(t) = \int_0^t \int_0^w f_1(s, x(0), y(0), x'(0), y'(0))dsdw + x'(0)t + x(0), \\ &y(t) = \int_0^t \int_0^w f_2(s, x(0), y(0), x'(0), y'(0))dsdw + y'(0)t + y(0). \end{aligned} \tag{11}$$

Since the functions $f_i(t, x(t), y(t), x'(t), y'(t))$, $i = 1, 2$ are continuous, there exist the following limits

$$\begin{aligned} x(t_1) &= \lim_{t\to t_1-0} x(t),\ x'(t_1) = \lim_{t\to t_1-0} x'(t), \\ y(t_1) &= \lim_{t\to t_1-0} y(t),\ y'(t_1) = \lim_{t\to t_1-0} y'(t). \end{aligned} \tag{12}$$

Now consider Eq. (4) for $t \in [t_1, t_2)$ with the initial conditions (12). Similarly, we have

$$\begin{aligned} x'(t) &= \int_{t_1}^{t} f_1(t, x(t_1), y(t_1), x'(t_1), y'(t_1))\mathrm{d}s + x'(t_1), \\ y'(t) &= \int_{t_1}^{t} f_2(s, x(t_1), y(t_1), x'(t_1), y'(t_1))\mathrm{d}s + y'(t_1) \end{aligned} \tag{13}$$

and

$$\begin{aligned} x(t) &= \int_{t_1}^{t}\int_{t_1}^{w} f_1(t, x(t_1), y(t_1), x'(t_1), y'(t_1))\mathrm{d}s\mathrm{d}w + x'(t_1)(t - t_1) + x(t_1), \\ y(t) &= \int_{t_1}^{t}\int_{t_1}^{w} f_2(s, x(t_1), y(t_1), x'(t_1), y'(t_1))\mathrm{d}s\mathrm{d}w + y'(t_1)(t - t_1) + y(t_1), \end{aligned} \tag{14}$$

where the following limits exist

$$\begin{aligned} x(t_2) &= \lim_{t\to t_2-0} x(t),\ x'(t_2) = \lim_{t\to t_2-0} x'(t), \\ y(t_2) &= \lim_{t\to t_2-0} y(t),\ y'(t_2) = \lim_{t\to t_2-0} y'(t). \end{aligned} \tag{15}$$

Let $t \in [t_{k-1}, t_k)$, $k = 2, \ldots, n$ and

$$\begin{aligned} x(t_{k-1}) &= \lim_{t\to t_{k-1}-0} x(t),\ x'(t_{k-1}) = \lim_{t\to t_{k-1}-0} x'(t), \\ y(t_{k-1}) &= \lim_{t\to t_{k-1}-0} y(t),\ y'(t_{k-1}) = \lim_{t\to t_{k-1}-0} y'(t). \end{aligned} \tag{16}$$

Then to equation (4) for $[t_{k-1}, t_k)$ with the initial conditions (16) we get

$$\begin{aligned} x'(t) &= \int_{t_k}^{t} f_1(s, x(t_k), y(t_k), x'(t_k), y'(t_k))\mathrm{d}s + x'(t_k), \\ y'(t) &= \int_{t_k}^{t} f_2(s, x(t_k), y(t_k), x'(t_k), y'(t_k))\mathrm{d}s + y'(t_k) \end{aligned} \tag{17}$$

and

$$\begin{aligned} x(t) &= \int_{t_k}^{t}\int_{t_k}^{w} f_1(s, x(t_k), y(t_k), x'(t_k), y'(t_k))\mathrm{d}s\mathrm{d}w + x'(t_k)(t-t_k) + x(t_k), \\ y(t) &= \int_{t_k}^{t}\int_{t_k}^{w} f_2(s, x(t_k), y(t_k), x'(t_k), y'(t_k))\mathrm{d}s\mathrm{d}w + y'(t_k)(t-t_k) + y(t_k). \end{aligned} \tag{18}$$

One can see that by construction the functions $x(t)$ and $y(t)$ defined by (17) are differentiable in $(0, 1)$ and continuous on $[0, 1]$, and $x''(t)$ and $y''(t)$ are differentiable in $(0, t_1)\bigcup(t_1, t_2)\bigcup\cdots\bigcup(t_{n-1}, 1)$ and one-sided limits at points $t_k, \quad k = 0, 1, \ldots, n$ exist. Clear that the vector function $(x(t), y(t))$ is unique solution of Eq. (4).

Proof of the Theorem 2. Since $(x_n(t), y_n(t))$ is solution of (4), for $t \in [t_k, t_{k+1})$ we have

$$x_n''(t) - f_1(t, x_n(t), y_n(t), x_n'(t), y_n'(t))$$

$$= f_1(t_k, x(t_k), y(t_k), x'(t_k), y'(t_k)) - f_1(t_k, x(t_k), y(t_k), x'(t_k), y'(t_k)),$$

$$y_n''(t) - f_2(t, x_n(t), y_n(t), x_n'(t), y_n'(t))$$

$$= f_2(t_k, x(t_k), y(t_k), x'(t_k), y'(t_k)) - f_2(t_k, x(t_k), y(t_k), x'(t_k), y'(t_k)).$$

Let

$$I(t) = f_i(t, x_n(t), y_n(t), x_n'(t), y_n'(t)) - f_i(t_k, x(t_k), y(t_k), x'(t_k), y'(t_k)),$$

then by (3)

$$I(t) = \sum_{0\le p+m+l+r\le 2} A^{(i)}_{\mathrm{pmlr}}(t)\Big(x^p(t)y^l(t)x'^m(t)y'^r(t) - x^p(t_k)y^l(t_k)x'^m(t_k)y'^r(t_k)\Big).$$

The function $I(t)$ can be evaluated as

$$\begin{aligned} |I(t)| \le A\Big(&|x^2(t) - x^2(t_k)| + |x'^2(t) - x'^2(t_k)| + |y^2(t) - y^2(t_k)| + |y'^2(t) - y'^2(t_k)|) \\ &+ |x(t)x'(t) - x(t_k)x'(t_k)| + |x(t)y(t) - x(t_k)y(t_k)| + |x(t)y'(t) - x(t_k)y'(t_k)| \\ &+ |y(t)y'(t) - y(t_k)y'(t_k)| + |(y(t)x'(t) - y(t_k)x'(t_k)| + |x'(t)y'(t) - x'(t_k)y'(t_k)| \\ &+ |x(t) - x(t_k)| + |x'(t) - x'(t_k)| + |y(t) - y(t_k)| + |y'(t) - y'(t_k)|\Big). \end{aligned}$$

A simple calculation gives

$$\begin{aligned}|I(t)| \le A\Big(&|x(t)-x(t_k)|(|x(t)|+|x(t_k)|+|x'(t)|+|x'(t_k)|+|y(t_k)|+|y'(t_k)|+1)\\ &+|y(t)-y(t_k)|(|y(t)|+|y(t_k)|+|x(t)|+|y'(t_k)|+|x'(t_k)|+1)\\ &+|x'(t)-x'(t_k)|(|x'(t)|+|x'(t_k)|+|x(t)|+|y(t)|+|y'(t_k)|+1)\\ &+|y'(t)-y'(t_k)|(|y'(t)|+|y'(t_k)|+|x(t)|+|y(t)|+|x'(t)|+1)\Big),\end{aligned} \tag{19}$$

where

$$A=\max\big(\max_{0\le p+m+l+r\le 2} A^{(1)}_{\text{pmlr}}(s), \max_{0\le p+m+l+r\le 2} A^{(2)}_{\text{pmlr}}(s)\big).$$

Using (5) from the last inequality we have

$$\begin{aligned}|I(t)| \le A(5C+1)\Big(&|x(t)-x(t_k)|+|y(t)-y(t_k)|\\ &+|x'(t)-x'(t_k)|+|y'(t)-y'(t_k)|\Big).\end{aligned} \tag{20}$$

Since

$$|\sum_{0\le p+m+l+r\le 2} x^p(t_k)y^l(t_k)x'^m(t_k)y'^r(t_k)A^{(1)}_{\text{pmlr}}(s)| \le 14AC^2+A$$

and

$$x(t)-x(t_k)=\int_{t_k}^{t}\int_{t_k}^{w}\sum_{0\le p+m+l+r\le 2} x^p(t_k)y^l(t_k)x'^m(t_k)y'^r(t_k)A^{(1)}_{\text{pmlr}}(s)\mathrm{d}s\mathrm{d}w + x'(t_k)(t-t_k),$$

we obtain

$$|x(t)-x(t_k)| \le (14AC^2+A)\frac{1}{n^2}+C\frac{1}{n} \quad \text{for} \quad t\in[t_{k-1},t_k). \tag{21}$$

By similar way we have

$$|y(t)-y(t_k)| \le (14AC^2+A)\frac{1}{n^2}+C\frac{1}{n} \quad \text{for} \quad t\in[t_{k-1},t_k). \tag{22}$$

$$|x'(t)-x'(t_k)| \le (14AC^2+A)\frac{1}{n} \quad \text{for} \quad t\in[t_{k-1},t_k). \tag{23}$$

$$|y'(t)-y'(t_k)| \le (14AC^2+A)\frac{1}{n} \quad \text{for} \quad t\in[t_{k-1},t_k). \tag{24}$$

According the inequalities (21)–(24) from (20) we have

$$|I(t)| \leq A(5C+1)((56AC^2+4A+4C)\frac{1}{n} \leq \frac{C_1}{n},$$

where $C_1 = A(5C+1)(56AC^2+4A+4C)$.

This claims the proof of the theorem.

4 Summary of Algorithms

Let the exact and approximate solutions at the point t be denoted by $x(t)$, $x_n(t)$ and $y(t)$, $y_n(t)$ respectively, then the absolute errors at the point AEx and AEy are defined as follows.

$$\begin{aligned} &\mathrm{AE}_x = |x(t_i) - x_n(t_i)|, \\ &\mathrm{AE}_y = |y(t_i) - y_n(t_i)|, \\ &i = 1, 2, \ldots, k. \end{aligned} \tag{25}$$

If an exact solution of the problem is not given or it is impossible to find the analytical solution, we can find the residual errors RE_x and RE_y at the point using an approximate solution. It is defined as:

$$\begin{aligned} &\mathrm{RE}_x = |x''_n(t_i) - f_1(t_i, x_n(t_i), y_n(t_i), x'_n(t_i), y'_n(t_i))|, \\ &\mathrm{RE}_y = |y''_n(t_i) - f_2(t_i, x_n(t_i), y_n(t_i), x'_n(t_i), y'_n(t_i))|, \\ &i = 1, 2, \ldots, k. \end{aligned} \tag{26}$$

Algorithms for solving the initial conditional problem is presented as follows:

Input: Initial conditions.

Step 1: Find $(x'(t), y'(t))$ and $(x(t), y(t))$ in the system of Eqs. (10) and (11) for $t \in [0, t_1)$;

Step 2: Find limits $x(t_1) = \lim_{t \to t_1 - 0} x(t)$, $y(t_1) = \lim_{t \to t_1 - 0} y(t)$, $x'(t_1) = \lim_{t \to t_1 - 0} x'(t)$, $y'(t_1) = \lim_{t \to t_1 - 0} y'(t)$.

Step 3: According to (16)–(18), we compute $(x'(t), y'(t))$ and $(x(t), y(t))$ for $[t_{k-1}, t_k)$ and we calculate the limits $x(t_k) = \lim_{t \to t_k - 0} x(t)$, $y(t_k) = \lim_{t \to t_k - 0} y(t)$, $(t_k) = \lim_{t \to t_k - 0} x'(t)$, $y'(t_k) = \lim_{t \to t_k - 0} y'(t)$ for $t \in [t_{k-1}, t_k)$, $k = 2, 3, \ldots, n$.

Step 4: We construct an approximate solution from (18).

Step 5: We draw the graph of the approximate solution and compare it with the graph of the exact solution by placing it on one coordinate.

Step 6: Determine the errors AE_x, AE_y (25) and RE_x, RE_y (26) using the absolute point error formulas.

Output: If the errors at the specified point are acceptable, then the for loop ends.

5 Numerical Results

In order to verify the effectiveness of the proposed methods, we consider several examples and give the absolute errors of approximate and exact solutions. To obtain an approximate solution of Eq. (1) from the Theorem 2, we need to calculate x_n and y_n, which has the explicit. In this case, when the functions $f_i(t, x(t), y(t), x'(t), y'(t))$, $i = 1, 2$ have exact antiderivatives, the numerical calculation of x_n and y_n is a simple.

We give a numerical method of calculation x_n and y_n for the case, when the variable coefficients $f_i(t, x(t), y(t), x'(t), y'(t)), i = 1, 2$ have no exact antiderivatives. For this case in the explicit of x_n and y_n we can change $f_i(t, x(t), y(t), x'(t), y'(t))$, $i = 1, 2$ by the step functions $x(t_k), y(t_k)$ for $t \in [t_k, t_{k+1})$.

Problem 1 Let the system of nonlinear equations of the second-order with the initial value have the following form

$$\begin{aligned}
&x''(t) + tx'(t) + x(t)y(t) = -t^4 - 3t^3 - t^2 - 3t, \quad t \in [0, 1] \\
&y''(t) + t^2y'(t) + (x(t) + 1)y(t) = t^4 + 5t^3 + 7t^2 + 6t + 6 \\
&x(0) = 1, \quad x'(0) = 0, \quad y(0) = 2, \; y'(0) = 3.
\end{aligned}$$

Note that the functions $f(t)$ and $g(t)$ are chosen so that the function $x(t) = t^2 + 1$, $y(t) = t^2 + 3t + 2$ is exact solution of this problem. In this case, the graphs of the approximate solution x_n,y_n and the exact solutions $x(t)$ and $y(t)$ of the system are compared in Fig. 1. The absolute errors at the points t_k between the approximate solution calculated by the proposed method and the exact solution of the problem are presented in Table 1.

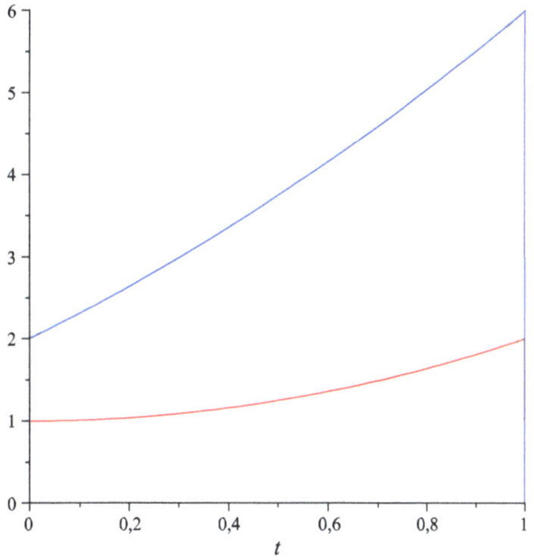

Fig. 1 Graphs of original (black color) and approximate (blue and red colors) solutions for Problem 1, respectively as $n = 10{,}000$

Table 1 Absolute errors of approximate and exact solutions for Problem 1

t	AE_x	AE_y
0.1	0	0
0.2	0	0
0.3	0	0
0.4	1.6×10^{-10}	0
0.5	9.75×10^{-10}	1.85×10^{-9}
0.6	2.02×10^{-9}	1.6×10^{-9}
0.7	3.36×10^{-9}	5×10^{-10}
0.8	2.16×10^{-9}	6.08×10^{-8}
0.9	3.53×10^{-9}	1.3085×10^{-7}

($n = 10{,}000$)

Table 2 Absolute errors of approximate and exact solutions for Problem 2

t	$AE_x n = 100$	$AE_x n = 1000$	$AE_x n = 10000$	$AE_y n = 100$	$AE_y n = 1000$	$AE_y n = 10000$
0.1	2.41×10^{-5}	2.48×10^{-6}	2.51×10^{-7}	7.51×10^{-7}	8.12×10^{-8}	1.48×10^{-8}
0.2	9.75×10^{-5}	9.89×10^{-6}	9.93×10^{-7}	6.27×10^{-6}	6.55×10^{-7}	6.32×10^{-8}
0.3	2.19×10^{-4}	2.21×10^{-5}	2.22×10^{-6}	2.12×10^{-5}	2.19×10^{-6}	2.17×10^{-7}
0.4	3.89×10^{-4}	3.92×10^{-5}	3.94×10^{-6}	5.03×10^{-5}	5.14×10^{-6}	5.07×10^{-7}
0.5	6.13×10^{-4}	6.17×10^{-5}	6.19×10^{-6}	9.88×10^{-5}	1.01×10^{-5}	1.00×10^{-6}
0.6	9.04×10^{-4}	9.10×10^{-5}	9.14×10^{-6}	1.74×10^{-4}	1.77×10^{-5}	1.76×10^{-6}
0.7	1.29×10^{-3}	1.30×10^{-4}	1.31×10^{-5}	2.92×10^{-4}	2.96×10^{-5}	2.96×10^{-6}
0.8	1.84×10^{-3}	1.87×10^{-4}	1.87×10^{-5}	4.83×10^{-4}	4.91×10^{-5}	4.92×10^{-6}
0.9	2.69×10^{-3}	2.74×10^{-4}	2.75×10^{-5}	8.14×10^{-4}	8.33×10^{-5}	8.35×10^{-6}

($n = \{100, 1000, 10{,}000\}$)

Since the exact solutions of this initial value problem are quadratic polynomials, the graphs converge even for small values of n (Table 2).

Problem 2 Let the system of differential equations of the second order with the initial value have the following form

$$\begin{aligned}
&x''(t) - 5t^3(x'(t))^2 + 2t^2 y(t)y'(t) - 4t^2 x(t)y(t) + 3t^3 y^2(t) + 2x(t) + 2 = f(t), t \in [0, 1] \\
&y''(t) - 4t^2(y'(t))^2 + 7t^3 x(t)x'(t) - 3t^3 x(t)y(t) + t^2 x^2(t) + 6y(t) - 1 = g(t), \\
&x(0) = 0, \quad x'(0) = 1, \quad y(0) = 1, \quad y'(0) = 0.
\end{aligned}$$

Table 3 Residual errors of approximate and exact solutions for Problem 2

t	$RE_x n = 100$	$RE_x n = 1000$	$RE_x n = 10000$	$RE_y n = 100$	$RE_y n = 1000$	$RE_y n = 10000$
0.1	1.5×10^{-10}	6.5×10^{-10}	5.0×10^{-11}	4×10^{-9}	3.0×10^{-9}	2.0×10^{-9}
0.2	1.1×10^{-10}	1.0×10^{-10}	5.0×10^{-10}	0	4×10^{-9}	0
0.3	3.0×10^{-10}	0	8.0×10^{-10}	4×10^{-9}	0	2.0×10^{-9}
0.4	4.0×10^{-10}	2.0×10^{-10}	1.0×10^{-10}	1.0×10^{-9}	2.0×10^{-9}	2.0×10^{-9}
0.5	2.0×10^{-10}	1.2×10^{-9}	2.0×10^{-10}	3.0×10^{-9}	3.0×10^{-9}	3.0×10^{-9}
0.6	6×10^{-10}	1.4×10^{-9}	4.0×10^{-10}	0	3.0×10^{-9}	1.00×10^{-9}
0.7	1.8×10^{-9}	3.2×10^{-9}	1.20×10^{-9}	3.0×10^{-9}	0	0
0.8	9×10^{-10}	9.0×10^{-10}	9.00×10^{-10}	3.0×10^{-9}	1.0×10^{-9}	1.00×10^{-9}
0.9	2.4×10^{-9}	6.0×10^{-10}	1.40×10^{-9}	1.0×10^{-9}	3.0×10^{-9}	4.00×10^{-9}

($n = \{100, 1000, 10{,}000\}$)

The exact solutions are given by $x(t) = \sin t$ and $y(t) = \cos t$, where $f(t) = -2t^3\cos^2 t - 3t^2 \sin 2t + \sin t + 2$, $g(t) = -3t^2 \sin^2 t + 2t^3 \sin 2t + 5\cos t - 1$. For this problem, the residual errors and absolute errors at the points t_k between the approximate solution and the exact solution are presented in Tables 2 and 3, respectively. In this case, the graphs of the approximate solution x_n, y_n and the exact solution $x(t)$, $y(t)$ of the system are compared in Fig. 2.

Problem 3 Let the system of differential equations of the second order with the initial value have the following form

$$\begin{aligned} &x''(t) - (y'(t))^2 + 2e^t x'(t)y'(t) - 4x^2(t) - 2y'(t) + 1 = e^t + 1, \quad t \in [0, 1],\\ &y''(t) - (x'(t))^2 - 2y(t)y'(t) + 4y^2(t) - 4y(t) + 3 = -e^{2t} + 3,\\ &x(0) = 1, \quad x'(0) = 1, \quad y(0) = 1, \quad y'(0) = 2. \end{aligned}$$

The exact solutions of this problem are $x(t) = e^t$ and $y(t) = e^{2t}$. Numerical results are tabulated in Tables 4 and 5. In this case, the graphs of the approximate solution x_n, y_n and the exact solution $x(t)$, $y(t)$ of the system are compared in Fig. 3.

Problem 4 Let the system of differential equations of the second order with the initial value have the following form

$$\begin{aligned} &x''(t) - (y'(t))^2 + 2x'(t) + e^t y'(t) - y^2(t) = 0, \quad t \in [0, 1]\\ &y''(t) - (x'(t))^2 - e^{2t}x'(t) + 3y'(t) - x^2(t) = 0,\\ &x(0) = 1, \quad x'(0) = 2, \quad y(0) = -1, \quad y'(0) = 0. \end{aligned}$$

Analytical solutions of this problem are not given and cannot be found by classical calculation. In this problem, the absolute errors and graphs of the points are found using the approximate solution of the system of equations. The results for the residual errors are presented in Table 6.

Graphs of the approximate solution of Problem 4 x_n, y_n are presented in Fig. 4.

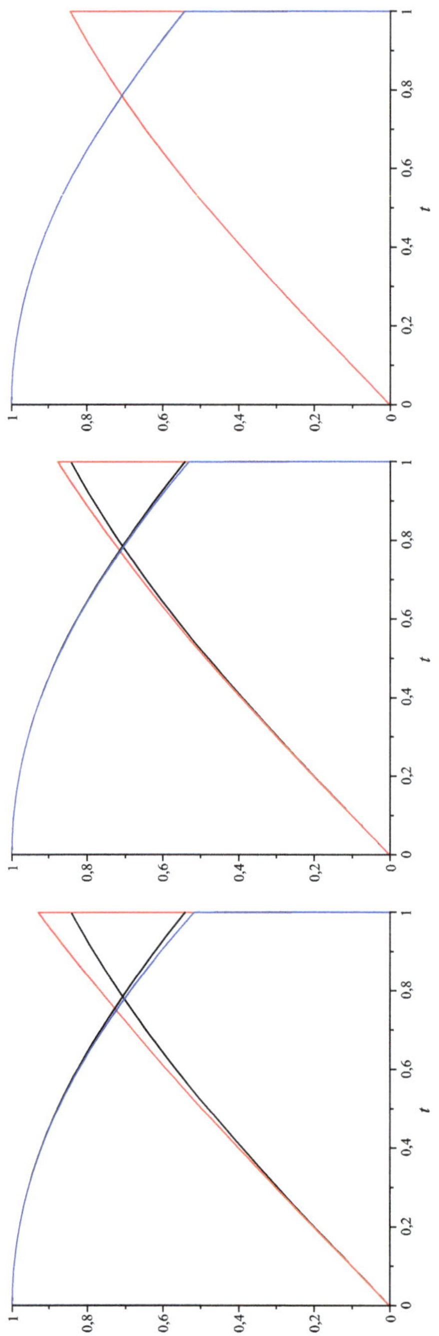

Fig. 2 Graphs of exact (black color) and approximate (red and blue colors) solutions for Problem 2, respectively as $n = 3$, $n = 10$ and $n \geq 100$

Table 4 Absolute errors of approximate and exact solutions for Problem 3 ($n = \{100, 1000, 10{,}000\}$)

t	$AE_x n = 100$	$AE_x n = 1000$	$AE_x n = 10000$	$AE_y n = 100$	$AE_y n = 1000$	$AE_y n = 10000$
0.1	4.57×10^{-5}	4.97×10^{-6}	7.70×10^{-8}	2.22×10^{-4}	2.32×10^{-5}	2.65×10^{-7}
0.2	2.92×10^{-4}	3.08×10^{-5}	3.85×10^{-7}	1.08×10^{-3}	1.11×10^{-4}	1.21×10^{-6}
0.3	9.20×10^{-4}	9.60×10^{-5}	1.39×10^{-6}	3.0×10^{-3}	3.10×10^{-4}	3.26×10^{-6}
0.4	2.15×10^{-3}	2.24×10^{-4}	3.86×10^{-6}	6.78×10^{-3}	7.03×10^{-4}	7.61×10^{-6}
0.5	4.33×10^{-3}	8.55×10^{-4}	9.27×10^{-6}	1.38×10^{-3}	1.45×10^{-4}	1.63×10^{-6}
0.6	8.10×10^{-3}	1.58×10^{-3}	2.12×10^{-5}	2.73×10^{-2}	2.88×10^{-3}	3.48×10^{-5}
0.7	1.47×10^{-2}	3.01×10^{-3}	5.06×10^{-5}	5.36×10^{-2}	5.76×10^{-3}	7.75×10^{-5}
0.8	2.73×10^{-2}	6.23×10^{-3}	1.32×10^{-4}	1.08×10^{-2}	1.20×10^{-2}	1.93×10^{-4}
0.9	5.38×10^{-2}	2.74×10^{-4}	4.12×10^{-4}	2.37×10^{-2}	2.77×10^{-2}	5.73×10^{-4}

Table 5 Residual errors of approximate and exact solutions for Problem 3 ($n = \{100, 1000, 10{,}000\}$)

t	$RE_x n = 100$	$RE_x n = 1000$	$RE_x n = 10{,}000$	$RE_y n = 100$	$RE_y n = 1000$	$RE_y n = 10{,}000$
0.1	1.1×10^{-9}	0	1.00×10^{-9}	2.0×10^{-9}	2.0×10^{-9}	0
0.2	1.5×10^{-8}	3.0×10^{-9}	3.00×10^{-9}	0	2.0×10^{-9}	3.00×10^{-9}
0.3	0	0	1.00×10^{-8}	1.0×10^{-9}	0	5.00×10^{-9}
0.4	1.0×10^{-8}	1.0×10^{-8}	1.00×10^{-8}	5.0×10^{-9}	0	2.00×10^{-9}
0.5	2.0×10^{-8}	2.0×10^{-8}	0	0	5.0×10^{-9}	0
0.6	2.0×10^{-8}	2.0×10^{-8}	1.00×10^{-8}	3.7×10^{-8}	8.0×10^{-9}	2.00×10^{-8}
0.7	1.0×10^{-8}	3.0×10^{-8}	8.00×10^{-8}	4.6×10^{-8}	8.0×10^{-9}	7.90×10^{-8}
0.8	1.5×10^{-7}	1.0×10^{-8}	0	8.9×10^{-8}	4.0×10^{-9}	6.00×10^{-9}
0.9	0	1.0×10^{-7}	1.00×10^{-7}	3.6×10^{-8}	3.6×10^{-8}	1.40×10^{-8}

Problem 5 Consider one more IVP [17]

$$\begin{aligned} &x''(t) = \sin h(x(t)), \\ &x(0) = 1, \ \ x'(0) = 0, \ \ t \in [0, 6]. \end{aligned}$$

Exact solutions of this problem are not given and cannot be found by classical calculation. In this problem, the residual error of the points are found using the approximate solution of the equation. The numerical values and absolute errors of the methods (piecewise constant argument method and the Folkner method given in [17]) are presented in Table 7.

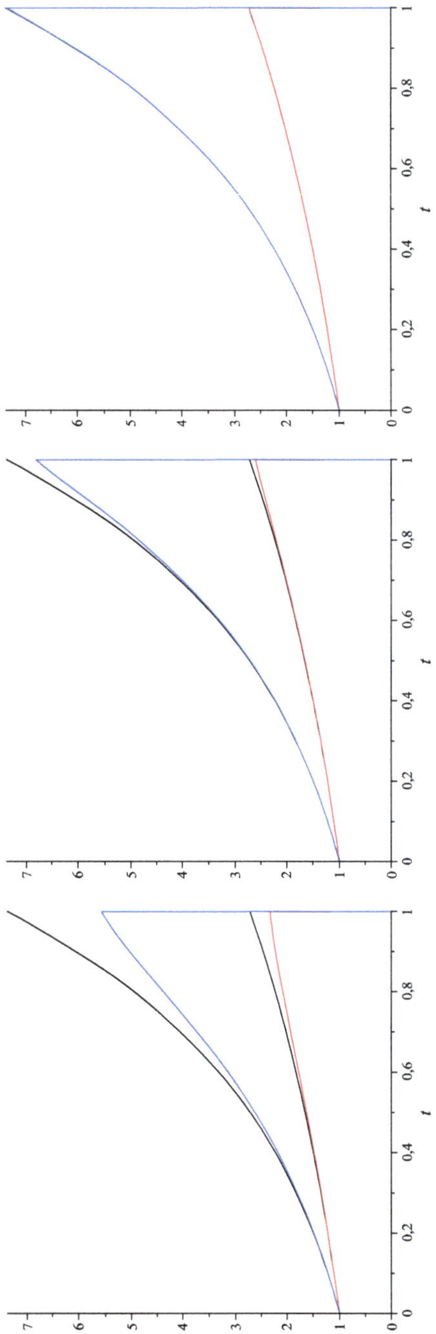

Fig. 3 Graphs of exact (black color) and approximate (blue and red colors) solutions for Problem 3, respectively as $n = 10$, $n = 100$ and $n \geq 1000$

Table 6 Residual errors of approximate solutions for Problem 4 (n = {100, 1000, 10,000})

t	$RE_x n = 100$	$RE_x n = 1000$	$RE_x n = 10,000$	$RE_y n = 100$	$RE_y n = 1000$	$RE_y n = 10,000$
0.1	1.0×10^{-10}	3.0×10^{-10}	1.06×10^{-10}	1.0×10^{-9}	2.0×10^{-9}	1.37×10^{-9}
0.2	2.0×10^{-9}	3.0×10^{-9}	1.86×10^{-9}	3.0×10^{-9}	3.0×10^{-9}	2.80×10^{-9}
0.3	0	3.0×10^{-9}	1.29×10^{-9}	2.00×10^{-9}	3.0×10^{-9}	1.14×10^{-9}
0.4	1.0×10^{-9}	1.0×10^{-9}	2.16×10^{-9}	6.0×10^{-9}	4.0×10^{-9}	3.95×10^{-9}
0.5	2.0×10^{-9}	0	7.06×10^{-9}	2.0×10^{-9}	5.0×10^{-9}	2.66×10^{-9}
0.6	1.0×10^{-9}	1.0×10^{-9}	4.22×10^{-9}	6.0×10^{-9}	1.0×10^{-9}	4.43×10^{-9}
0.7	1.0×10^{-9}	3.0×10^{-9}	1.13×10^{-9}	2.0×10^{-9}	6.0×10^{-9}	1.24×10^{-9}
0.8	1.0×10^{-9}	2.0×10^{-9}	1.21×10^{-9}	2.0×10^{-9}	5.0×10^{-9}	3.30×10^{-9}
0.9	1.0×10^{-9}	2.0×10^{-9}	1.10×10^{-9}	4.0×10^{-9}	9.0×10^{-9}	1.66×10^{-9}

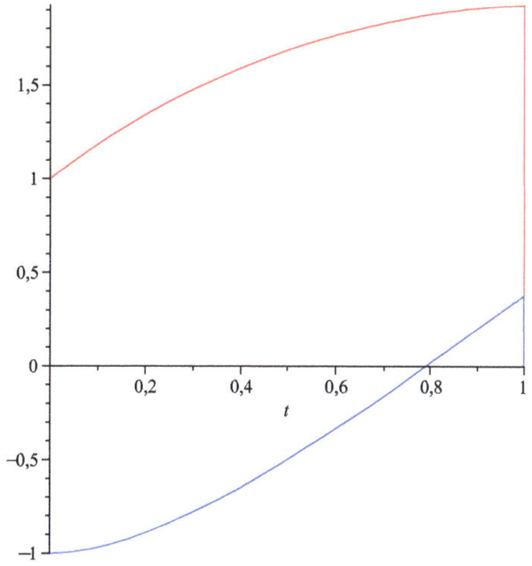

Fig. 4 Graphs of approximate (blue and red colors) solutions for Problem 4, $n = 1000$

Table 7 Accuracy comparison of [17] for Problem 5

Method	ABS.ERR.
VOVS	15.23×10^{-3}
STOR(6)	2.91×10^{-5}
FALK(6)	2.16×10^{-5}
VOVS	5.25×10^{-4}
STOR(8)	6.92×10^{-6}
FALK(8)	1.48×10^{-7}
RE_x (our method)	7.30×10^{-6}

6 Conclusion

In this paper, an efficient numerical algorithm based on the use of the method of constant arguments is proposed to solve the approximate solution of the system of nonlinear second-order differential equations with an initial value. First, the differential equation with piecewise constant arguments corresponding to the considering equation has been constructed. The definition of the solution of ordinary differential equations with a piecewise constant5 argument is introduced as a piecewise smooth function with respect to a positive integer. This piecewise smooth solution is proved to be an approximate solution of the considered initial value problem for a system of second-order differential equations for adults. Numerical results are performed to demonstrate the effectiveness of the proposed methods, and the errors between the approximate and exact solutions are evaluated. The presented method has great advantages in solving single initial value problems due to its simplicity and high accuracy.

Acknowledgements We would like to thank the anonymous reviewers for their constructive comments that have greatly contributed to improving the manuscript.

Disclosure of Interests. We declare that we have no known competing financial interests or personal relationships that could have appeared to influence the work reported in this paper.

References

1. Jator, S.N., King, K.L., et al.: Integrating oscillatory general second-order initial value problems using a block hybrid method of order 11. Math. Probl. Eng. **2018** (2018)
2. Kwari, L.J., Sunday, J., Ndam, J.N., Shokri, A., Wang, Y.: On the simulations of second-order oscillatory problems with applications to physical systems. Axioms **12**(3), 282 (2023)
3. Jator, S.N., Swindell, S., French, R.: Trigonometrically fitted block Numerov type method for $y'' = f(x, y, y')$. Numer. Algorithms **62**, 13–26 (2013)
4. Sunday, J., Shokri, A., Kwanamu, J.A., Nonlaopon, K.: Numerical integration of stiff differential systems using non-fixed step-size strategy. Symmetry **14**(8), 1575 (2022)
5. Franco, J.M.: A class of explicit two-step hybrid methods for second-order IVPs. J. Comput. Appl. Math. **187**(1), 41–57 (2006)
6. Abdulganiy, R.I., Wen, S., Feng, Y., Zhang, W., Tang, N.: Adapted block hybrid method for the numerical solution of Duffing equations and related problems. AIMS Math. (2021). American Institute of Mathematical Sciences (AIMS)
7. Sunday, J., Shokri, A., Marian, D.: Variable step hybrid block method for the approximation of Kepler problem. Fractal Fract. **6**(6), 343 (2022)
8. Allogmany, R., Ismail, F.: Direct solution of $u'' = f(t, u, u')$ using three point block method of order eight with applications. J. King Saud Univ.-Sci. **33**(2), 101337 (2021)
9. Ramos, H., Rufai, M.A.: Third derivative modification of k-step block Falkner methods for the numerical solution of second order initial-value problems. Appl. Math. Comput. **333**, 231–245 (2018). https://doi.org/10.1016/j.amc.2018.03.098
10. Li, J., Deng, S., Wang, X.: Multi-step Nyström methods for general second-order initial value problems $y''(t) = f(t, y(t), y'(t))$. Int. J. Comput. Math. **96**(6), 1254–1277 (2019)

11. Li, J., Wang, X.: Multi-step Runge–Kutta–Nyström methods for special second-order initial value problems. Appl. Numer. Math. **113**, 54–70 (2017)
12. Jia, L., Lio, W., Yang, X.: Numerical method for solving uncertain spring vibration equation. Appl. Math. Comput. **337**, 428–441 (2018)
13. Aldosary, S.F., Ahmed, R.: Stability and bifurcation analysis of a discrete Leslie predator-prey system via piecewise constant argument method. AIMS Math. **9**, 4684–4706 (2024)
14. Muminov, M.I., Radjabov, T.A.: Existence conditions for 2-periodic solutions to a non-homogeneous differential equation with piecewise constant argument. Ex. Count. Examples **5**, 100145 (2024)
15. Muminov, M.I., Radjabov, T.A.: On existence conditions for periodic solutions to a differential equation with constant argument. Nanosyst. Phys. Chem. Math. **13**(5), 491–497 (2022)
16. Muminov, M.I., Jumaev, Z.Z.: Exact periodic solutions of second-order differential equations with piecewise constant arguments. Adv. Math. Sci. J. **10**(9), 3113–3128 (2021)
17. Vigo-Aguiar, J., Ramos, H.: Variable stepsize implementation of multistep methods for $y'' = f(x, y, y')$. J. Comput. Appl. Math. **192**(1), 114–131 (2006)
18. Muminov, M.I.: On the method of finding periodic solutions of second-order neutral differential equations with piecewise constant arguments. Adv. Differ. Equ. **2017**, 1–17 (2017)
19. Muminov, M.I., Murid, H.M.: Existence conditions for periodic solutions of second-order neutral delay differential equations with piecewise constant arguments. Open Math. **18**(1), 93–105 (2020)

Performance Measure of Malaysian International Airports Using Data Envelopment Analysis

Anis Izzati Mohamad Iskandar, Mashitah Maisarah Zaili, Zuraida Alwadood, and Norlenda Mohd Noor

Abstract This study addresses the problem of limited research on the performance efficiency of Malaysian airports, with most studies focusing on airports of other countries. To initiate such performances analysis, this study aims to fill this gap by employing Data Envelopment Analysis (DEA) model to calculate the relative efficiency of Malaysian international airports and determine the ideal value of inputs and outputs to maximize the performance of the inefficient airports. DEA is a mathematical performance measure technique that can evaluate the relative efficiency of decision-making units (DMUs). The DMU in this study are eight selected Malaysian international airports. The input data are the number of runways, runway length, check-in counters, boarding gates, while the output data are passenger movement and aircraft movement. The result has revealed varying levels of relative efficiency among the airports, with DMU1, DMU7, and DMU8 achieving the highest efficiency scores. The findings suggest rooms for improvement for other five international airports. Improvement strategies were assessed by adjusting factors of input and output values. By implementing the proposed improvement strategies, the relative efficiency level of the DMU have increased significantly. This study provides valuable insights for airport authorities to enhance operational efficiency, contributing to the development and improvement of Malaysian international airports.

Keywords Efficiency · Malaysian airports · Performance measure

1 Introduction

According to Malaysia Airports Holdings Berhad (MAHB) Annual Report [1], the increased proportion of international passengers for both Malaysia and overseas operations is responsible for the increased profitability. Passenger traffic increased by 6%

A. I. M. Iskandar · M. M. Zaili · Z. Alwadood (✉) · N. M. Noor
Faculty of Computer and Mathematical Sciences, Universiti Teknologi MARA, Shah Alam, Selangor, Malaysia
e-mail: zuraida794@uitm.edu.my

A. Shuib et al. (eds.), *Mathematical Modeling and Numerical Analysis*, Springer Proceedings in Mathematics & Statistics 510,
https://doi.org/10.1007/978-981-96-9350-4_8

has contributed to the growth in revenue and earnings. As the number of passengers continues to rise exponentially, performance efficiency is currently used as one of the necessities to ensure that the anticipated level of efficiency is encountered.

Operational efficiency assesses the ability of an organization to utilize its resources and deliver outputs in the most efficient manner [2]. It refers to the overall effectiveness and productivity of an organization in utilizing its resources to deliver outputs. Technical efficiency is a component of operational efficiency, which specifically focuses on the organization's ability to achieve maximum output from a given set of inputs. On the other hand, relative efficiency refers to a measure used to compare the efficiency of units by calculating their efficiency scores relative to each other. This study evaluated airport performance based on relative efficiency.

In this study, the efficiency of airports in Malaysia is measured using Data Envelopment Analysis (DEA). The DEA model which was formulated by Charnes et al. [3] involves creating a non-parametric piece-wise frontier over the data using a linear programming (LP) method. The efficiency metrics are computed in relation to this frontier. This method calculates the relative efficiency of decision-making units (DMUs) using multiple inputs and multiple outputs. The DEA model is also very helpful for identifying relative inefficiencies so that performance can be improved [4]. Many research papers used the DEA models to calculate the operational efficiency and productivity growth.

Various studies have attempted to apply DEA in the airline sector. Gutierrez and Lozano [5] used DEA to investigate the technical efficiencies among 21 airports in Europe. Stichhauerova and Pelloneova [6] measured the relative technical efficiencies of the 27 most important airports in Germany. Tsironis et al. [7] examined the technical efficiency of 14 regional airports in Greek in 2016, just before their privatization. Efficiency measurement is a vital aspect of the management control of every organization. It can be used as a basis for improving the performance of a facility or as a guide for making informed decisions.

A study conducted by Sopadang and Suwanwong [8] has utilized the technique of DEA to evaluate airport performance. The study involved 19 airports, including Singapore Changi Airport (Singapore), Incheon International Airport (South Korea), Hong Kong International Airport (Hong Kong), Tokyo International Airport (Japan), and a few others. Other study done by Huynh et al. [9] has employed DEA-Slack Based Measurement (DEA-SBM) to measure the efficiency of nine major airports in South East Asia including Changi Airport (Singapore), Suvarnabhumi Airport (Thailand), Kuala Lumpur Airport (Malaysia), Soekarno Hatta Airport (Indonesia), Ninoy Aquino Airport (Philippines), Noi Bai Airport, Tan Son Nhat Airport (Vietnam), Guangzhou Baiyun Airport (China), and Hong Kong Airport (Hong Kong). From the findings, Kuala Lumpur Airport turned out to be the most inefficient airport.

Many studies consider the movement of passengers, aircraft, and cargo to be the most appropriate variables to use as outputs. As a result, they are found in most works investigating airport efficiency [4, 5, 10–17]. Passenger movement is the most used output and has grown in importance over time. Furthermore, the number of aircraft movements is significant because landing and parking fees generate revenue for the airport. Thus, the movement of passengers and aircraft reflects business results and is used as an output.

While previous studies have extensively analyzed the efficiency of airports in other countries, there is a scarcity of research on Malaysian airports. Therefore, this study aims to fill the gap by employing the DEA model to assess the Malaysian international airport efficiencies. DEA is one of the most suitable methods to compute airports' performance due to its ability to assess the effect of many factors [7].

Sopadang and Suwanwong [8] and Huynh et al. [9] have reported that Penang International Airport and Kuala Lumpur International Airport, respectively, are inefficient airports. The management might under looked the elements that contributed to airport efficiency and whether resources are being used to their full potential. They may be unaware of how their services compare to one another; thus, they are unable to determine whether their business is operating satisfactorily. Therefore, the objectives of this study are to calculate the relative efficiency of Malaysia international airports and determine the ideal value of inputs and outputs to maximize the performance of the inefficient airports. The result is expected to provide ideas to the airport management in upgrading service quality, leading to a more solid and reliable airport experience for passengers.

There are over 60 airports in Malaysia, but this study focused on eight international airports. The dataset used in this research was based on Airports Statistics 2019 from the MAHB corporate website and the official site of the Ministry of Transport Malaysia [18]. Due to the COVID-19 pandemic, the travel restrictions were enforced domestically and internationally in 2020. While the aviation industry began transitioning to the "new normal" in 2021, travel restrictions continued to impact the industry and Malaysia international borders were reopened in 2022 [19]. Therefore, due to data availability, this study utilized the 2019 dataset in the mathematical modeling. This study makes an academic contribution to the research field because of the scarcity of a detailed study of Malaysian airports in the literature. Moreover, this study can provide managerial benefits of benchmarking for the airport performance improvement to airport authorities.

2 Methodology

Based on the literature reviews, this study has selected four input variables and two output variables. The secondary data used for this study was gathered from the 2020 Malaysia Transportation Statistics on the Ministry of Transport Malaysia official portal and Airports Statistics 2019. The data is collected for use in modeling the model sets and parameters. The choice of the input and output factors are made based on the gap analysis done on the related studies, as well as the data availability from the sources.

There are four factors selected to be the input data in the DEA model which are the number of runways, length of the runway, number of check-in counter and number of boarding gates. The number of runways is important for faster landing and take-off of an aircraft and reducing the number of line-ups at the airport. The length of the runway is the total runway length which should be capable of handling any size and type of aircraft. The number of check-in counters cater the process of checking-in and ensures that passengers do not have to wait so long to check-in. Having multiple check-in counters could help to alleviate passenger congestion. The number of boarding gates allows airlines to park their aircraft directly in front of the boarding bridge gate. Therefore, passengers can easily board the aircraft. Table 1 displayed the data for the selected input and output variables for eight Malaysian international airports.

Apart from the input factors, there are two factors selected to be the output in this study. The first output is the passenger movements which refer to the total number of passengers arriving or departing from the airport. The other output is the aircraft movements which refer to the total number of planes for landings and take-offs to and from the airport.

The mathematical model used in DEA is based on Charnes et al. [3]. The DEA model is used to measure the efficiency of the DMUs, which are the eight selected airports. In this analysis, four inputs and two outputs were used to evaluate the efficiency of selected Malaysian airports. This model consists of sets, parameters, decision variables, objective functions, and constraints. Therefore, the mathematical model that is used in this study can be expressed as follows:

Sets:

DMUs, $j = \{1, 2, 3, 4, 5, 6, 7, 8\}$
Inputs, $h = \{1, 2, 3, 4\}$
Outputs, $i = \{1, 2\}$.

Parameters:

x_{hj} = the value of input h produced by DMU j

Table 1 Data for inputs and outputs for selected Malaysian airports for the year 2019

DMU	Inputs				Outputs	
	No. of runway	Length of runway	No. of check-in counter	No. of boarding gates	Passenger movement	Aircraft movement
DMU1	3	3993	242	27	62,336,469	407,315
DMU2	1	3354	53	11	8,331,291	78,400
DMU3	1	3780	64	12	9,445,494	82,125
DMU4	1	3780	36	9	5,956,141	53,610
DMU5	1	3813	24	8	2,946,150	25,673
DMU6	1	2135	6	4	135,569	2502
DMU7	1	3782	14	12	2,259,595	52,082
DMU8	1	3800	16	5	4,270,144	58,313

Source Malaysia Airports Holdings Berhad [1], Ministry of Transport Malaysia [18]

y_{ij} = the value of output i produced by DMU j.

Decision variables:

u_h = weight obtained for input h
v_i = weight obtained for output i
E_j = the relative efficiency of jth DMU.

The objective function is aimed to maximize the efficiency of DMU j, represented by:

$$\text{Maximize } E_j = \frac{\sum_{i=1}^{2} v_i y_{ij}}{\sum_{h=1}^{4} u_h x_{hj}} \tag{1}$$

Subject to:

$$\frac{\sum_{i=1}^{2} v_i y_{ij}}{\sum_{h=1}^{4} u_h x_{hj}} \leq 1 \tag{2}$$

The objective function (1) corresponds to the maximum efficiency score acquired by DMU$_j$. Constraint (2) is the range of relative efficiency E_j where it must be between zero and one. Charnes et al. [3] stated that the above model can be rewritten into the linear programming model shown below using transformations.

$$\text{Maximize } E = \sum_{i=1}^{2} v_i y_{ij} \tag{3}$$

Subject to

$$\sum_{h=1}^{4} u_h x_{hj} = 1 \tag{4}$$

$$\sum_{i=1}^{2} v_i y_{ij} - \sum_{h=1}^{4} u_h x_{hj} \leq 0, \quad j = 1, 2, \ldots, 8, \ h = 1, 2, 3, 4, \ i = 1, 2 \tag{5}$$

where h and i denote the number of inputs and outputs, respectively, and j is the index for DMUs. The objective function (3) is to maximize the outputs based on the numerator in Eq. (1). Constraint (4) is the denominator which is the inputs in Eq. (1) must be equal to unity one. Constraint (5) is the difference between the numerator and denominator, which must be less than zero.

3 Result and Discussion

By using QM for Windows software, the weight for each input and output is generated. After each weight is obtained, the efficiency of each airport is evaluated by substituting the weight value generated into Eq. (1). The calculation of efficiency using DEA demonstrates a DMU's ability to utilize its inputs optimally to generate the maximum possible outputs based on the available technology or production.

Table 2 shows the relative efficiency of all selected international airports. It can be shown that three airports have a full score of efficiency of 100%, while another five airports did not show good efficiency with the percentage of range between 11 and 71%.

DMU1, DMU7, and DMU8 have achieved 100% efficiency by maximizing the use of their inputs and outputs. The existence of three runways in DMU1 provides for more efficient aircraft scheduling and management, decreasing congestion and delays. The length of its runways allows for a wide range of aircraft sizes, improving connection and passenger flow. The large number of check-in counters ensures efficient passenger processing, lowering wait times and increasing overall efficiency. DMU7 has recorded a good efficiency by successfully utilizing its single runway and securely accepting a wide range of aircraft sizes. Apart from this, DMU8 has also attained high efficiency by managing its single runway efficiently and effectively. The airport optimizes passenger processing and gate utilization with 16 check-in counters and five boarding gates, contributing to its great efficiency in handling passenger movements and aircraft operations. The efficient use of inputs such as runways, check-in counters, and boarding gates, combined with excellent control of outputs such as passenger movement and aircraft movement, enables these airports to obtain 100% efficiency.

DMU2 which gained 71% efficiency has a single runway with a length of only 3354 m. This allows a limited range of aircraft sizes and restricted handling facilities during take-offs and landings. It has only 11 boarding gates which may limit its ability to handle numerous airline operations at the same time, affecting turnaround times and capacity utilization. Besides this, DMU3 and DMU4 have recorded an efficiency rating of 66% and 64%, respectively. Both airports have high passenger movements and aircraft movements, which reflect the high volume of people serviced and the number of aircraft operations at the airports. High passenger traffic involves efficient check-in, luggage processing, and passenger services, whereas considerable aircraft traffic necessitates effective runway operations, scheduling, and airspace management. The unsatisfactory efficiency rating indicates that there is a room for improvement in operational processes and overall performance of DMU3 and DMU4.

Table 2 Relative efficiency for eight DMUs

	DMU1	DMU2	DMU3	DMU4	DMU5	DMU6	DMU7	DMU8
Relative efficiency (%)	100	71	66	64	47	11	100	100

DMU5 only achieved 47% of efficiency rating, indicating a crucial need for improvement strategies. With only eight boarding gates, DMU5 must effectively manages gate utilization to maintain smooth boarding and disembarkation processes. The considerate number of passenger movement highlights the importance of effective check-in, baggage handling, and passenger services. The airport of DMU6 received a poor efficiency rating of 11%, indicates the urgent need for improvement strategies in its performance and resource utilization. DMU6 has only six check-in counters, indicating a low capacity for passenger processing. With only four boarding gates, it may encounter difficulties coordinating simultaneous flight operations and passenger movement, potentially resulting in delays and inefficiencies. The volume of people served and the number of aircraft movement at the airport are 569 and 2502 aircraft, respectively. These numbers shows that demand for air travel is relatively low, which may limit the airport's revenue potential and impede investment in infrastructure and service enhancements. This issue can also indirectly pose an impact on the airport's overall efficiency and attractiveness.

An improvement strategy should be planned so that the efficiency level of the inefficient airports could be raised. Improvement was done by determining the ideal value of the number of runways, length of the runway, number of checks in counter, number of boarding gates, number of passenger movements and aircraft movement by either increasing or decreasing the values of inputs and outputs. El-Mahgary and Lahdelma [20] model is applied to get the ideal values of inputs and outputs to increase up to almost 100% or achieved 100% efficiency.

The improvement model to obtain the suggested values of inputs and outputs is,

$$x_{kE'} = x_{kM}\lambda_{kM} + x_{kQ}\lambda_{kQ} \tag{6}$$

where

- $x_{kE'}$ is the revised value of input or output k of the less efficient DMU
- x_{kM} is the value of input or output k of reference set DMU M
- λ_{kM} is the dual weight of the input or output k of reference set DMU M
- x_{kQ} is the value of the input or output k of reference set DMU Q
- λ_{kQ} is the dual weight of the input or output k of reference set DMU Q.

Reference set is a set taken from the dual weight value of two DMUs with highest level of efficiency, namely DMU M and DMU Q. They are comprised predominantly of airports with higher efficiencies and serve as a basis for enhancing the efficiency of airports with lower efficiencies. Equation (6) is used to calculate the dual weight for the less efficient DMU. The dual weight obtained from each reference set later is used as an indicator to evaluate the new suggested values of inputs and outputs for the DMU with less efficiency.

Table 3 provides a comprehensive list of current, suggested, and practical values for inputs and outputs of all DMUs which are the airports under study. The "current value" row displays the existing data, while the "suggested value" row provides the recommended values obtained by solving the El-Mahgary and Lahdelma [20] model and recommended by the improvement strategy. The "practical value" row lists the most feasible and reasonable values between the current and suggested values to ensure the efficiency and practicality of airport operations.

The first column of input shows the current, suggested, and practical values of the airport number of runway. For DMU2, DMU3, and DMU4, all the current value equals the suggested value, which is one airport runway. However, the suggested values for DMU5 and DMU6 are recorded as zero. Since this value is regarded as

Table 3 List of current, suggested, and practical values of inputs and outputs of inefficient DMUs

		Input				Output	
		No. of runway	Length of runways (m)	No. of check-in counter	No. of boarding gates	Passenger movement	Aircraft movement
DMU2	Current value	1	3354	53	11	8,331,291	78,400
	Suggested value	1	1894	32	5	8,329,464	66,155
	Practical value	1	3354	32	5	8,329,464	66,155
DMU3	Current value	1	3780	64	12	9,445,494	82,125
	Suggested value	1	1498	37	5	9,446,328	69,421
	Practical value	1	3780	37	5	9,446,328	69,421
DMU4	Current value	1	3780	3	9	5,956,141	53,610
	Suggested value	1	1925	23	4	5,900,675	51,900
	Practical value	1	3780	23	4	5,900,675	51,900
DMU5	Current value	1	3813	24	8	2,946,150	25,673
	Suggested value	0	1654	13	3	3,298,667	33,992
	Practical value	1	3813	13	3	3,298,667	33,992
DMU6	Current value	1	2135	6	4	135,569	2502
	Suggested value	0	175	1	0	135,557	2503
	Practical value	1	2135	1	1	135,557	2503

not feasible for an airport, therefore it is practical to maintain the current number of runways, which is one.

The second column of input in Table 3 shows the length of runways, in meter, for each airport. The suggested value for the length of runways for all airports is far smaller than the current value. However, reducing the length of runways seems to be impractical as it is not only jeopardizing the safety of moving aircraft, but also demanding a high construction cost. Therefore, the suggested value of runway length for each of the DMU it is not recommended, and the practical value should remain the current value of runway length.

For the number of check-in counters, the values suggested by the improvement strategy seems feasible and reasonable to carry out. DMU2, DMU3, DMU5, and DMU6 were suggested to reduce the number of check-in counter, except for DMU4 which put forward a larger number of counters. As the suggested values are acceptable, therefore the practical values follow the suggested ones.

The suggested number of boarding gates in the improvement strategies are all approximately halved except for DMU6. The reduction number of boarding gates would be able to reduce the operating cost of the airport. Nevertheless, the suggested value of zero number of boarding gates for DMU6 is neither practical nor feasible. As all other DMU were suggested to reduce the number of boarding gates, keeping the same number of current values of four gates at DMU6 may not be necessary. Therefore, the practical value of boarding gate was recommended to be one.

All current values and suggested values for both output values, namely passenger movement and aircraft movement have shown only a small variation. Some suggested values are greater than the current values and vice versa. Therefore, all the suggested values are taken as the practical values for these outputs.

Table 4 shows the results of the efficiencies of the DMUs after improvement strategies were implemented. The new efficiencies were computed based on the "practical values" stated in Table 3. DMU2 shows a perfect score of 100% for efficiency after the values of inputs and outputs are revised. Meanwhile, DMU3, DMU4, and DMU5 recorded a satisfactory result when the new efficiencies are nearly 100%. For DMU6, even though the new efficiency is not very satisfactory, but it shows a great increase in the number, from 11 to 56%. The low percentage is expected as the new efficiency was calculated based on one out of four suggested values of input given by the improvement strategy. That is, due to impracticality, only the number of check-in counters follow the suggested value of input.

Table 4 Relative efficiencies of DMUs after improvement strategies

	Current efficiencies (%)	New efficiencies (%)
DMU2	71	100
DMU3	66	99
DMU4	64	98
DMU5	47	96
DMU6	11	56

4 Conclusion

This study has successfully achieved its objectives of calculating the relative efficiency of Malaysia's international airports and determining the ideal value of inputs to maximize their efficiency. The findings from this study provide valuable insights into the efficiency levels of the selected airports. The efficiency of international airports in Malaysia was assessed, and improvements were proposed to enhance their efficiency levels. The top three efficient airports, which are DMU1, DMU7, and DMU8 served as benchmarks for the other airports. Based on the analysis, improvements were suggested for other DMUs. These improvements have included adjusting factors such as the number of check-in counters, the number of boarding gates, passenger movement, and aircraft movement. By optimizing these factors, the airports aimed to improve resource allocation, reduce congestion, and enhance operational efficiency. After implementing the proposed improvements, the airports' efficiency levels increased significantly with DMU2 achieving 100%, DMU3 reaching 99%, DMU4 operating at 98%, DMU5 improving to 96% and finally, DMU6 increasing to 56% efficiency.

For future studies, it is recommended that the performance measurement to be expanded to include airports throughout Malaysia, including not only international but also domestic and regional airports. A comparative analysis of multiple Malaysian airports would be useful for benchmarking and identifying best practices. Future study could also enhance this analysis to incorporate a broader perspective of inputs and outputs. For instance, inputs or outputs could include terminal capacity, baggage handling systems, passenger services, and aircraft punctuality to provide a more comprehensive evaluation of airport performance.

Acknowledgements The authors would like to thank all parties who are involved in this case study. A special appreciation for College of Computing, Informatics and Mathematics, Universiti Teknologi MARA (UiTM) Shah Alam for the financial support on the publication of this paper.

References

1. Malaysia Airports Holdings Berhad (MAHB) Homepage. Annual Report 2019. https://www.malaysiaairports.com.my/sites/corporate/files/2020-06/MAHB-Annual-Report-2019_0.pdf. Accessed 13 Feb 2024
2. Bukh, P.N., Nielsen, C., Gormsen, P.: Financial ratios and the predictability of efficiency in bankruptcy prediction models. Int. J. Account. Inf. Manag. **20**(3), 217–230 (2012)
3. Charnes, A., Cooper, W.W., Rhodes, E.: Measuring the efficiency of decision-making units. Eur. J. Oper. Res. **2**(6), 429–444 (1978)
4. Bezić, H., Šegota, A., Vojvodić, K.: Evaluating airport efficiency using data envelopment analysis. Adv. Bus.-Related Sci. Res. J. **1**(1), 55–66 (2010)
5. Gutierrez, E., Lozano, S.: Efficiency assessment and output maximization possibilities of European small and medium-sized airports. Res. Transp. Econ. **56**, 3–14 (2016)
6. Stichhauerova, E., Pelloneova, N.: An efficiency assessment of selected German airports using the DEA model. J. Competitiveness **11**(1), 135–151 (2019)

7. Tsironis, L.K., Toskas-Tasios, M.L., Madas, M.A.: Measuring the efficiency of Greek regional airports prior to privatization using data envelopment analysis. Int. J. Econ. Manag. Syst. **6**, 547–559 (2021)
8. Sopadang, A., Suwanwong, T.: The analysis of airport operational performance case study: Chiang Mai International Airport, Thailand. In: Proceedings of the 2016 International Conference on Industrial Engineering and Operations Management. Kuala Lumpur, Malaysia (2016)
9. Huynh, T.M., Kim, G., Ha, H.K.: Comparative analysis of efficiency for major Southeast Asia airports: a two-stage approach. J. Air Transp. Manag. **89** (2020)
10. Ulutas, B.H.: Efficiency analysis of airports in Turkey considering environmental effects. Cent. Eur. Rev. Econ. Manag. **2**(1), 65–78 (2018)
11. Wanke, P.F.: Physical infrastructure and flight consolidation efficiency drivers in Brazilian airports: a two-stage network-DEA approach. J. Air Transp. Manag. **31**, 1–5 (2013)
12. Lai, P.L., Potter, A., Beynon, M., Beresford, A.: Evaluating the efficiency performance of airports using an integrated AHP/DEA-AR technique. Transp. Policy **42**, 75–85 (2015)
13. Lin, Z.F., Choo, Y.Y., Oum, T.H.: Efficiency benchmarking of North American airports: comparative results of productivity index, data envelopment analysis and stochastic frontier analysis. J. Transp. Res. Forum **52**(1), 47–67 (2013)
14. Rapee, P., Peng, K.C.: Measuring efficiency of main airports in Thailand by an integrated DEA-AHP method. Adv. Mater. Res. 2203–2206 (2014)
15. Nwaogbe, O.R., Ogwude, I.C., Omoke, V., Barros, C.P.: An assessment of productivity and efficiency in Nigerian Airports using data envelopment analysis. In: Proceedings of 19th Air Transport Research Society (ATRS) World Conference, Singapore (2015)
16. Ripoll-Zarraga, A.E., Mar-Molinero, C.: Exploring the reasons for efficiency in Spanish airports. Transp. Policy **99**, 186–202 (2020)
17. Ozsoy, V.S., Orkcu, H.H.: Structural and operational management of Turkish airports: a bootstrap data envelopment analysis of efficiency. Util. Policy **69** (2021)
18. Ministry of Transport Malaysia: Malaysia Transportation Statistics 2020. https://www.mot.gov.my/en/Statistik%20Tahunan%20Pengangkutan/Transport%20Statistics%20Malaysia%202020.pdf. Accessed 05 June 2024
19. Malaysia Airports Holdings Berhad (MAHB) Homepage. Annual Report 2021. https://mahb.listedcompany.com/misc/ar/mahb_ar2021.pdf. Accessed 13 Feb 2024
20. El-Mahgary, S., Lahdelma, R.: Data envelopment analysis: visualizing the results. Eur. J. Oper. Res. **83**(3), 700–710 (1995)

Ranking Success Criteria of Open Distance Learning Using Interval Neutrosophic MARCOS

Noor Azzah Awang, Wan Rosanisah Wan Mohd, Norsyahida Zulkifli, and Lazim Abdullah

Abstract In decision-making and modeling uncertainty, fuzzy numbers have been extensively used due to their ability to represent imprecision and vagueness. However, in many real-world applications, decision-makers (DMs) face not only uncertainty but also degrees of indeterminacy and inconsistency that fuzzy numbers may not adequately capture. This study employs interval neutrosophic sets (INS), an extension of neutrosophic sets, to provide a more comprehensive framework for addressing such situations by incorporating truth, indeterminacy, and falsity values into the decision-making method known as Measurement of Alternatives and Ranking according to Compromise Solution (MARCOS). The proposed model is applied to examine learning success criteria in an open distance learning (ODL) environment. The study aims to investigate success criteria for learning and the type of e-learning systems that contribute to successful learning outcomes in online distance education. Specifically, it seeks to identify the criteria and alternatives that best support success in ODL using the interval neutrosophic MARCOS method. The selection of the best criteria for ODL success is based on six criteria: student characteristics, instructor characteristics, learning environment, instructional design, information technology, and the level of collaboration. The results of this study confirm that asynchronous learning is the most effective system for open distance learners. Furthermore, it is

N. A. Awang (✉)
Faculty of Computer and Mathematical Sciences, Universiti Teknologi MARA, Shah Alam, Selangor, Malaysia
e-mail: azzahawang@uitm.edu.my

W. R. W. Mohd
Faculty of Artificial Intelligence, Universiti Teknologi Malaysia, Kuala Lumpur, Selangor, Malaysia

N. Zulkifli
Department of Mathematical Sciences, Faculty of Science, Universiti Teknologi Malaysia, Johor Bahru, Johor, Malaysia

L. Abdullah
Management Science Research Group, Faculty of Computer and Mathematics, Universiti Malaysia Terengganu, Kuala Nerus, Terengganu, Malaysia

A. Shuib et al. (eds.), *Mathematical Modeling and Numerical Analysis*,
Springer Proceedings in Mathematics & Statistics 510,
https://doi.org/10.1007/978-981-96-9350-4_9

recommended as the optimal method, particularly for those seeking the best option for success in ODL.

Keywords COVID-19 · MARCOS · Multi-criteria decision-making (MCDM) · Neutrosophic set · Online distance learning

1 Introduction

In real-world decision-making, decision-makers (DMs) often face challenges in evaluating alternatives using real numbers due to the inherent complexity and uncertainty of such problems. Traditional decision-making models, which rely on precise numerical values, often fall short of capturing the nuances and intricacies of these complex scenarios. Fuzzy numbers provide a compelling alternative by enabling the representation of imprecise and ambiguous information, thereby improving the decision-making process. The concept of fuzzy sets (FS) was pioneered by Zadeh [1], where FS is defined by a single membership degree within the closed interval [0, 1]. Since its introduction, numerous advancements have been made to enhance the FS framework including intuitionistic fuzzy sets (IFS) [2], interval-valued fuzzy sets (IVFS) [3], Pythagorean fuzzy sets (PFS) [4, 5], among others. While fuzzy numbers are effective in managing uncertainty by representing data through degrees of membership, they are limited in their ability to explicitly address indeterminacy and inconsistency in information [6]. This limitation can result in oversimplified models and suboptimal decisions, especially in complex environments where data is not only uncertain but also ambiguous and inconsistent. These shortcomings highlight the need for more advanced approaches that can account for both the uncertainty and the multifaceted nature of real-world decision-making scenarios.

Thus, Smarandache [7] proposed a neutrosophic set, extending the concept of FS and IFS by incorporating three independent degrees which are the truth membership degree (T), indeterminacy-membership degree (I), and falsity-membership degree (F). These sets are well-suited for capturing incomplete, indeterminate, and inconsistent information, making them a powerful tool in decision-making scenarios where uncertainty is prevalent. Despite their theoretical robustness, neutrosophic sets face challenges in practical applications due to their non-standard intervals. The degrees T, I, and F are independent and can lie anywhere within the interval [0, 1], leading to complexity in computational procedures. To address the computational challenges, Wang et al. [8] developed single-valued neutrosophic sets (SVNS). In an SVNS, each element is characterized by single values for T, I, and F, simplifying the representation and computational processes. Building on SVNS, Wang et al. [9] further introduced interval neutrosophic sets (INS), where each membership degree is represented as an interval rather than a single value. This extension allows for a more flexible and comprehensive representation of uncertainty. Due to its advantages and efficiency, INS has widely been extended and applied in decision-making problems.

Multi-criteria decision-making (MCDM) is a valuable tool for identifying the best options in complex situations. When combined with fuzzy theory, MCDM becomes even more effective for making reliable judgments across various domains. The integration of fuzzy MCDM approaches enables more precise identification of optimal solutions, particularly in scenarios where evaluating numerous factors is a challenging and intricate process [10]. Over the years, several well-known fuzzy MCDM methods have been proposed to address MCDM problems. Among these are fuzzy Analytic Hierarchy Process (AHP) [11], fuzzy Technique for Order of Preference by Similarity to Ideal Solution (TOPSIS) [12], fuzzy Decision-Making Trial and Evaluation Laboratory (DEMATEL) [13], fuzzy VlseKriterijumska Optimizacija i Kompromisno Resenje (VIKOR) [14], and fuzzy Measurement Alternatives and Ranking according to Compromise Solution (MARCOS) [15].

Fuzzy MARCOS is particularly useful in complex decision-making environments characterized by uncertainty and vagueness. Notable applications include supply chain management, risk assessment, healthcare, project management, education, and financial decision-making. However, since the onset of the COVID-19 pandemic, decision-making across various sectors has become increasingly complex due to heightened uncertainty, rapid changes, and conflicting criteria. The pandemic has also accelerated the adoption of advanced decision-making frameworks in education and remote learning. To minimize face-to-face interaction between lecturers and students, higher education institutions worldwide adopted open distance learning (ODL).

Despite its widespread use, ODL has faced significant challenges in underdeveloped countries, where limited internet access, financial constraints, lack of face-to-face interaction, delayed response times, and the absence of traditional classroom socialization have hindered its effectiveness. Furthermore, while fuzzy MARCOS offers valuable insights, it struggles to adequately address uncertainty and incomplete information. To overcome these limitations, interval neutrosophic numbers which consider truth, indeterminacy, and falsity membership are incorporated into the MARCOS technique to better represent uncertainty in human beliefs and expert judgments.

This study extends fuzzy MARCOS by integrating interval neutrosophic numbers, replacing crisp values with interval neutrosophic numbers to examine key success criteria for learning in ODL environments. The proposed interval neutrosophic MARCOS method is a novel ranking algorithm designed to rationally prioritize options in group decision-making. By leveraging this advanced framework, the method provides a comprehensive approach for processing complex data based on group expert evaluations, thereby enhancing the effectiveness of decision-making in uncertain and multifaceted scenarios.

To achieve this goal, the remaining sections of this paper are structured as follows: Sect. 2 describes the methodology. Subsequently, Sect. 3 briefly explains the implementation of interval neutrosophic MARCOS, followed by Sect. 4, which presents the results obtained and discusses them. Finally, this paper concludes with conclusions and recommendations. Subsequent paragraphs, however, are indented.

2 Methodology

MARCOS is an MCDM method used to evaluate and rank alternatives based on multiple criteria. The MARCOS method was introduced by Stević et al. [16] as a novel MCDM approach which developed to address the need for a more balanced decision-making model. Its purpose is to generate robust and realistic rankings of alternatives in complex decision scenarios. Traditional MCDM methods, such as TOPSIS, VIKOR, and AHP, often have limitations related to normalization techniques, sensitivity to criteria weights, and the inability to handle both ideal and anti-ideal solutions comprehensively. This developed method aims to identify a compromise solution that balances multiple criteria by taking into account both the best and worst scenarios. This section demonstrates the general step-by-step of the MARCOS method which consists of nine steps.

Step 1: Obtain the DMs preferences.

Decision-makers are experts who play a crucial role in making important decisions. According to Taherdoost and Madanchian [17], certain decision-making methods involve multiple decision-makers in their case studies. In this problem, $\text{DM} = \{\text{DM}_1, \text{DM}_2, \ldots, \text{DM}_k\}$ represents the decision-makers.

Step 2: Determine the criteria and weightage of DMs.

The criteria, $C = \{C_1, C_2, \ldots, C_N\}$ and the weightage of DMs, $W = \{W_1, W_2, \ldots, W_q\}$ for critical success criteria are determined, ensuring that the total weight sums to 1. Table 1 shows the number of years of decision-makers' experience and Table 2 provides the linguistic terms in INS that are used to transform the decision-makers judgments.

Step 3: Construct INS.

Table 1 Number of years of DMs' experience

Years of experience	DM
< 3	1
4–6	2
7–10	3
> 10	4

Table 2 Linguistic terms of INS [17]

Linguistic terms	Interval neutrosophic set
Most non-preferred (MN)	(0.1, 0.2), (0.5, 0.6), (0.7, 0.8)
Non-preferred (NP)	(0.2, 0.4), (0.5, 0.6), (0.5, 0.6)
Neutral (N)	(0.4, 0.6), (0.4, 0.5), (0.3, 0.4)
Preferred (P)	(0.6, 0.8), (0.3, 0.4), (0.2, 0.4)
Most preferred (MP)	(0.7, 0.9), (0.2, 0.3), (0.1, 0.2)

Step 4: Aggregate all the INS.

The purpose of aggregation is to combine all the input individual data into single representative one. All the INS decisions are aggregated by applying Interval Neutrosophic Weighted Aggregate (INWA) operators to decision-making.

$$\mathrm{INWA}_w(A_1, A_2, \ldots, A_n) = \left\langle \begin{array}{c} \left[1 - \prod_{i=1}^{n}\left(1 - \inf T_{A_i}\right)^{w_j}, 1 - \prod_{i=1}^{n}\left(1 - \sup T_{A_i}\right)^{w_j}\right], \\ \left[\prod_{i=1}^{n} \inf I_{A_i}^{w_j}, \prod_{i=1}^{n} \sup I_{A_i}^{w_j}\right], \left[\prod_{i=1}^{n} \inf F_{A_i}^{w_j}, \prod_{i=1}^{n} \sup F_{A_i}^{w_j}\right] \end{array} \right\rangle \quad (1)$$

Let $A_i = \langle T_i, I_i, F_i \rangle$ $(i = 1, 2, \ldots, n)$ be collection of INS and $W = (w_1, w_2, \ldots, w_n)$ is the weight vector of A_i $(i = 1, 2, \ldots, n)$ with $w_j \in [0, 1]$ and $\sum_{j=1}^{n} w_{j=1}$.

Step 5: Normalize the decision-making values of different criteria.

Normalization is the process of neutralizing the impact of different types of criteria. Normalization of the extended initial matrix X. The elements of the normalized matrix are obtained by applying Eqs. (2) and (3).

$$n_{ij} = \left(n_{ij}^T, n_{ij}^L, n_{ij}^F\right) = \left(\frac{x_{id}^T}{x_{ij}^F}, \frac{x_{id}^T}{x_{ij}^L}, \frac{x_{id}^T}{x_{ij}^T}\right) \text{ if } j \in C \quad (2)$$

$$n_{ij} = \left(n_{ij}^T, n_{ij}^L, n_{ij}^F\right) = \left(\frac{x_{id}^T}{x_{ij}^F}, \frac{x_{id}^L}{x_{ij}^F}, \frac{x_{id}^F}{x_{ij}^F}\right) \text{ if } j \in B, \quad (3)$$

where elements x_{ij} and x_{id} represents the elements of the matrix X.

Step 6: Calculate the sum of the elements of the weight matrix, S_i.

The S_i $(i = 1, 2, 3 \ldots, m)$ matrix implies the sum of values by row including anti-ideal and ideal solutions by using the following expression:

$$S_i = \sum_{i=1}^{n} v_{ij} \quad (4)$$

$$v = \left(v_{ij}^T, v_{ij}^I, v_{ij}^F\right) = n_{ij} \otimes w_j = \left(n_{ij}^T \times w_j^T, n_{ij}^I \times w_j^I, n_{ij}^F \times w_j^F\right) \quad (5)$$

where v_{ij} is from $n_{ij} \times w_j$.

Step 7: Calculate the deneutrosophication to obtain a crisp value.

$$D(\tilde{\tilde{x}}_j) = \left(\frac{T_j^L + T_j^U}{2}\right) + \left(1 - \frac{I_j^L + I_j^U}{2}\right) * \left(I_j^U\right) + \left(\frac{F_j^L + F_j^U}{2}\right) * \left(1 - F_j^U\right), \tag{6}$$

where $\tilde{\tilde{x}}_j = \left\langle \left[T_j^L, T_j^U\right], \left[I_j^L, I_j^U\right], \left[F_j^L, F_j^U\right] \right\rangle$ is a collection of interval-valued neutrosophic numbers were $j = 1, 2, \ldots, k$ and k is the number of DMs.

Step 8: Computing and determination of utility degree, matrix T_i, new number of D and utility functions.

$$K_i^- = \frac{S_i}{S_{\text{aai}}} \tag{7}$$

$$K_i^+ = \frac{S_i}{S_{\text{ai}}} \tag{8}$$

$$T_i = K_i^- \oplus K_i^+ \tag{9}$$

$$D = \max_i T_i \tag{10}$$

$$f(K_i) = \frac{K_i^+ + K_i^-}{1 + \frac{1 - f(K_i^+)}{f(K_i^+)} + \frac{1 - f(K_i^-)}{f(K_i^-)}}, \tag{11}$$

where $f\left(k_i^-\right)$ represent the utility function in relation to anti-ideal (AAI), while $f\left(k_i^+\right)$ represent the utility function in relation to the ideal (AI) solution. The utility function was determined by Eqs. (12) and (13).

$$f\left(K_i^-\right) = \frac{K_i^+}{D} \tag{12}$$

$$f\left(K_i^+\right) = \frac{K_i^-}{D}. \tag{13}$$

Step 9: Ranking the criteria and alternatives based on criteria weight.

3 Implementation

In light of the widespread use of ODL and the emerging trend toward e-learning, educators must choose an appropriate medium to ensure students can easily access all learning materials. Regarding these issues, numerous online education programs has been modified, and numerous online learning sites have also made their services available for free [18–20]. Thus, this project was conducted for UiTM Shah Alam lecturers to find out generally suitable medium for online classes to improve the effectiveness of online teaching and learning experience. Lectures most frequently used the following e-learning platforms:

A_1: Blended learning
A_2: Asynchronous learning
A_3: Synchronous learning.

Each platform may have different features, which differentiate them from other platforms.

Step 1: Obtain the DMs preferences.

Five DMs, comprising lecturers from the College of Computing, Informatics, and Media (KPPIM), UiTM Shah Alam, who are experts in teaching, are selected to address the problem. The DMs were selected randomly, and this set of DMs is referred as $\text{DM} = \{\text{DM}_1, \text{DM}_2, \text{DM}_3, \text{DM}_4, \text{DM}_5\}$.

Step 2: Identify the decision criteria.

The DMs are required to determine the criteria weights. The set of criteria known as $C = \{C_1, C_2, C_3, C_4, C_5, C_6\}$.

C_1: Student characteristics
C_2: Instructor characteristics
C_3: Learning environment
C_4: Instructional environment
C_5: Information Technology
C_6: Level of collaboration.

Step 3: Determine the criteria.

Each of experts determine the criteria based on their perspectives, experience and knowledge of utilized online learning platform. Expert's preference on the criteria's in shows in Table 3.

After that, transform linguistic variables in Table 3 to the INS and represent in the matrix form.

Table 3 Decision-makers' preferences based on criteria

	DM_1	DM_2	DM_3	DM_4	DM_5
C_1	MP	P	P	N	MP
C_2	P	N	P	MP	MP
C_3	MP	N	P	MP	P
C_4	MP	P	P	MP	N
C_5	P	P	P	MP	P
C_6	N	N	P	MP	N

$$X = \begin{pmatrix} x_{11} & x_{12} & x_{13} & x_{14} & x_{15} \\ x_{21} & x_{22} & x_{23} & x_{24} & x_{25} \\ x_{31} & x_{32} & x_{33} & x_{34} & x_{35} \\ x_{41} & x_{42} & x_{43} & x_{44} & x_{45} \\ x_{51} & x_{52} & x_{53} & x_{54} & x_{55} \\ x_{61} & x_{62} & x_{63} & x_{64} & x_{65} \end{pmatrix}$$

Step 4: Perform the aggregation of the matrix of criteria.

To combine the input from each expert into a single representative using INWA operators, the data from each DM must first be aggregated. Using Eq. (1), the DM's opinion can be aggregated; for instance, for MP, the element x_{11} can be obtained as follows:

$$x_{11} = \left\langle \begin{bmatrix} 1 - \begin{pmatrix} (1-0.7)^{0.1111} \times (1-0.6)^{0.2222} \times (1-0.6)^{0.2222} \\ \times (1-0.4)^{0.2222} \times (1-0.7)^{0.2222} \end{pmatrix}, \\ 1 - \begin{pmatrix} (1-0.9)^{0.1111} \times (1-0.8)^{0.2222} \times (1-0.8)^{0.2222} \\ \times (1-0.6)^{0.2222} \times (1-0.9)^{0.2222} \end{pmatrix} \end{bmatrix}, \right.$$

$$\begin{bmatrix} \begin{pmatrix} (0.2)^{0.1111} \times (0.3)^{0.2222} \times (0.3)^{0.2222} \\ \times (0.4)^{0.2222} \times (0.2)^{0.2222} \end{pmatrix}, \\ \begin{pmatrix} (0.3)^{0.1111} \times (0.4)^{0.2222} \times (0.4)^{0.2222} \\ \times (0.5)^{0.2222} \times (0.3)^{0.2222} \end{pmatrix} \end{bmatrix},$$

$$\left. \begin{bmatrix} \begin{pmatrix} (0.1)^{0.1111} \times (0.2)^{0.2222} \times (0.2)^{0.2222} \\ \times (0.3)^{0.2222} \times (0.4)^{0.2222} \end{pmatrix}, \\ \begin{pmatrix} (0.2)^{0.1111} \times (0.4)^{0.2222} \times (0.4)^{0.2222} \\ \times (0.4)^{0.2222} \times (0.2)^{0.2222} \end{pmatrix} \end{bmatrix} \right\rangle$$

$$= \langle [0.60231, 0.814825], [0.279376, 0.3819], [0.173707, 0.31748] \rangle.$$

The rest of the elements are calculated similarly. After performing the aggregation, the results are shown in Table 4.

Step 5: The crisp aggregate matrix of criteria is obtained by deneutrosophication.

Deneutrosophication is the process to transform the interval-valued neutrosophic number to the crisp number. Equation (6) is used to obtain crisp value of each criterion. For instance, for the element x_{11} where it refers to C_1 and DM1.

$$C_1 = \left(\frac{0.60231 + 0.814825}{2}\right) + \left(1 - \frac{0.279376 + 0.3819}{2}\right)$$

Table 4 Value of T_i

T_i	Values
T_1	1.99162474
T_2	1.97265674
T_3	1.99162474
T_4	1.99162474
T_5	2.04154681
T_6	1.96011046

$$* (0.3819) + \left(\frac{0.173707 + 0.31748}{2}\right) * (1 - 0.31748)$$
$$= 1.097729$$

The rest of the criteria are calculated in a similar way.

Step 6: Normalize the decision matrix using the aggregated matrix.

In this step, each number in a row of the aggregate matrix is divided by the largest number in that row. Normalized matrix denoted as n can be achieved using Eqs. (2) and (3).

$$\begin{aligned} n_1 &= \left(\left(\frac{0.60231}{0.814825}, \frac{0.814825}{0.814825}\right), \left(\frac{0.279376}{0.814825}, \frac{0.3819}{0.814825}\right), \right. \\ &\quad \left. \left(\frac{0.173707}{0.814825}, \frac{0.31748}{0.814825}\right)\right) \\ &= ((0.739189, 1), (0.342866, 0.468689), (0.213184, 0.38963)) \\ n_6 &= \left(\left(\frac{0.529971}{0.748016}, \frac{0.748016}{0.748016}\right), \left(\frac{0.321662}{0.748016}, \frac{0.424751}{0.748016}\right), \right. \\ &\quad \left. \left(\frac{0.214765}{0.748016}, \frac{0.342898}{0.748016}\right)\right) \\ &= ((0.708503, 1), (0.430021, 0.567837), (0.287113, 0.45841)). \end{aligned}$$

Step 7: Calculate the weights of the normalized decision matrix.

In step 5, each number is multiplied by the weight assigned to each criterion. The values of the weighted normalized matrix show are obtained using Eq. (5).

$$\begin{aligned} v_1 &= ((0.739189 \times 0.168253), (1 \times 0.168253)), \\ &\quad ((0.342866 \times 0.168253), (0.468689 \times 0.168253)), \\ &\quad ((0.213184 \times 0.168253), (0.38963 \times 0.168253)) \\ &= (0.124371, 0.168253), (0.057688, 0.078859), (0.035869, 0.065557). \end{aligned}$$

Step 8: Calculation of the utility degree of criteria C_i, utility function $f(C_i)$ using Eqs. (7)–(13) by replacing K to C as below.

$$C_i^- = \frac{SC_i}{SC_{\text{aai}}}$$

$$C_i^+ = \frac{SC_i}{SC_{\text{ai}}}$$

$$f(C_i) = \frac{C_i^+ + C_i^-}{1 + \frac{1-f(C_i^+)}{f(C_i^+)} + \frac{1-f(C_i^-)}{f(C_i^-)}}$$

where $f(C_i^-) = \frac{C_i^+}{C_i^+ + C_i^-}, f(C_i^+) = \frac{C_i^-}{C_i^+ + C_i^-}$

$$C_1^- = \frac{0.53059641}{0.52220057} = 1.01607781$$

$$C_1^+ = \frac{0.53059641}{0.54389634} = 0.975547.$$

The other elements were also calculated similarly.

Using Eq. (9), the value T_i is calculated as in Table 4.

Next, it is essential to calculate the new value of D using Eq. (10), $D = \max_i T_i$ and $D = 2.0415$ is obtained which will result in the crisp number. Utility functions in relation to ideal (AAI), $f\left(C_i^+\right)$ and anti-ideal (AI), $f\left(C_i^-\right)$ solution are determined by applying Eqs. (12) and (13).

$$f\left(C_1^+\right) = \frac{1.01607781}{2.041547} = 0.4977$$

$$f\left(C_1^-\right) = \frac{0.975547}{2.041547} = 0.477847.$$

Determination of utility functions of criteria $f(C_i)$. The utility functions of criteria are defined by Eq. (11).

$$f(C_1) = \frac{C_i^+ + C_i^-}{1 + \frac{1-f(C_i^+)}{f(C_i^+)} + \frac{1-f(C_i^-)}{f(C_i^-)}}$$

$$= \frac{1.01607781 + 0.975547}{1 + \frac{1-0.4977}{0.4977} + \frac{1-0.477847}{0.477847}} = 0.642053.$$

Next, we use similar steps and formulas to calculate the alternatives based on criteria.

Step 9: Ranking the criteria and the alternatives.

Table 5 Final result and ranking for criteria using interval neutrosophic MARCOS

	$f(C_i^-)$	$f(C_i^+)$	C_i^-	C_i^+	$f(C_i)$	Rank
C_1	0.477	0.497	1.016	0.975	0.642	2
C_2	0.473	0.493	1.006	0.966	0.627	5
C_3	0.477	0.497	1.016	0.975	0.642	2
C_4	0.477	0.497	1.016	0.975	0.642	2
C_5	0.489	0.510	1.041	1	0.680	1
C_6	0.470	0.489	1	0.96	0.618	6

Table 6 Final result and ranking for alternatives using interval neutrosophic MARCOS

	$f(A_i^-)$	$f(A_i^+)$	A_i^-	A_i^+	$f(A_i)$	Rank
A_1	0.008	0.086	1	0.093	0.008	3
A_2	0.008	0.087	1.014	0.095	0.008	2
A_3	0.086	0.914	10.64	1	0.992	1

The calculation for the final two steps and the results obtained using the interval neutrosophic MARCOS method are shown in Tables 5 and 6.

4 Results

The goal of this study was to determine the learning success criteria and types of learning system during ODL. The weightage of each criterion has been set by five experts from KPPIM of UiTM Shah Alam who are experienced in ODL were form as a sample in the study. The interval neutrosophic MARCOS technique ranks the best criteria and alternative depending on the DMs preferences.

In this study, six criteria are utilized to evaluate and select the most effective ODL systems. These criteria are as follows: C_1 (student characteristics), C_2 (instructor characteristics), C_3 (learning environment), C_4 (instructional design), C_5 (information technology), and C_6 (level of collaboration). The weightage assigned to each criterion is illustrated in Fig. 1.

The criteria are ranked based on all data that has been collected and computed. The mathematical result is visualized using a bar chart, as seen in Figs. 1 and 2. The criterion's ranking is shown at the end of each bar. Based on the calculation, we can conclude that information technology is the most important criterion in the success of open distance learning.

This survey identified three alternatives in total. Blended learning is designated as A_1, Synchronous learning as A_2 and asynchronous learning as A_3. The percentage of each alternative is shown in Fig. 3 and the corresponding ranking is illustrated in Fig. 4.

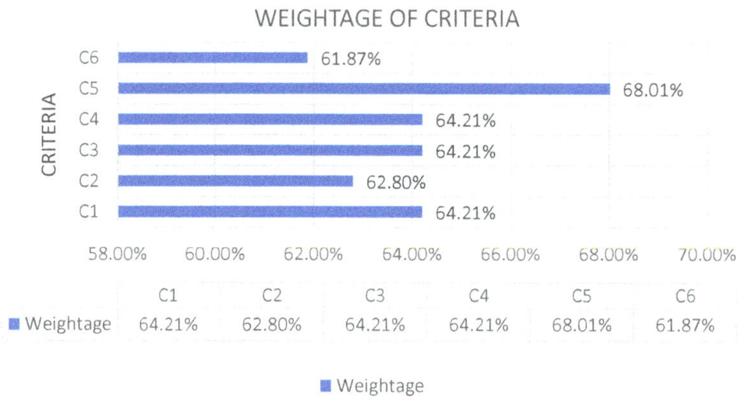

Fig. 1 Weightage of criteria

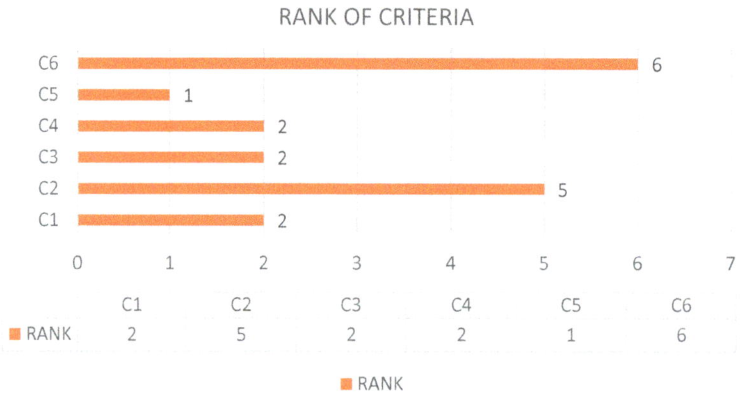

Fig. 2 Rank of criteria

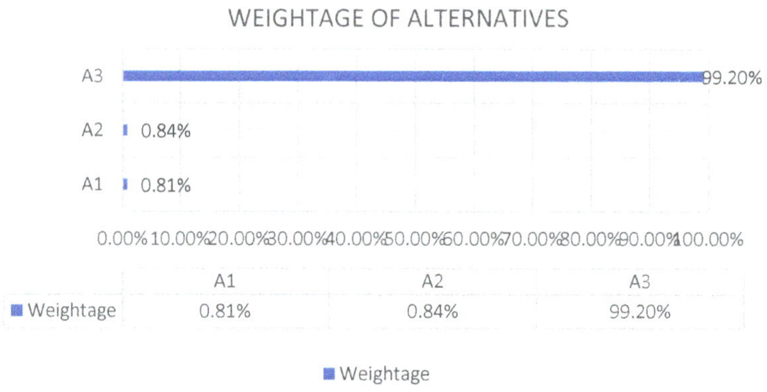

Fig. 3 Weightage of alternatives

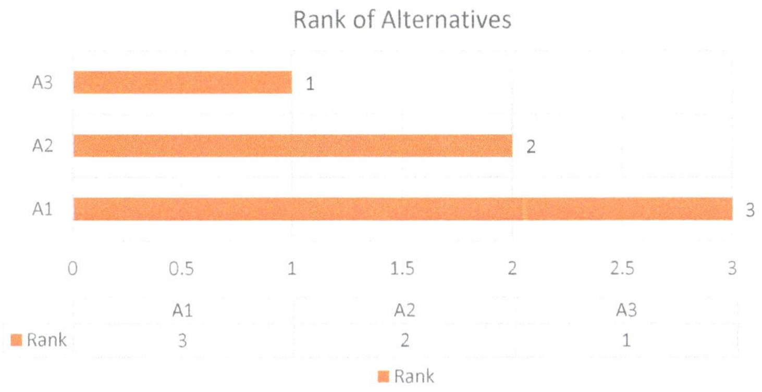

Fig. 4 Rank of alternatives

The outcome of the calculation of the values of similar steps as ranking the criteria is depicted by the bar chart. The ranking for the alternatives based on criteria is stated at the end of each bar. As a result, we can conclude that asynchronous learning allows students to access learning materials and complete assignments at their own pace and preferred time. This flexibility is particularly beneficial for individuals with busy schedules, working professionals, or those in different time zones, as they can fit learning around their commitments.

5 Conclusion

In conclusion, this study identifies the optimal criteria for open distance learning by applying the interval neutrosophic MARCOS method. It incorporates indeterminacy, inconsistency, and incompleteness by utilizing linguistic variables and decision matrices. The research aimed to investigate the effectiveness of this innovative approach and identify the key criteria that contribute to successful learning outcomes in an open distance learning environment. The study aims to determine which criteria and alternatives are the best for success in the study during open distance learning using interval neutrosophic MARCOS. The selection of the most effective success criteria for open distance learning is determined by six criteria: student characteristics, instructor characteristics, learning environment, instructional design, information technology, and the level of collaboration, along with three alternatives: blended learning, synchronous learning, and asynchronous learning.

On the other hand, the ranking of open distance learning systems can be established using the interval neutrosophic MARCOS method. Based on the findings of this study, it can be concluded that asynchronous learning is the most effective system for open distance learning. Asynchronous learning, enabled by information technology, offers a flexible, accessible, personalized, and scalable open distance learning system. By utilizing the strengths of asynchronous learning and tapping into the capabilities of information technology, educational institutions can design dynamic and impactful learning experiences that enable learners to reach their educational objectives, irrespective of time, location, or individual circumstances.

Therefore, based on our findings and discussions, decision-makers can effectively apply the interval neutrosophic MARCOS method by following these steps to evaluate alternatives and criteria. This approach serves as an efficient and recommended method, especially for those seeking to make optimal choices and decisions.

Acknowledgements We would like to express our appreciation and record our gratitude to the School of Mathematical Sciences. Faculty of Computer and Mathematical Sciences, Universiti Teknologi MARA for their support in conducting this research. This study was funded by Fundamental Research Grant Scheme, FRGS/1/2022/STG06/UITM/02/17, 600-RMC/FRGS 5/3 (126/2022), Ministry of Higher Education Malaysia, and MYRA LPHD Grant from Universiti Teknologi MARA (UiTM), 600-RMC/GPM LPHD 5/3 (139/2021).

References

1. Zadeh, L.A.: Fuzzy sets. Inf. Control. **8**, 338–353 (1965)
2. Atanassov, K.T.: Intuitionistic fuzzy sets. Fuzzy Sets Syst. **20**, 87–96 (1986)
3. Gorzałczany, M.B.: A method of inference in approximate reasoning based on interval-valued fuzzy sets. Fuzzy Sets Syst. **21**(1), 1–17 (1987)
4. Yager, R.R.: Pythagorean membership grades in multicriteria decision making. IEEE Trans. Fuzzy Syst. **22**(4), 958–965 (2014)
5. Yager, R.R., Abassov, A.M.: Pythagorean membership grades, complex numbers, and decision-making. Int. J. Intell. Syst. **28**(5), 436–452 (2013)
6. Anvari-Sotoudeh, A.A.: Critical review on theoretical drawbacks and mathematical incorrect assumptions in fuzzy OR methods: review from 2010 to 2020. Appl. Soft Comput. **93**, 106354 (2020)
7. Smarandache, F.: Neutrosophy: Neutrosophic Probability, Set, and Logic, pp. 1–141. American Research Press, Rehoboth (1998)
8. Wang, H., Smarandache, F., Zhang, Y., Sunderraman, R.: Single valued neutrosophic set. Multisp. Multistruct. IV **4**, 410–413 (2010)
9. Wang, H., Smarandache, F., Zhang, Y., Sunderraman, R.: Interval Neutrosophic Sets and Logic: Theory and Applications in Computing. Hexis, Phoenix, AZ (2005)
10. Kaya, I., Çolak, M., Terzi, F.A.: Comprehensive review of fuzzy multi criteria decision making methodologies for energy policy making. Energy Strategy Rev. **24**, 207–228 (2019)
11. Emrouznejad, A., Ho, W.: Fuzzy Analytic Hierarchy Process. Taylor & Francis, CRC Press (2022)
12. Baharin, N.H., Rashidi, N.F., Mahad, N.F.: Manager selection using fuzzy TOPSIS method. J. Phys. Conf. Ser. **1988**, 012057 (2021)
13. López-Ospina, H., Pardo, D., Rojas, A., Barros-Castro, R., Palicio, K., Quezada, L.: A revisited fuzzy DEMATEL and optimization method for strategy map design under the BSC framework: selection of objectives and relationships. Soft. Comput. **26**, 6619–6644 (2022)
14. Talib, A.: Fuzzy VIKOR approach to evaluate the information security policies and analyze the content of press agencies in gulf countries. J. Inf. Secur. **11**, 189–200 (2020)
15. Zolfani, S.H., Gorener, A., Toker, K.: A hybrid fuzzy MCDM approach for prioritizing the solutions of resource recovery business model adoption to overcome its barriers in emerging economies. J. Clean. Prod. **413**, 137362 (2023)

16. Stević, Z., Pamučar, D., Puška, A., Chatterjee, P.: Sustainable supplier selection in healthcare industries using a new MCDM method: Measurement of Alternatives and Ranking according to COmpromise Solution (MARCOS). Comput. Ind. Eng. **140**, 106231 (2020)
17. Taherdoost, H., Madanchian, M.: Multi-criteria decision making (MCDM) methods and concepts. Encyclopedia **3**, 77–87 (2023)
18. Ferri, F., Grifoni, P., Guzzo, T.: Online learning and emergency remote teaching: opportunities and challenges in emergency situations. Societies **10**, 1–18 (2020)
19. Mukhtar, K., Javed, K., Arooj, M., Sethi, A.: Advantages, limitations and recommendations for online learning during COVID-19 pandemic era. Pak. J. Med. Sci. **36**, S27–S31 (2020)
20. El-Sakran, A., Salman, R., Alzaatreh, A.: Impacts of emergency remote teaching on college students amid COVID-19 in the UAE. Int. J. Environ. Res. Public Health **19**, 2979 (2022)

Optimizing Workforce Scheduling Using Binary Programming Model for a Veterinary Clinic

Nur Farhana Mohd Asri, Norlenda Mohd Noor, and Zuraida Alwadood

Abstract A veterinary clinic is a medical facility that treats sick or wounded animals. Many veterinary clinics have been set up in Malaysia to assist animal enthusiasts in treating their pets. Regular check-ups for pets, emergency, and urgent treatment, as well as boarding and grooming, are all available at the clinic. Many of the services need many staff to handle. Thus, to manage many staff and ensure that each position is well assigned, a good workforce schedule must be created. Therefore, this study modified a Binary Programming (BP) model to represent the real-life problem of workforce scheduling for a veterinary clinic. The study aimed to minimize the weekly number of staff in any working shifts to produce the optimal schedule in the veterinary clinic. This weekly schedule is created to determine the task for each staff shift, while considering the rest day in between working shifts, the minimum number of employees for each shift, among other consideration. This paper focused on two sub-case studies, and they are based on the data collected from a local veterinary clinic. The modified BP model was then solved using Open Solver software. To determine the best working schedule of veterinary staff, the results obtained by the two case studies were compared based on the shortcomings. The findings indicate that the new model could be used to offer a flexible working schedule that is both reliable and effective; however, there are also some rooms for improvement.

Keywords Workforce scheduling · Binary linear programming

1 Introduction

Scheduling is an active field of discrete optimization that plays an important role in the manufacturing and service industries. Scheduling is described as allocating resources to tasks throughout time to optimize a restricted performance index. Generally, labor

N. F. M. Asri · N. M. Noor (✉) · Z. Alwadood
Faculty of Computer and Mathematical Sciences, Universiti Teknologi MARA, Shah Alam, Selangor, Malaysia
e-mail: norlenda@uitm.edu.my

A. Shuib et al. (eds.), *Mathematical Modeling and Numerical Analysis*,
Springer Proceedings in Mathematics & Statistics 510,
https://doi.org/10.1007/978-981-96-9350-4_10

or workforce scheduling is the process of assigning an adequate number of workers to tasks on each workday [1]. Hospitals, retail stores, contact centers, cleaning, home care, guarding, manufacturing, and transportation of products are some of the industries that use employee rostering and shift generation [2]. Nowadays, in the challenging era due to the technology development and epidemic phase, creating an effective schedule is strongly needed.

Labor or workforce scheduling is a complicated task because it involves many aspects, including meeting labor requirements, management norms, government regulations, and also the human capabilities which is the most important factor [3]. When the organization grows larger and the number of employees increases, the complexity of scheduling all employees while adhering to law, regulations, and worker preferences will also increase [4]. Workforce scheduling problems are generally divided into three categories which are shift scheduling, days off scheduling, and tour scheduling [5]. Shift scheduling determines each employee's work and refreshment hours per day. Next, days off scheduling defines every employee's workday and off days each week or multiple-week work cycle. Meanwhile, tour scheduling is defined as a combination between shift and day off scheduling in determining each employee's workday hours and off days each week or multiple-week work cycle.

Ineffective scheduling may necessitate the hiring of temporary employees, raising costs, while schedules that do not comply with all legal requirements may result in fines and employee displeasure [6]. The crisis of a significant number of workers suffering from workplace stress is a serious worldwide issue that affects both local and international countries [7]. This is due to excessive working hours, insufficient break or refreshment time, and imbalance salary. The result of a work process is determined by which workers are allocated with which duties. However, optimum assignments based on employees' qualifications might occasionally result in an unequal and imbalanced job allocation. According to [8], equitable workload allocation without taking into account the employee's qualifications may result in inefficient employee-task matching, which leads to poor performance. Thus, adequate and appropriate labor scheduling should be effectively structured to ensure job satisfaction among all employees.

Various studies have attempted to apply mathematical model in producing effective work schedule. Among mathematical model being used are Linear Programming, LP by [9] and [10]. Yange et al. [9] developed a web to manage employee schedules at a veterinary hospital using LP model and implemented a meta-heuristic approach which is Tabu Search Algorithm. Hasan and Arefin [10], on the other hand, studied the application of LP to the shift scheduling problem on police officers at the Clark County Sheriff's Department. The study successfully obtained the minimum value of the model objective function which indicates that there must be 19 number of officers required to satisfy all shift requirements.

Alwadood et al. [11] modified the Binary Integer Programming (BIP) model to obtain the appropriate number of workers for each shift in each housekeeping sector by minimizing the number of personnel on duty for each position throughout all shifts each day. Result shows that the new schedule necessitates 140 working shifts every week, in contrast to the previous schedule of 216 working shifts for the same timeframe.

Shin et al. [5] developed an ILP model in optimizing the workforce schedule for the security staff at University Tun Hussein Onn Malaysia (UTHM). The study aimed to maximize the job satisfaction of the security staff by allowing them to select their desired shift and day off while preserving the university's rules in consideration. The results indicate that the value of overall satisfaction of security personnel concerning work shifts and days off has been effectively increased from 228.33 to 394.33 and successfully satisfied all the constraints that have been set. Permatasari et al. [12] also developed an ILP model to represent aircraft maintenance manpower shift that can be utilized to identify the best personnel allocation and shift planning. These two criteria strongly affect the operating cost; therefore, they are considered to meet the target of minimum total cost. As for the finding, the model had succeeded in achieving its objective as there was a 30% reduction in operational and manpower costs. Another study by Herawati et al. [3] focused on minimizing the violation of workers' preference fulfillment by implementing ILP and using the exact method approach to obtain the optimal solution. As a result, the optimal schedule was achieved with the physical shift workloads were distributed evenly among the employees.

Integer Goal Programming, IGP model by Mahmud et al. [7] have contributed in enhancing the workforce scheduling in electronic industrial company fields. The goal of the study was to enhance employee work scheduling concerning the company's staff requirements and staff preferences. The schedule produced by this study is effective in producing an ideal solution with respect to different work limits and expenses, balancing workloads, and controlling idle days among all technical staff members, without any changes in total working hours.

Mixed Integer Linear Programming, MILP model by Akyurt et al. [13] has determined the optimal shift workforce planning of pilots in an airline company by minimizing the total number of existing pilots and newly employed pilots. The goal of the study was to simulate a workforce-planning problem for pilot in an airline firm, such as captain and first officer, and to come up with an efficient strategy that maximizes the use of minimum workforce needs. During the planning horizon, the model considered various variables such as hiring pilots with various skill types, resignations, retirements, pilot vacations, and transitions between different skills. As a result, it shows that a firm should use transitions approach instead of hiring new employees and this is a better human resource planning decision. Patidar et al. [14] formulated and implemented two optimization models, Pure Integer (PI) and Mixed-Integer Linear Programming (MILP) to solve staff scheduling problems within the aircraft maintenance workforce. The study aimed to minimize the employee labor costs based on the tour cost. Determining the tour cost is a complex operation, and the airline company must have set the minimum salary for its employees. Results for both models were compared, and it turned out to generate the same optimal solution. The study has succeeded to minimize the tour cost to Rs.10,000.

Several reasons, a visit to veterinary clinic might be stressful for the customers and pets [15]. Therefore, a good service provided by the clinic can at least reduce the customers' stress. The services provided should also be commensurate with the price charged. Undoubtedly, without efficient work management, the quality of services offered to customers will be affected. Hence, this study focuses on optimizing the workforce schedule for employees at a local veterinary clinic. It is developed based on the number of staff, shift, working days, and the department where the staff are positioned. The Binary Programming (BP) model is employed to optimize staff scheduling. Once the optimal schedule is determined, staff receive balanced task allocations, thereby improving work quality and enabling higher service standards.

2 Methodology

The Binary Programming, BP model is based on Paramadevan [16]. This model was designed where each veterinary staff is allowed only one shift per day in specific department within the time horizon. The model formulation including the symbol of subscript, parameters, and decision variables of the binary are listed below:

Sets and Parameters:

- I Set of staffs, $I = \{1, 2, \ldots, i\}$.
- J Set of shifts, $J = \{1, 2, \ldots, j\}$.
- K Set of working days, $K = \{1, 2, \ldots, k\}$.
- L Set of departments, $L = \{1, 2, \ldots, l\}$.
- A_{jk} The minimum number of shifts in a week.
- B_{jk} The maximum number of shifts in a week.
- C_{ik} The minimum number of staff in a day.
- D_{ijk} The minimum number of staff working in morning shift per day.
- E_{ijk} The minimum number of staff working in afternoon shift per day.
- x_{ijkl} Staff (i) is assigned for the shift-type (j) on a day (k) at a department (l).

Decision variables:

$$x_{ijkl} = \begin{cases} 1, & \text{if staff } i \text{ is assigned to shift } j \text{ on day } k \text{ at department l} \\ 0, & \text{otherwise} \end{cases}.$$

The objective function is aimed to minimize the number of staff working at any shifts in a week:

$$\text{Min}, Z = \sum\sum\sum\sum x_{ijkl}, \quad \forall i \in I, j \in J, k \in K, l \in L. \tag{1}$$

Subject to:

$$\sum_{i=1}^{I} x_{ijkl} \le A_{jk}, \quad \forall i \in I, j \in J, k \in K, l \in L, \tag{2}$$

$$\sum_{i=1}^{I} x_{ijkl} = B_{jk}, \quad \forall j \in J, k \in K, l \in L, \tag{3}$$

$$\sum_{i=1}^{I} x_{ijkl} \leq C_{ik}, \quad \forall j \in J, k \in K, l \in L, \tag{4}$$

$$\sum_{i=1}^{I} x_{i1kl} \leq D_{ijk}, \quad j = 1, \forall k \in K, l \in L, \tag{5}$$

$$\sum_{i=1}^{I} x_{i2kl} \leq E_{ijk}, \quad j = 2, \forall k \in K, l \in L. \tag{6}$$

The objective function (1) corresponds to the minimum the number of staff working at any shifts in a week. Constraint (2) explained that staff shall be assigned according to the number of employees required in any shifts in a day except off days. Next, Constraint (3) ensures that a staff is required to work for a maximum number of shifts in a week. Constraint (4) restricts the number of staff needs to be assigned to work in specific department, in a shift for each day. Constraint (5) specifies the number of staff that should be allocated in morning shift, meanwhile Constraint (6), indicates the number of staff should be placed in afternoon shift for each day in a week.

3 Results and Discussion

This paper consists of two sub-cases which are Case Study I and Case Study II. The mathematical model for both case studies was constructed to address the problems in veterinary clinics. There are four pre-defined departments in this model. Department 1 refers to the veterinarian department, Department 2 refers to the veterinarian assistant department, and Department 3 and Department 4 refer to the receptionist department and the dispensary department, respectively. The difference between these two case studies is the job assignment of veterinarian assistants. In Case Study I, veterinarian assistants are assigned to Department 2, Department 3, and Department 4. This is because veterinarian assistants can perform the tasks in these three departments besides assisting veterinarians in the treatment room. Meanwhile, in Case Study II, veterinarian assistants were only assigned to Department 2, whereby they are only required to assist the veterinarian in the treatment room. The two outcomes from the case studies are compared and analyzed.

3.1 Case Study 1

For Case Study I, four veterinarians are placed under Department 1, and four clinical assistant veterinarians are categorized under Department 2, Department 3, and Department 4. And five clinic assistants are placed under Department 3 and Department 4. Table 1 shows the summary of results obtained based on model from case study I.

Table 1 shows the summary of the results generated and obtained from the Open Solver software for Case Study I. The number "1" indicates that veterinary staff i will be on duty on certain shift at any day in a predetermined department, while "0" means otherwise. For example, on day 1 in department 1, veterinary staff 1 will be working on afternoon shift, while veterinary staff 2 and 4 will be working on the morning shift.

The total in the last column of Table 1 is referring to the optimum value of the objective function. The original number of shifts available at the clinic in a week is 182 shifts and by using the model, it shows that only 65 shifts have been utilized. The available shifts of 182 is obtained from the number of shifts per day, number of days per week and total number of staffs and departments. Based on the generated result, all staff were assigned to at least one shift per day with two off days in a week. The total number of shifts per shift is also perfectly adequate which is five shifts per week.

For the veterinarian department, it indicates that there are two veterinarians in the morning shift and one veterinarian in the afternoon shift except for Day 6. Day 6 shows that only one veterinarian has been assigned to the morning shift. Nevertheless, this is acceptable since a predetermined constraint of this model only ensures at least two veterinarians should be allocated in the morning shift.

It is difficult to ensure that all veterinarian assistants are equally assigned into every available department. Therefore, the model constraint was set to restrict that there are at least two veterinarian assistants who must be assigned in the morning shift and one veterinarian assistant in the afternoon shift for Department 2. Meanwhile, for Department 3 and Department 4, it is sufficient to assign only one veterinarian assistant in the morning shift and one in the afternoon shift. However, the entire existing veterinarian assistant in the clinic could not accommodate the duties in Department 2. For example, on Day 2, Day 3, Day 4, and Day 6, no veterinarian assistant is assigned to Department 2 and this is due to the shortage of staff in the clinic. Therefore, this working schedule is not considered as an optimal schedule for veterinary staff.

Next, for Department 3 and Department 4, it shows that the assignment of tasks in the department is considered complete and satisfactory. This is because there are five clinic assistants being assigned and the number is sufficient to handle all the tasks in the department. Even though there is no clinic assistant was assigned to work in Department 4 on Day 3, there is one veterinary assistant staff who can do the clinic assistant's tasks, which is Staff 7 in the morning shift and Staff 5 in the afternoon shift.

Table1 Summary of results obtained based on model from case study I

D(l)	V(i)	Day (k)/Shift (j)														Total
		1		2		3		4		5		6		7		
		M	A	M	A	M	A	M	A	M	A	M	A	M	A	
1	1	0	1	0	1	1	0	1	0	0	0	0	0	1	0	9
	2	1	0	1	0	0	0	0	1	0	1	0	1	0	0	10
	3	0	0	0	0	0	1	1	0	1	0	1	0	0	1	11
	4	1	0	1	0	1	0	0	0	1	0	0	0	1	0	12
2	5	0	0	0	0	0	0	0	0	0	1	0	0	0	0	13
	6	1	0	0	0	0	0	0	0	0	0	0	0	0	0	9
	7	0	0	0	0	0	0	0	0	0	0	0	0	0	1	10
	8	0	0	0	0	0	0	0	0	0	0	0	0	1	0	11
3	5	0	0	0	0	0	0	0	0	0	0	0	0	0	1	12
	6	0	0	0	0	0	1	0	0	0	0	0	0	1	0	13
	7	0	0	1	0	0	0	0	0	1	0	0	1	0	0	9
	8	0	0	0	1	0	0	0	1	0	0	0	0	0	0	10
4	5	0	0	1	0	0	1	0	0	0	0	0	1	0	0	11
	6	0	0	0	0	0	0	0	1	1	0	0	0	0	0	12
	7	0	0	0	0	1	0	0	0	0	0	0	0	0	0	13
	8	1	0	0	0	0	0	0	0	0	1	0	0	0	0	9
3	9	0	0	1	0	1	0	1	0	0	0	0	0	0	0	10
	10	1	0	0	0	0	0	0	0	0	0	1	0	0	1	11
	11	0	0	0	0	0	0	0	0	0	1	1	0	0	0	12
	12	0	0	0	1	0	0	0	0	0	0	0	0	1	0	13
	13	0	1	0	0	0	1	0	1	0	0	0	0	0	0	9
4	9	0	0	0	0	0	0	0	0	0	0	0	1	0	1	10
	10	0	0	0	1	0	0	1	0	0	0	0	0	0	0	11
	11	1	0	0	0	0	0	0	1	0	0	1	0	1	0	12
	12	0	1	0	0	0	0	0	1	0	0	1	0	0	0	13
	13	0	0	1	0	0	0	0	0	1	0	0	0	0	0	9
Total		6	3	6	4	4	4	4	6	4	4	5	4	6	5	65

Note [M] Morning; [A] Afternoon; [D] Department; [V] Veterinary staff

3.2 *Case Study II*

For Case Study II, four veterinarians are placed under Department 1, four veterinarian assistants are categorized under Department 2 and five more clinic assistant are placed under Department 3 and Department 4. Table 2 shows the summary of results obtained based on model from Case Study II.

Table 2 Summary of results obtained based on model from Case Study II

D(*l*)	*V*(*i*)	Day (*k*)/Shift (*j*)														Total
		1		2		3		4		5		6		7		
		M	A	M	A	M	A	M	A	M	A	M	A	M	A	
1	1	0	1	0	1	1	0	1	0	0	0	0	0	1	0	5
	2	1	0	1	0	0	0	0	1	0	1	0	1	0	0	5
	3	0	0	0	0	0	1	1	0	1	0	1	0	0	1	5
	4	1	0	1	0	1	0	0	0	1	0	0	0	1	0	5
2	5	1	0	1	0	1	0	1	0	0	1	0	0	0	0	5
	6	0	1	0	0	1	0	1	0	0	0	0	1	1	0	5
	7	1	0	0	1	0	0	0	0	1	0	1	0	0	1	5
	8	0	0	1	0	0	1	0	1	0	0	1	0	1	0	5
3	9	0	0	0	0	1	0	0	0	0	0	0	1	1	0	3
	10	0	1	0	0	0	0	0	0	0	0	0	0	0	0	1
	11	0	0	1	0	0	0	1	0	1	0	1	0	0	0	4
	12	1	0	0	0	0	0	0	1	0	1	0	0	0	1	4
	13	0	0	0	0	0	1	0	0	0	0	0	0	0	0	1
4	9	0	0	0	0	0	0	0	1	1	0	0	0	0	0	2
	10	0	0	0	1	1	0	0	0	0	0	0	1	0	1	4
	11	0	0	0	0	0	0	0	0	0	0	0	0	1	0	1
	12	0	0	1	0	0	0	0	0	0	0	0	0	0	0	1
	13	1	0	0	0	0	0	1	0	0	1	1	0	0	0	4
Total																65

Note [M] Morning; [A] Afternoon; [D] Department; [V] Veterinary Staff

Table 2 represents the summary of the results generated and obtained from the Open Solver software based on a model from Case Study II. Similar to the result from a Case Study I, the objective shows that 65 shifts have been utilized from 182 shifts and all the constraints were satisfied. Based on the results obtained, all staff were assigned to at least one shift per day and the total number of shifts per staff is also five in a week, which satisfied the model constraint.

Another similarity with Case Study I, except for Day 6, Department 1 shows that two veterinarians were assigned to morning shift and one veterinarian in afternoon shift. On Day 6, just one veterinarian is assigned to the morning shift.

Since it is challenging to guarantee that all veterinarian assistants are evenly distributed throughout Department 2, Department 3, and Department 4, therefore, the model constraint has specified that veterinary assistants will only remain in Department 2 and do not have to be assigned with other duties from other departments. They should only assist the veterinarian in the treatment room. For Department 2, the model constraint has restricted that there must be at least two veterinarian assistants in the morning shift and one in the afternoon shift. This restriction is satisfied

Table 3 Summary of differences model and results obtained based on case study I and case study II

Differences	Case study I	Case study II
Satisfy all constraint	Yes	Yes
Veterinarian assistant department	Veterinary assistants are placed in Department 2, Department 3 and Department 4	Veterinary assistants are only placed in Department 2

except for Day 5. In Day 5, there is only one veterinary staff who will assist the veterinarian in the morning shift. This is acceptable because the veterinarian assistants who work in the afternoon shift will start work at 1 pm. So, they will be able to assist the veterinarian when needed.

In contrast to the Case Study I, if there were any clinic assistants who were not assigned to Department 3 and Department 4 on a particular day, there will be no replacement from veterinary assistants to carry out duties in those departments. This scenario can be seen on Day 1 and Day 3 where no clinic assistants are assigned in the afternoon shift in Department 4. Similarly, there are no clinic assistants which are assigned in the afternoon shift in Department 3 on Day 2.

3.3 The Comparison of Results Between Case Study I and Case Study II

Table 3 shows the summary of differences model and results obtained based on Case Study I and Case Study II.

All the model constraints are satisfied for both case studies. The difference between both case studies is whether veterinary assistants are allowed to work in only department 2 or they could be placed in either Departments 2, 3, or 4.

4 Conclusion

Over the years, veterinary clinic managements are still using manual approach in developing veterinary staff schedule. This has resulted in a time-consuming process and the flexibility of the result in many aspects is quite limited. The weakness in generating a good schedule will affect the work performance of staff as well as disrupt the daily operations of the clinic. Therefore, this study has proposed a quick and systematic mechanism to produce staff working schedule. The mathematical model proposed in this study can reduce the time in developing staff schedules and at the same time overcome the clinic workforce problems such as an unbalanced assignment of tasks among staff and overtime work.

Acknowledgements The authors would like to thank the Kolej Pengajian Pengkomputeran, Informatik dan Matematik, UiTM Shah Alam, for the unconditional support given while completing this study.

References

1. Al-rawi, O.Y.M., Mukherjee, T.: Application of linear programming in optimizing labour scheduling. J. Math. Financ. **9**, 272–285 (2019)
2. Nurmi, K., Kyngäs, N., Kyngäs, J.: Workforce optimization: the general task-based shift generation problem. IAENG Int. J. Appl. Math. **49**(4), 1–8 (2019)
3. Herawati, A., Yuniartha, D.R., Purnama, I.L.I., Dewi, L.: Shift scheduling model considering workload and worker's preference for security department. In: International Conference on Industrial and System Engineering (IConISE) 2017, p. 337 (2018)
4. Soriano, J., Jalao, E.R., Martinez, I.A.: Integrated employee scheduling with known employee demand, including breaks, overtime, and employee preferences. J. Ind. Eng. Manag. **13**(3), 451–463 (2020)
5. Shin, Y.A., Razali, S.N.A.M., Sie, L.K.: Optimized preference of security staff scheduling using integer linear programming approach. 3–12 (2019)
6. Kletzander, L., Musliu, N.: Solving the general employee scheduling problem. In: PATAT 2018—Proceedings of the 12th International Conference on the Practice and Theory of Automated Timetabling, pp. 469–472 (2018)
7. Mahmud, N., Hafawati, S., Syafina, I.: Optimization of workforce scheduling using integer goal programming approach. Jurnal Intelek **13**(2), 27–36 (2018)
8. Kandemir, C., Handley, H.A.H.: Work process improvement through simulation optimization of task assignment and mental workload. Comput. Math. Organ. Theory **25**(4), 389–427 (2019)
9. Yange, T.S., Onyekware, O., Abdulmuminu, Y.M.: A duty scheduler for veterinary teaching hospitals using tabu search algorithm. J. Comput. Sci. Appl. **8**(1), 21–29 (2020)
10. Hasan, M.M., Arefin, M.R.: Application of linear programming in scheduling problem. Dhaka Univ. J. Sci. **65**(2), 145–150 (2017)
11. Alwadood, Z., Noor, N.M., Mainor, N.A.: An optimization model for hotel housekeeping personnel scheduling in pandemic outbreak. Discov. Math. **43**(2), 83–92 (2021)
12. Permatasari, C.I., Yuniaristanto, Sutopo, W., Hisjam, M.: Aircraft maintenance manpower shift planning with multiple aircraft maintenance licenced. In: IOP Conference Series: Materials Science and Engineering, vol. 495, pp. 1–8 (2019)
13. Akyurt, İZ., Kuvvetli, Y., Deveci, M., Garg, H., Yuzsever, M.: A new mathematical model for determining optimal workforce planning of pilots in an airline company. Complex Intell. Syst. **8**(1), 429–441 (2022)
14. Patidar, M., Choudhary, S., Bhardwaj, R.: The solution of staff scheduling problem in an aircraft using linear programming. Mater. Today Proc. **47**, 6826–6829 (2020)
15. Lind, A.K., Hydbring-Sandberg, E., Forkman, B., Keeling, L.J.: Assessing stress in dogs during a visit to the veterinary clinic: correlations between dog behavior in standardized tests and assessments by veterinary staff and owners. J. Vet.Y Behav. Clin. Appl. Res. **17**, 24–31 (2017)
16. Paramadevan, P.: Linear programming approach for doctors scheduling in a hospital. J. Sci. **8**(2), 12 (2017)

Solving SEIR Model: Application of Multistep Block Method

Nur Tasnem Jaaffar and Zanariah Abdul Majid

Abstract The Susceptible-Exposed-Infectious-Recovered (SEIR) model serves as a pivotal tool in simulating infectious diseases like COVID-19, offering a structured framework to analyze how diseases spread through populations over time. It allows researchers and policymakers to estimate key epidemiological parameters such as transmission rates, incubation periods, and recovery rates, crucial for designing effective public health interventions. This study implements multistep block method with predictor and corrector scheme to approximate the population representing individuals who are susceptible to the disease, those that exposed to the disease, those who has infected, and also those who has recovered from the disease. The multistep method is derived by using Lagrange interpolation and the two-points solutions are obtained simultaneously in block method. These two solutions are then iterated by using predictor-corrector scheme to improve the accuracy of the approximate solutions. The proposed method then tested to solve the SEIR model and compared with the established previous methods to observe its accuracy performance. The factors contributed to its computational time required such as total steps and total function calls also discussed and compared with previous method. Overall, the research contributes a robust computational framework for SEIR modeling, advancing capabilities to simulate and understand disease spread dynamics effectively.

Keywords SEIR Model · Multistep method · Block method

N. T. Jaaffar (✉)
Faculty of Computer and Mathematical Sciences, Universiti Teknologi MARA (UiTM) Shah Alam, Shah Alam, Selangor, Malaysia
e-mail: tasnemjaaffar@gmail.com

Z. A. Majid
Department of Mathematics, Faculty of Science, Universiti Putra Malaysia, UPM Serdang, Selangor, Malaysia

A. Shuib et al. (eds.), *Mathematical Modeling and Numerical Analysis*,
Springer Proceedings in Mathematics & Statistics 510,
https://doi.org/10.1007/978-981-96-9350-4_11

1 Introduction

The SEIR (Susceptible-Exposed-Infectious-Recovered) model is a widely used mathematical framework for examining the spread of infectious diseases like COVID-19. It segments the population into categories: those at risk of infection, those exposed to the pathogen, those who are infectious, and those who have recovered or are no longer affected. Various numerical and approximation techniques have been employed to solve the SEIR model's differential equations and accurately simulate disease transmission.

Researchers have applied different methods to solve the SEIR model. For example, Rustan and Handayani [19] used the Euler method to address the differential equations in a modified SEIR model for COVID-19 in Indonesia. Aakash et al. [1] employed both the Euler method and the fourth-order Runge-Kutta method to manage the complexity of the SEIR model's equations. Recently, Palma et al. [18] revealed that simulations using both the fourth- and fifth-order Runge-Kutta methods generate the same patterns as those from the ODE45 algorithm in Matlab software. Additionally, Mungkasi [16] introduced a multistage method, along with successive approximation and variational iteration methods, to derive semi-analytical solutions for the SEIR model. The application of the Mickens method, a non-standard finite difference method, was shown to preserve the dynamic properties of the SEIR model while being easy to implement [10].

In terms of predictor-corrector techniques, Farooqi et al. [7] applied this approach to a non-standard finite difference method, while Gao et al. [8] utilized a modified predictor-corrector method to solve a fractional-order SIR model. Recently, Adebayo et al. [20] employed the Adams-Bashforth-Moulton method with a predictor-corrector approach to solve an SIRS model, while Arif et al. [2] incorporated a third-order explicit predictor-corrector scheme combined with a neural network to solve and simulate the SEIR model. However, to our knowledge, no research has yet focused on using a multistep block method with a predictor-corrector approach for the SEIR model.

Selecting appropriate numerical algorithms for differential equations is essential for faster computational times. Recent studies have aimed to develop computational methods that are more cost-effective compared to traditional approaches, emphasizing the importance of computational efficiency in solving ordinary differential equations (ODEs) [5]. For instance, the modified Euler method has been shown to significantly accelerate moderately stiff chemical kinetics in reactive-flow simulations using GPUs, compared to conventional methods [17]. This underscores the impact of method selection on computational performance, particularly when handling large numbers of independent ODE systems effectively.

The computational time required to solve the SEIR model can be considerable, particularly for large populations, as the model must be simulated over numerous time steps to capture the full epidemic trajectory [4]. Therefore, efficient computational methods are crucial for rapid analysis and forecasting during fast-moving epidemics. Slow simulations can hinder the ability to quickly evaluate interventions like lockdowns and vaccinations.

In conclusion, developing an accurate and time-efficient numerical method for solving the SEIR model is critical for predicting disease dynamics and guiding public health decisions. This study aims to enhance both the precision and computational speed of solutions by focusing on a multistep block method with a predictor-corrector approach. The predictor step provides an initial estimate, which is then refined by the corrector step, leading to improved accuracy. Additionally, the two-point solutions obtained simultaneously using the block method reduce the total number of iteration steps needed to complete the algorithm, thereby leading to faster computational times. This approach seeks to balance computational efficiency with the accuracy necessary for reliable SEIR model simulations.

2 SEIR Model

The nonlinear SEIR model of the COVID-19 can be represented as follows [9]:

$$
\begin{aligned}
S'(t) &= -\beta\frac{S(t)I(t)}{N} - \frac{Z}{N}S(t) + (\rho_I + \rho E) + \left(\frac{e_I}{N} + \frac{e_E}{N}\right)S(t) + \upsilon N(t) - \mu S(t),\\
E'(t) &= -\beta\frac{S(t)I(t)}{N} + \frac{Z}{N}S(t) - \alpha E(t) - \left(\frac{e_I}{N} + \frac{e_E}{N}\right)E(t) - \mu E(t) - \sigma E(t),\\
I'(t) &= \alpha E(t) - \gamma I(t) - \left(\frac{e_I}{N} + \frac{e_E}{N}\right)I(t) - \mu I(t),\\
R'(t) &= \gamma I(t) - \mu R(t) + \sigma E(t),
\end{aligned} \tag{1}
$$

subject to initial conditions:

$$
S(0) = S_0, \quad E(0) = E_0, \quad I(0) = I_0, \quad R(0) = R_0,
$$

where the $S(t)$, $E(t)$, $I(t)$, and $R(t)$ are the individuals that susceptible, exposed, infected, and recovered, respectively with t is the time in this situation. The total population is given by $N(t) = S(t) + E(t) + I(t) + R(t)$. The parameters involved are β, the transmission rate for infected people, α, the infection rate for exposed people, γ, the average number of people currently infected, μ, the natural death rate, υ, the natural birth rate, σ, the recovery rate, Z, the infection rate in standard situation, ρ_I and ρ_E are the average daily number of international and domestic passengers, respectively that arrived by air, e_I and e_E are the average daily number of international and domestic passengers, respectively that departed by air.

3 Methodology

Supposed that the SEIR system above can be written as below:

$$\begin{aligned} S'(t) &= u\left(t, S(t), E(t), I(t), R(t)\right), \\ E'(t) &= u\left(t, S(t), E(t), I(t), R(t)\right), \\ I'(t) &= u\left(t, S(t), E(t), I(t), R(t)\right), \\ R'(t) &= u\left(t, S(t), E(t), I(t), R(t)\right), \end{aligned} \tag{2}$$

where $t \in [a, b]$. For all the upcoming procedures, we shall assume for $S'(t)$ only since the procedures are repeated for other differential equations, $E'(t)$, $I'(t)$, and $R'(t)$.

In this study, we convert the continuous interval $[a, b]$ into the discrete points of $\{t_i | i = 0, 1, 2, 3, \ldots, M\}$ where $M = \dfrac{b-a}{H}$ and H is the step size used. The points S_{i+1} and S_{i+2} within each block are determined simultaneously at positions t_{i+1} and t_{i+2}. Each block is selected to include interpolation points arranged in a diagonal fourth-order pattern adopted from Jaaffar et al. [12]: $\{t_{i+1}, t_i, t_{i-1}, t_{i-2}\}$ for the first point and $\{t_{i+2}, t_{i+1}, t_i, t_{i-1}\}$ for the second point.

The integration of $S'(t)$ is obtained for the first and second points along the interval $[t_i, t_{i+1}]$ and $[t_i, t_{i+2}]$ respectively.
First point:

$$\int_{t_i}^{t_{i+1}} S'(t)dt = \int_{t_i}^{t_{i+1}} u\left(t, S(t), E(t), I(t), R(t)\right)dt$$

$$S_{i+1} = S_i + \int_{t_i}^{t_{i+1}} u\left(t, S(t), E(t), I(t), R(t)\right)dt \tag{3}$$

Second point:

$$\int_{t_i}^{t_{i+2}} S'(t)dt = \int_{t_i}^{t_{i+2}} u\left(t, S(t), E(t), I(t), R(t)\right)dt$$

$$S_{i+2} = S_i + \int_{t_i}^{t_{i+2}} u\left(t, S(t), E(t), I(t), R(t)\right)dt. \tag{4}$$

where $u\,(t, S(t), E(t), I(t), R(t))\,dt$ in (3) and (4) are approximated by the Lagrange polynomial of degree 3 for both first point and second point.

Consequently, the corrector formula for the fourth-order two-point multistep block method for the SEIR model (BM4-SEIR) is derived as follows:

$$
\begin{aligned}
S_{i+1} &= S_i + \frac{H}{24}(9u_{i+1} + 19u_i - 5u_{i-1} + u_{i-2}) \\
S_{i+2} &= S_i + \frac{H}{3}(u_{i+2} + 4u_{i+1} + u_i) \\
E_{i+1} &= E_i + \frac{H}{24}(9u_{i+1} + 19u_i - 5u_{i-1} + u_{i-2}) \\
E_{i+2} &= E_i + \frac{H}{3}(u_{i+2} + 4u_{i+1} + u_i) \\
I_{i+1} &= I_i + \frac{H}{24}(9u_{i+1} + 19u_i - 5u_{i-1} + u_{i-2}) \\
I_{i+2} &= I_i + \frac{H}{3}(u_{i+2} + 4u_{i+1} + u_i) \\
R_{i+1} &= R_i + \frac{H}{24}(9u_{i+1} + 19u_i - 5u_{i-1} + u_{i-2}) \\
R_{i+2} &= R_i + \frac{H}{3}(u_{i+2} + 4u_{i+1} + u_i)
\end{aligned}
\tag{5}
$$

The method employs a predictor-corrector (PECE) approach, where the predictor formula is one order lower than the corrector formula. The predictor formula is a third-order two-point multistep block method (BM3), as demonstrated below:

$$
\begin{aligned}
S_{i+1} &= S_i + \frac{H}{12}(23u_i - 16u_{i-1} + 5u_{i-2}) \\
S_{i+2} &= S_i + \frac{H}{3}(19u_i - 20u_{i-1} + 7u_{i-2}) \\
E_{i+1} &= E_i + \frac{H}{12}(23u_i - 16u_{i-1} + 5u_{i-2}) \\
E_{i+2} &= E_i + \frac{H}{3}(19u_i - 20u_{i-1} + 7u_{i-2}) \\
I_{i+1} &= I_i + \frac{H}{12}(23u_i - 16u_{i-1} + 5u_{i-2}) \\
I_{i+2} &= I_i + \frac{H}{3}(19u_i - 20u_{i-1} + 7u_{i-2}) \\
R_{i+1} &= R_i + \frac{H}{12}(23u_i - 16u_{i-1} + 5u_{i-2}) \\
R_{i+2} &= R_i + \frac{H}{3}(19u_i - 20u_{i-1} + 7u_{i-2})
\end{aligned}
\tag{6}
$$

This proposed method requires backward values since it is a multistep method therefore, Euler order 1 and Modified Euler order 2 (PECE) methods are used to approximate the starting values needed. Euler method order 1 and Modified Euler method order 2 are chose to match with BM3 that have order 3 so that no order gap exist since it can affect the accuracy of overall performance of our proposed method.

4 Analysis of Method

First, by converting the continuous interval $[a, b]$ into a discrete set, we obtain the points $\{t_i | i = 0, 1, 2, ..., (b - a)/H\}$, where H is the specified step size. Let S_i denote an estimate of the theoretical solution at t_i. When a numerical method to find the sequence $\{S_i\}$ is expressed through a linear relationship between S_{i+j} and u_{i+j} for $j = 0, 1, ..., k$, it is known as a linear multistep method (LMM) of step number k. The general LMM for first-order differential equations in (2) can be formulated as follows:

$$\sum_{j=0}^{k} A_j S_{i+j} = H \sum_{j=0}^{k} B_j u_{i+j}, \tag{7}$$

where $u_{i+j} = S'_{i+j}$ for $j = 0, 1, 2, .., k$. A_j and B_j are constants. Additionally, throughout this paper, we assume that $A_k = 1$.

The linear difference operator associated with the LMM (7) is defined as follows:

$$L\big[S(t); H\big] = \sum_{j=0}^{k} \big[A_j S(t + jH) - HB_j S'(t + jH)\big], \tag{8}$$

given that $S(t)$ represents a theoretical solution that is continuously differentiable over the interval $[a, b]$. By performing a Taylor series expansion for $S(t + jH)$ and $S'(t + jH)$ around t, and subsequently merging the terms, we can eventually derive

$$L\big[S(t); H\big] = C_0 S(t) + C_1 H S'(t) + \cdots + C_p H^p S^{(p)}(t)$$

where

$$\begin{aligned} C_0 &= \sum_{j=0}^{k} A_j, \\ C_1 &= \sum_{j=0}^{k} (j A_j - B_j), \\ &\vdots \\ C_p &= \sum_{j=0}^{k} \left(\frac{j^p}{p!} A_j - \frac{j^{p-1}}{(p-1)!} B_j \right), \quad p = 2, 3, ... \end{aligned} \tag{9}$$

where p is the method's order. The truncation error, T_{i+k} is considered local under the assumption that no previous truncation errors have occurred, which is known as the localizing assumption. Suppose that

$$S_{i+j} = S(t_{i+j}), \quad j = 0, 1, ..., k-1$$

where all preceding values are exact. Let $\hat{S_{i+k}}$ represent the approximate solution of the LMM (7) at t_{i+k} when the localizing assumption is applied [11].

From (8)

$$\sum_{j=0}^{k} A_j S(t_i + jH) = H \sum_{j=0}^{k} B_j u\big(t_i + jH, S(t_{i+j}), E(t_{i+j}), I(t_{i+j}), R(t_{i+j})\big) + L[S(t_i); H] \tag{10}$$

where $S(t)$ is assumed to represent the exact solution of the problem. The approximate solution, $\hat{S_{i+k}}$ generated by the method (7) fulfills

$$\hat{S_{i+k}} + \sum_{j=0}^{k-1} A_j S_{i+j} = H \sum_{j=0}^{k-1} B_j u(t_{i+j}, S_{i+j}, E_{i+j}, I_{i+j}, R_{i+j}) + H B_k u(t_{i+k}, \hat{S_{i+k}}, \hat{E_{i+k}}, \hat{I_{i+k}}, \hat{R_{i+k}}), \tag{11}$$

given that $A_k = 1$. Subtracting (11) from (10) and applying the localizing assumption leads to

$$\begin{aligned} S(t_{i+k}) - \hat{S_{i+k}} = {} & H B_k \Big[u\big(t_{i+k}, S(t_{i+k}), E(t_{i+k}), I(t_{i+k}), R(t_{i+k})\big) - u\big(t_{i+k}, \hat{S_{i+k}}, \hat{E_{i+k}}, \hat{I_{i+k}}, \hat{R_{i+k}}\big) \Big] \\ & + T_{i+k}. \end{aligned} \tag{12}$$

Applying the mean value theorem on the right hand side,

$$\begin{aligned} & u\big(t_{i+k}, S(t_{i+k}), E(t_{i+k}), I(t_{i+k}), R(t_{i+k})\big) - u(t_{i+k}, \hat{S_{i+k}}, \hat{E_{i+k}}, \hat{I_{i+k}}, \hat{R_{i+k}}) = [S(t_{i+k}) - \hat{S_{i+k}}] \frac{\delta u}{\delta S}(\eta_{i+k}) \\ & + [E(t_{i+k}) - \hat{E_{i+k}}] \frac{\delta u}{\delta E}(\eta_{i+k}) + [I(t_{i+k}) - \hat{I_{i+k}}] \frac{\delta u}{\delta I}(\eta_{i+k}) + [R(t_{i+k}) - \hat{R_{i+k}}] \frac{\delta u}{\delta R}(\eta_{i+k}) \end{aligned}$$

where η_{i+k} is in the interval whose endpoints are $\big(t_{i+k}, S(t_{i+k}), E(t_{i+k}), I(t_{i+k}), R(t_{i+k})\big)$ and $(t_{i+k}, \hat{S_{i+k}}, \hat{E_{i+k}}, \hat{I_{i+k}}, \hat{R_{i+k}})$. Thus,

$$\begin{aligned} S(t_{i+k}) - \hat{S_{i+k}} = H B_k \Big[& [S(t_{i+k}) - \hat{S_{i+k}}] \frac{\delta u}{\delta S}(\eta_{i+k}) + [E(t_{i+k}) - \hat{E_{i+k}}] \frac{\delta u}{\delta E}(\eta_{i+k}) \\ & + [I(t_{i+k}) - \hat{I_{i+k}}] \frac{\delta u}{\delta I}(\eta_{i+k}) + [R(t_{i+k}) - \hat{R_{i+k}}] \frac{\delta u}{\delta R}(\eta_{i+k}) \Big] + T_{i+k}. \end{aligned}$$

Hence, when H approaching to zero, the local truncation error (LTE), T_{i+k} is then

$$T_{i+k} = S(t_{i+k}) - \hat{S_{i+k}}.$$

Based on definition,

Definition 1 (Lambert [14]) *The LTE of the method (7) at t_{i+k}, denoted by T_{i+k} is defined by*

$$T_{i+k} = L\big[S(t_i);\, H\big],$$

where exact solution $S(t) \in C'[a, b]$. If we assume that $S(t) \in C^{p+1}[a, b]$, where p is the order of the method, therefore,

$$T_{i+k} = S(t_{i+k}) - \hat{S_{i+k}} = C_{p+1} H^{p+1} S^{(p+1)}(t_i) + O(H^{p+2}). \tag{13}$$

From the LTE, the consistency of the method also can be shown as in definition below

Definition 2 (Lambert [14]) *The LMM (7) is said to be consistent if the LTE, T_{i+k} satisfies*

$$\lim_{H \to 0} \frac{1}{H} T_{i+k} = 0.$$

According to Lambert [15] and Baharum et al. [3], the non-vanishing coefficient of the error constant obtained from (9) is determined as follows:

$$C_0 = C_1 = C_2 = C_3 = C_4 = [0 \quad 0]^T,$$
$$C_5 = \begin{bmatrix} -\frac{19}{720} & -\frac{1}{90} \end{bmatrix}^T.$$

Thus, based on (13) and **Definition 2**,

$$\lim_{H \to 0} \frac{1}{H} \left[\begin{bmatrix} -\frac{19}{720} & -\frac{1}{90} \end{bmatrix}^T H^5 S^{(5)}(t_i) + O(H^6) \right] = 0.$$

This has proven that our proposed method is consistent based on **Definition 2**. The significance of consistency lies in its ability to regulate the magnitude of the Local Truncation Error (LTE) obtained at each computational stage, thus can improve the accuracy of our proposed method if we chose smaller step size. Both consistency and LTE concepts are fundamental in the analysis and development of effective multistep methods for solving SEIR model.

Meanwhile, based on definition of order below

Definition 3 (Lambert [14]) *A linear multistep method of order p is said to have error constant C_{p+1} given by* (9).

Therefore, our proposed method is order four since error constant $C_{p+1} = C_5$. The order of a numerical method is important as it reflects the accuracy and efficiency of the method in solving mathematical problems. Higher-order approximations offer even greater efficiency improvements. Therefore, applying high-order schemes is expected to significantly reduce computational costs [6].

5 Implementation of Method

The implementation of the methodology above is written in C programming by following algorithm below:

5.1 Algorithm of BM4-SEIR

1. **Start**
2. **Initialize parameters and initial conditions**:
    - Set t_0, t_n,N,S_0, E_0, I_0, R_0, β, α, γ, μ,v,σ,Z,ρ_I,ρ_E,e_I,e_E,H
    - Create time array t from t_0 to t_n with step size H
3. **Use Euler and Modified Euler to get the first few steps (i = 0, 1, 2)**:
    - For $i = 0$ to 2:
        - Calculate $u\left(S_i, E_i, I_i, R_i\right)$ using SEIR derivatives
        - Predictor step:

          Estimate $\left(S_{i+1}, E_{i+1}, I_{i+1}, R_{i+1}\right)$ predictor using Euler method order 1
        - Corrector step:

          Calculate $u\left(S_{i+1}, E_{i+1}, I_{i+1}, R_{i+1}\right)$ using SEIR derivatives at $\left(S_{i+1}, E_{i+1}, I_{i+1}, R_{i+1}\right)$ predictor obtained previously
          Update $\left(S_{i+1}, E_{i+1}, I_{i+1}, R_{i+1}\right)$ using Modified Euler order 2.
4. **Apply BM4-SEIR method for subsequent steps** ($i >= 3$):
    - For $i = 3$ to $n - 2$:
        - Calculate $u\left(S_i, E_i, I_i, R_i\right)$ using SEIR derivatives
        - Calculate $u\left(S_{i-1}, E_{i-1}, I_{i-1}, R_{i-1}\right)$ using SEIR derivatives
        - Calculate $u\left(S_{i-2}, E_{i-2}, I_{i-2}, R_{i-2}\right)$ using SEIR derivatives
        - Predictor step:

          Estimate $\left(S_{i+1}, E_{i+1}, I_{i+1}, R_{i+1}\right)$ predictor using two-point block method order 3 (BM3) in (6)
        - Corrector step:

Calculate $u\,(S_{i+1}, E_{i+1}, I_{i+1}, R_{i+1})$ using SEIR derivatives at $(S_{i+1}, E_{i+1}, I_{i+1}, R_{i+1})$ predictor obtained previously
Update $(S_{i+1}, E_{i+1}, I_{i+1}, R_{i+1})$ using two-point block method order 4 (BM4) in (5)

5. **Obtain the results**:
    - Obtain S, E, I, R over time
6. **End**

Since we also want to observe the computational performance of our proposed method thus, we will compare our method with the established one-step method which is Runge-Kutta order 4 (RK4). Therefore, both methods are run in the same computer.

6 Numerical Results

For $t \in [0, 20]$, the initial conditions are given by:

$$S(0) = 2500, E(0) = 1, I(0) = 1, R(0) = 0,$$

$$N(0) = S(0) + E(0) + I(0) + R(0) = 2502.$$

Meanwhile, the parameters are as follows:

$$\beta = 0.8, \alpha = 0.75, \sigma = 0.1, \gamma = 0.05, \upsilon = 0.009/N,$$

$$\mu = 0.01, Z = 0.001, \rho_I = \rho_E = 0.15, e_I = 0.01, e_E = 0.03.$$

We implemented the BM4-SEIR method with $H = 1/4$. We then compared the approximate solution obtained to those from the Taylor Collocation Method (TCM) [13] and Runge-Kutta order 4 (RK4), as shown in Tables 1 and 2. The comparison of computational performance between BM4-SEIR and RK4 is presented in Table 3 (Figs. 1, 2, 3 and 4).

7 Discussions

Tables 1 and 2 show that our proposed block multistep method (BM4-SEIR) effectively solves the SEIR model, as the approximate solutions obtained are comparable

Table 1 Comparison of approximate solutions for susceptible (S) and exposed (E)

t (days)	S (Susceptible)			E (Exposed)		
	RK4	TCM	BM4-SEIR	RK4	TCM	BM4-SEIR
0	2500	2500	2500	1	1	1
2	2448	2448	2448	2	2	2
4	2394	2394	2394	4	4	4
6	2334	2334	2334	8	8	8
8	2260	2260	2260	17	17	17
10	2157	2157	2157	34	34	34
12	2000	2000	2000	66	66	66
14	1754	1754	1754	119	119	119
16	1398	1398	1397	184	184	184
18	967	967	966	230	230	230
20	567	567	567	219	221	219

Table 2 Comparison of approximate solutions for infected (I) and recovered (R)

t (days)	I (Infected)			R (Recovered)		
	RK4	TCM	BM4-SEIR	RK4	TCM	BM4-SEIR
0	1	1	1	0	0	0
2	3	3	3	0	0	0
4	6	6	6	1	1	1
6	13	13	13	3	3	3
8	28	28	28	7	7	7
10	59	59	60	16	16	16
12	122	122	122	34	34	34
14	237	237	237	69	69	69
16	425	425	425	129	129	130
18	675	675	675	223	223	223
20	923	923	924	343	343	344

Table 3 Comparison of computational time and its factors for RK4 and BM4-SEIR

	RK4	BM4-SEIR
Total function calls	1280	644
Total step of iterations	80	41
Time taken (seconds)	0.000202	0.000188

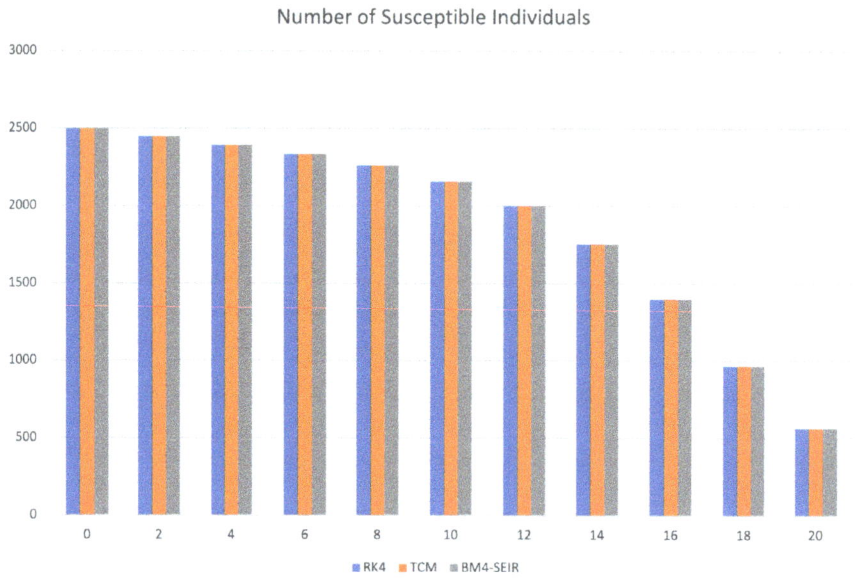

Fig. 1 Simulations of approximate solutions for susceptible (S) individuals for RK4, TCM, and BM4-SEIR

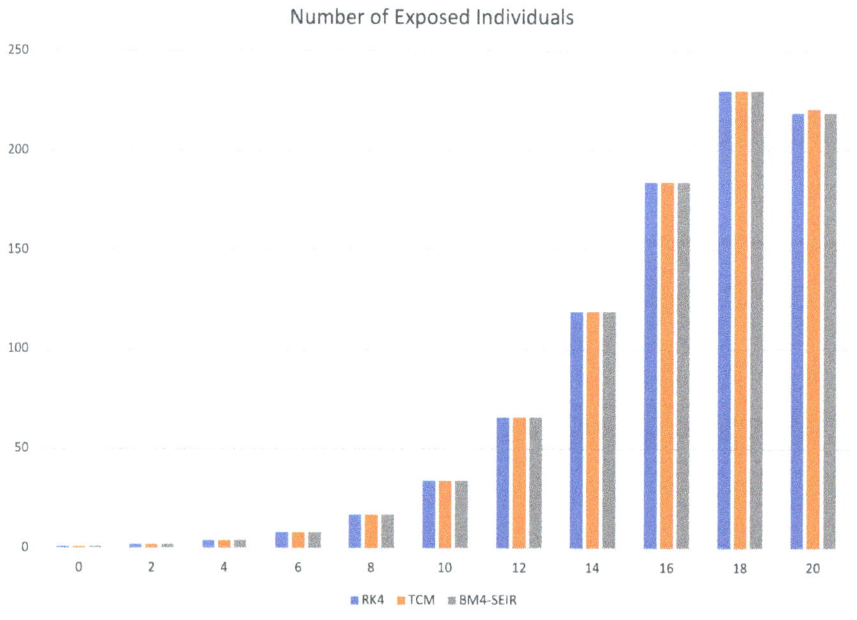

Fig. 2 Simulations of approximate solutions for exposed (E) individuals for RK4, TCM, and BM4-SEIR

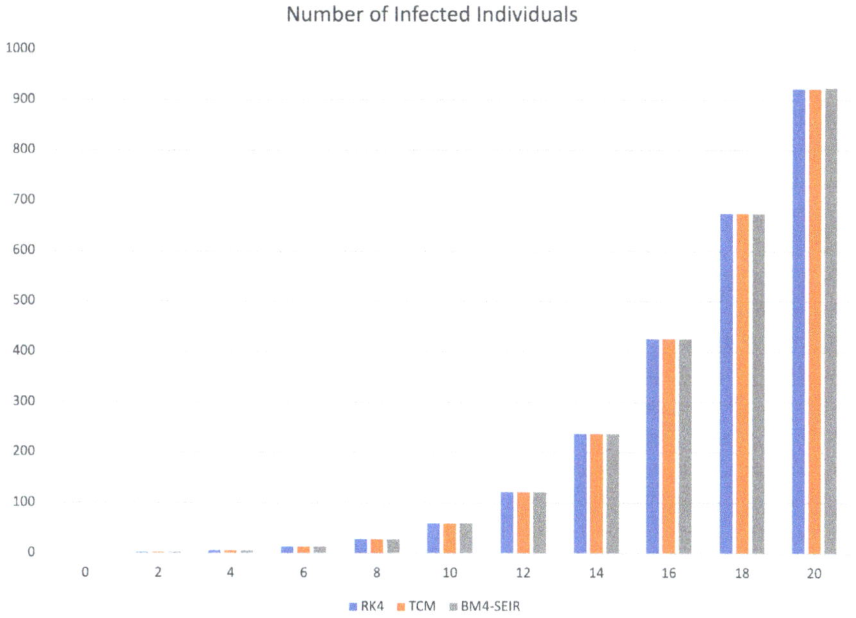

Fig. 3 Simulations of approximate solutions for infected (I) individuals for RK4, TCM, and BM4-SEIR

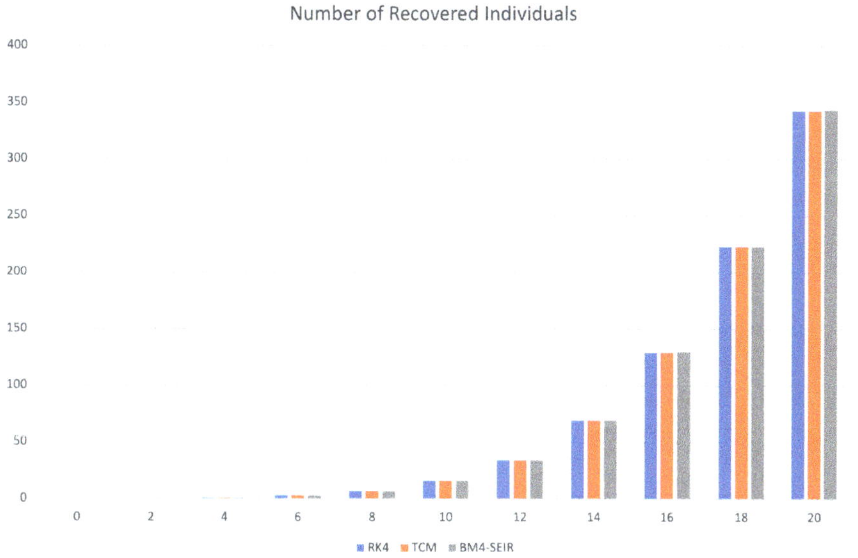

Fig. 4 Simulations of approximate solutions for recovered (R) individuals for RK4, TCM, and BM4-SEIR

and almost identical to those generated by previous methods, such as RK4 and TCM. Accuracy in solving the SEIR model is crucial due to its significant implications for public health, policy-making, economic stability, and overall societal well-being. Reliable and accurate predictions enable effective responses to infectious disease outbreaks, helping to save lives, manage healthcare resources efficiently, and minimize the socio-economic impact of epidemics.

Beyond accuracy, our method also shows advantages over the one-step method RK4 in terms of the total number of iteration steps and function calls required to complete the process, as these factors can influence the computational time needed. As shown in Table 3, our method requires fewer function calls and iteration steps, resulting in faster execution times. Since RK4 is a one-step method that requires four function evaluations per iteration step, it leads to a higher total number of function calls compared to our method. Furthermore, RK4 is not a block method, which results in double the number of iteration steps compared to our two-point block method that approximates two solutions simultaneously per iteration. These two factors significantly influence the overall computational time required to run the program. An algorithm with faster computational times is essential, as many scientific and engineering problems require large-scale simulations with millions or even billions of variables, ultimately saving computational or operational costs.

8 Conclusion

In conclusion, our proposed method, BM4-SEIR, is feasible for solving infectious disease models by providing comparable accuracy and better computational efficiency when compared to previously established methods.

Acknowledgements All authors are thankful for each other's contributions to this paper.

References

1. Aakash, M., Gunasundari, C., Al-Mdallal, Q.: Mathematical modeling and simulation of SEIR model for COVID-19 outbreak: a case study of Trivandrum. Front. Appl. Math. Stat. **9** (2023). https://doi.org/10.3389/fams.2023.1124897
2. Arif, M.S., Abodayeh, K., Nawaz, Y.: A third-order two stage numerical scheme and neural network simulations for SEIR epidemic model: a numerical study. Emerg. Sci. J. **8**(1), 326–340 (2024)
3. Baharum, N.A., Majid, Z.A., Senu, N.: Numerical method for Volterra Integro-differential equation with implicit multistep block method (2018)
4. Balsa, C., Lopes, I., Guarda, T., Rufino, J.: Computational simulation of the COVID-19 epidemic with the SEIR stochastic model. Comput. Math. Organ. Theory 1–19 (2021). https://doi.org/10.1007/s10588-021-09327-y
5. Cockayne, J., Oates, C., Sullivan, T., Girolami, M.: Bayesian probabilistic numerical methods. Siam Rev. **61**(3), 756–789 (2019). https://doi.org/10.1137/17m1139357

6. Ekaterinaris, J.A.: High-order accurate, low numerical diffusion methods for aerodynamics. Prog. Aerosp. Sci. **41**(3–4), 192–300 (2005)
7. Farooqi, A., Ahmad, R., Farooqi, R., Alharbi, S., Baleanu, D., Rafiq, M., Khan, I., Ahmad, M.: An accurate predictor-corrector-type nonstandard finite difference scheme for an SEIR epidemic model. J. Math. **2020**(1) (2020)
8. Gao, W., Veeresha, P., Cattani, C., Baishya, C., Baskonus, H.M.: Modified predictor–corrector method for the numerical solution of a fractional-order sir model with 2019-nCoV. Fractal Fract. **6**(2) (2022)
9. Harir, A., Melliani, S., Harfi, H.E., Chadli, L.S.: Variational iteration method and differential transformation method for solving the SEIR epidemic model. Int. J. Differ. Equ. **2020**(1) (2020)
10. Hoang, M.T., Ehrhardt, M.: A dynamically consistent nonstandard finite difference scheme for a generalized SEIR epidemic model. J. Differ. Equ. Appl. **30**(4), 409–434 (2024). https://doi.org/10.1080/10236198.2023.2291151
11. Jaaffar, N.T.: Solving Higher Order Delay Differential Equations with Boundary Conditions using Multistep Block Method. Ph.D. thesis, Universiti Putra Malaysia, Malaysia (2021)
12. Jaaffar, N.T., Abdullah, N., Pozin, M.A.A.: Numerical technique of solving boundary delay differential equations. Appl. Math. Comput. Intell. (AMCI) **12**(3), 1–19 (2023)
13. Laib, H., Bellour, A., Boulmerka, A.: Taylor collocation method for a system of nonlinear Volterra delay Integro-differential equations with application to covid-19 epidemic. Int. J. Comput. Math. **99**(4), 852–876 (2022)
14. Lambert, J.D.: Numerical methods for Ordinary Differential Systems, vol. 146. Wiley, New York (1991)
15. Lambert, J.D.: Computational Methods in Ordinary Differential Equations. Wiley (1973)
16. Mungkasi, S.: Successive approximation, variational iteration, and multistage-analytical methods for a SEIR model of infectious disease involving vaccination strategy. Commun. Biomath. Sci. **3**(2), 114–126 (2021). https://doi.org/10.5614/cbms.2020.3.2.3
17. Niemeyer, K., Sung, C.: GPU-based parallel integration of large numbers of independent ODE systems, pp. 159–182 (2014). https://doi.org/10.1007/978-3-319-06548-9_8
18. Palma, D.I., Mungkasi, S.: Fourth and fifth-order Runge-Kutta methods for solving a susceptible-exposed-infected-recovered mathematical model of the spread of COVID-19. In: AIP Conference Proceedings. vol. 3074. AIP Publishing (2024)
19. Rustan, R., Handayani, L.: The outbreak's modeling of coronavirus (COVID-19) using the modified SEIR model in Indonesia. Spektra Jurnal Fisika Dan Aplikasinya **5**(1), 61–68 (2020). https://doi.org/10.21009/spektra.051.07
20. Salaudeen, A., Sathasivam, S., Nam, C.J., Hang, T.Y.: Modelling the early outbreak of COVID-19 disease in Malaysia using sirs model with 4-step Adams-Uashforth-Moulton predictor-corrector method. Appl. Math. Comput. Intell. (AMCI) **13**(2), 121–136 (2024)

Credit Score Prediction Through Hybrid Machine Learning Models

Ruslan Murzagaliyev, Bizhigit Sagidolla, and Shirali Kadyrov

Abstract Assessing credit risk and ensuring financial stability in lending institutions hinge on accurate credit default prediction. This study evaluates the efficacy of diverse machine learning models in predicting credit default using a dataset comprising varied financial attributes. Initially, we present the outcomes of five baseline models—random forest, XGBoost, kNN, logistic regression, and neural network—in terms of precision, recall, $F1$-score, and overall accuracy. We subsequently investigate the impact of principal component analysis (PCA) and hyperparameter tuning via GridSearchCV on model performance. Our results demonstrate notable enhancements in model performance, particularly in accurately identifying non-default instances, upon integrating PCA and hyperparameter optimization. Furthermore, we underscore the adeptness of the multilayer perceptron (MLP) neural network model, which features two hidden layers with 64 and 32 neurons respectively, in capturing intricate data patterns. This research contributes to advancing credit default prediction methods and underscores the significance of employing comprehensive modeling techniques to bolster predictive accuracy in financial risk assessment.

Keywords Banking sector · Credit risk assessment · Data preprocessing · Machine learning algorithms · Hybrid models with kNN · Predictive modeling · Risk management

R. Murzagaliyev · B. Sagidolla (✉)
SDU University, City of Kaskelen, Almaty Region, Kazakhstan
e-mail: bizhigit.sagidolla@sdu.edu.kz
URL: https://sdu.edu.kz/

S. Kadyrov
Oxus University, Tashkent, Uzbekistan
URL: https://oxusuniversity.uz/

A. Shuib et al. (eds.), *Mathematical Modeling and Numerical Analysis*,
Springer Proceedings in Mathematics & Statistics 510,
https://doi.org/10.1007/978-981-96-9350-4_12

1 Introduction

The dynamic nature of finance has led to a recognition of the interconnectedness of risks, prompting a call for a more integrated approach to risk management [6]. This is particularly crucial for financial institutions, which face a myriad of risks in both product-market and capital-market transactions. The need for a holistic risk management strategy is further emphasized by the increasingly complex business environment and threat landscape [16]. However, the implementation of such a strategy is not without its challenges, including the human tendency of greed and the need for prophylactic measures. Despite the advancements in risk measurement and the application of quantitative techniques, a deep understanding of the nature of risk and the importance of internal controls remain paramount.

Credit risk is a significant concern in the financial sector, with the potential for widespread impact [5]. It is particularly crucial for lenders to assess this risk, as it can affect their profitability and stability. Managing credit risk is essential for financial stability, especially in the face of economic uncertainties. This involves maintaining credit risk exposure within acceptable parameters and considering its relationship with other risks [10]. Effective credit risk management is therefore a critical component of overall risk management in the banking sector.

This paper explores credit risk prediction, employing cutting-edge machine learning techniques to enhance predictive capabilities and refine risk management strategies. By leveraging real-world data from a banking institution, we aim to demonstrate the practical applicability and efficacy of our model. Through this effort, we aim to offer insights that enable financial entities to navigate the complexities of credit risk with greater precision and foresight in an era characterized by rapid change and interconnectivity.

The objectives of this paper are twofold: firstly, to illuminate the multifaceted nature of credit risk and its implications for financial institutions, and secondly, to demonstrate the potential of advanced machine learning algorithms in augmenting credit risk prediction and management practices. By combining theoretical knowledge with empirical research, we hope to advance the conversation about risk management and open the door for more effective and flexible risk-reduction tactics in the banking industry.

2 Literature Review

The body of literature on credit risk prediction is both vast and varied, reflecting the evolution of methodologies from traditional statistical models to advanced machine learning algorithms. Zhiyong Li, Jonathan Crook and Galina Andreeva introduced a novel approach using Malmquist data envelopment analysis, which assesses corporate efficiency over time, providing a more nuanced understanding of financial distress predictors than static models. Their research highlights the utility of this method in adjusting the efficiency frontier dynamically, offering a robust framework for predicting financial distress with greater accuracy [13].

As the financial landscape has grown in complexity, the focus has shifted toward more nuanced and data-driven approaches. By creating an improved hybrid ensemble machine learning model and using RS-MultiBoosting to predict SMEs' credit risk in supply chain finance, You Zhu, Li Zhou, Chi Xiea, Gang-Jin Wanga, and Truong V. Nguyen have contributed to this evolution. They have shown significant improvements in forecasting accuracy through the integration of multiple data-driven techniques [20].

Xolani Dastile, Turgay Celik and Moshe Potsane provide a comprehensive comparison of statistical and machine learning models used in credit scoring, highlighting recent trends and performance evaluations [7].

Beyond traditional statistical methods, recent decades have seen the rise of machine learning techniques in credit risk assessment. Amir E. Khandani, Adlar J. Kim and Andrew W. Lo demonstrate the application of support vector machines (SVM) and neural networks, highlighting their potential to capture complex nonlinear relationships in data [11]. Similarly, Stefan Lessmann, Bart Baesens, Hsin-Vonn Seow and Lyn C. Thomas conduct a comprehensive comparison of over 40 machine learning techniques, including random forests, gradient boosting machines, and deep learning models, concluding that ensemble methods often outperform single-model approaches in predictive accuracy [12]. Xueheng Qiu, Le Zhang, Ye Ren, P. N. Suganthan and Gehan Amaratunga discuss the use of ensemble deep learning techniques in time series forecasting, including applications to financial markets for risk prediction [18]. The most nonlinear models significantly outperformed logistic regression in a ground-breaking study by Kristina Bluwstein, Marcus Buckmann, Andreas Joseph, Sujit Kapadia, and Ozgur Simsek that developed early warning models for financial crisis prediction using machine learning techniques applied to macrofinancial data from 17 countries between 1870 and 2016. Key economic drivers, such as credit growth and the slope of the yield curve, were identified using a novel framework based on Shapley values to elucidate their nonlinear relationships with crisis risk [3].

The study of Mohammad Mahbobi, Salman Kimiagari and Marriappan Vasudevan presents an innovative algorithm designed to enhance predictive accuracy for imbalanced datasets, particularly in the context of predicting default payments, by utilizing a combination of machine learning classifiers including DNN, SVM, kNN, and ANN alongside sophisticated oversampling and undersampling strategies like SMOTE and ALL-kNN. Notably, the implementation of the SVM classifier with the ALL-kNN sampling method demonstrated exceptional performance, achieving an accuracy of 98.6% and the lowest cross entropy loss of 0.028, thereby offering significant improvements in credit risk assessment by effectively addressing the challenges posed by imbalanced data [15].

Recent studies further augment this view. Tobias Berg, Valentin Burg, Ana Gombovic, and Manju Puri explore the use of digital footprints as an alternative data source in credit scoring, demonstrating significant improvements in predictive power [2]. Meanwhile, Zan Huang, Hsinchun Chen, Chia-Jung Hsu, Wun-Hwa Chen, and Soushan Wu examine the efficacy of neural networks and deep learning frameworks in credit rating analysis and risk assessment, respectively, offering new insights into their practical applications in the finance sector [8]. In line with the current demands for more explainable AI in finance, Feng Zhao, Yan Lu, Xinning Li, Lina Wang, Yingjie Song, Deming Fan, Caiming Zhang, and Xiaobo Chen propose a novel deep learning framework for credit risk assessment [19]. This framework addresses both predictive accuracy and model interpretability.

Each credit risk prediction model carries inherent strengths and weaknesses. Traditional models, such as the Z-score and logistic regression, benefit from their interpretability and ease of use, making them accessible to a wide range of financial practitioners. However, these models may fail to capture the full complexity of modern financial datasets, lacking the flexibility to adapt to nonlinear patterns and interactions [11]. On the other hand, machine learning models excel in handling large and complex datasets, offering superior predictive performance [12]. Yet, their "black box" nature poses significant challenges in terms of interpretability and transparency, which are crucial for regulatory compliance and stakeholder trust. The research of Tisa Maria Antony and B. Sathish Kumar highlighted the effectiveness of the decision tree algorithm, which outperformed others such as linear regression, kNN, SVR, RF, XGB, and MLP, demonstrating the lowest mean absolute error (MSE) of 0.1637 and root mean squared error (RMSE) of 0.2158, thereby providing a promising tool for more accurate credit risk prediction [1].

Despite the advancements in credit risk prediction methodologies, several gaps remain in the literature. Models that successfully strike a balance between interpretability and predictive accuracy are necessary to enable financial institutions to anticipate risk and comprehend the underlying causes of those forecasts [4]. Additionally, the integration of alternative data sources (e.g., social media activity, mobile phone usage) into credit risk models remains underexplored, potentially offering richer insights into borrower behavior but also raising concerns regarding privacy and data quality [9]. Wei-Yang Lin, Ya-Han Hu, and Chih-Fong Tsai provide critical insights into the potential of machine learning in financial crisis prediction and the use of Bayesian analysis in credit scoring, highlighting the ongoing need for innovative approaches that cater to an increasingly complex financial environment [14]. There is a lack of comprehensive studies on the application of advanced machine learning techniques, such as deep learning, in the context of credit risk prediction across different markets and borrower segments [17].

This paper aims to bridge these gaps by presenting a novel credit risk prediction model that combines the strengths of machine learning's predictive power with enhanced interpretability features. By incorporating alternative data sources and employing advanced analytics, we seek to provide a more holistic view of borrower risk profiles. Furthermore, our research contributes to the burgeoning field of machine learning in finance by empirically validating the performance of deep learning models in credit risk assessment, offering insights into their applicability across diverse financial contexts.

3 Methodology

3.1 Dataset and Preprocessing

The dataset comprises 1,000 rows (representing individual records) and 84 variables, capturing various aspects of individuals' financial behavior and status. These variables include measures of income, savings, debt, and expenditure across different categories such as clothing, education, entertainment, fines, gambling, groceries, health, housing, taxes, travel, utilities, and overall expenditure.

The `INCOME` variable has a mean value of $52,490 with a standard deviation of $26,390, ranging from $5000 to $200,000. The `SAVINGS` variable has an average of $14,562, with values spanning from $1000 to $50,000. Debt (`DEBT`) varies widely, with a mean of $11,234 and a standard deviation of $8987, ranging from $500 to $50,000.

The dataset includes several ratios that represent relationships between income, savings, debt, and expenditures, such as:

- `R_SAVINGS_INCOME` (mean: 0.28)
- `R_DEBT_INCOME` (mean: 0.22)
- `R_DEBT_SAVINGS` (mean: 0.77)

Expenditures are broken down into categories:

- Clothing expenditures over the past year (`T_CLOTHING_12`) have a mean of $1,200.
- Education expenditures (`T_EDUCATION_12`) average $3,500.
- Health expenditures (`T_HEALTH_12`) have a mean of $2,000.
- Overall expenditure (`T_EXPENDITURE_12`) averages $24,000.

Categorical features indicate the presence of specific financial products or obligations:

- 62% of individuals have some form of debt (`CAT_DEBT`).
- 17% have a mortgage (`CAT_MORTGAGE`).
- 99% have a savings account (`CAT_SAVINGS_ACCOUNT`).
- 15% have dependents (`CAT_DEPENDENTS`).

The dataset also includes a `CREDIT_SCORE` variable with scores ranging from 300 to 800, with a mean of 587 and a standard deviation of 63.

The dependent variable `DEFAULT` indicates if the customer has defaulted (1) or not (0). In the dataset, 28.4% of the individuals have defaulted.

For model training and performance evaluation, the dataset was split into training and test sets with a 70% to 30% ratio. The distribution of defaulted (1) and not defaulted (0) cases is illustrated in the pie chart in Fig. 1a. For a more detailed

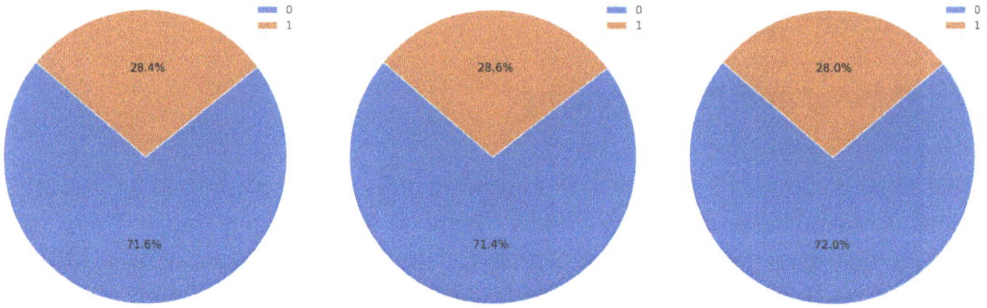

(a) All data (b) Train data (c) Test data

Fig. 1 Distribution of labels indicating default (1) or non-default (0) in the dependent variable

explanation and access to the dataset, please refer to the Kaggle page dedicated to this dataset.[1] There are no missing data.

The Credit Score dataset underwent initial preprocessing steps.

The "CAT_GAMBLING" feature was encoded to numerical values, with "High" being assigned 2, "Low" being assigned 1, and "No" being assigned 0. For subsequent model training, the dataset was split into independent variables (features) represented by a matrix and the dependent variable "DEFAULT". Standardization of features was performed using the StandardScaler from Scikit-learn. This process ensures that all features have a mean of 0 and a standard deviation of 1, thereby preventing any particular feature from dominating the model training process due to its scale.

3.2 Standard Machine Learning Algorithms

To provide a comprehensive evaluation and benchmark against traditional methods, we also consider several standard machine learning algorithms:

Logistic Regression: A logistic regression model serves as the baseline for comparison. It is trained using standardized features, and its performance is evaluated using accuracy, classification report, and confusion matrix. Mathematically, logistic regression can be represented as:

$$P(Y = 1|X) = \frac{1}{1 + e^{-(\beta_0 + \beta_1 X_1 + \cdots + \beta_p X_p)}} \quad (1)$$

where $P(Y = 1|X)$ is the probability of the positive class given the features X, β_0 is the intercept term, $\beta_1, \ldots, \beta_p$ are the coefficients, and X_1, ..., X_p are the features.

K-Nearest Neighbors (kNN): kNN algorithm is used with the value of $k = 3$ that is selected based on the hyperparameter that yields the best accuracy on the validation

[1] https://www.kaggle.com/datasets/conorsully1/credit-score/data.

set. The performance of the kNN model is assessed using accuracy, classification report, and confusion matrix. The kNN algorithm classifies a sample by a majority vote of its k nearest neighbors in the feature space. Mathematically, the prediction for a new sample x is determined by:

$$\hat{y} = \arg\max_{y_i} \sum_{i=1}^{k} I(y_i = y) \tag{2}$$

where $\hat{y}$ is the predicted class label for x, y_i are the class labels of the k nearest neighbors of x, and $I(\cdot)$ is the indicator function that returns 1 if its argument is true and 0 otherwise.

The distance metric used to find the nearest neighbors can vary, with the Euclidean distance being a common choice. For two samples x and x' with p features, the Euclidean distance is calculated as:

$$d(x, x') = \sqrt{\sum_{i=1}^{p} (x_i - x'_i)^2} \tag{3}$$

where x_i and x'_i are the feature values of x and x' respectively for the ith feature.

In our implementation, features are scaled to ensure each feature contributes equally to the distance calculation.

Random Forest: A random forest classifier comprising 100 decision trees is trained without feature scaling. Evaluation metrics are calculated to assess its performance. Random Forest combines multiple decision trees to improve generalization and reduce overfitting.

A random forest is an ensemble learning technique that works by building a large number of decision trees during training and producing a class that represents the mean prediction (regression) or mode of the classes (classification) of the individual trees. Mathematically, the prediction of a random forest $F(x)$ for a sample x is determined by:

$$F(x) = \frac{1}{N} \sum_{i=1}^{N} T_i(x) \tag{4}$$

where N is the number of trees in the forest, and $T_i(x)$ is the prediction of the i-th decision tree for sample x.

Each decision tree in the random forest is trained using a subset of the training data, and a subset of the features. This introduces randomness into the training process, which helps to reduce overfitting. Additionally, during prediction, each tree's output is combined through averaging (for regression) or voting (for classification), leading to more robust predictions. The splitting criterion in each decision tree is typically based on reducing impurity, such as Gini impurity or entropy for classification, and mean squared error for regression.

XGBoost: An XGBoost classifier is trained and evaluated using accuracy as the metric. Extreme gradient boosting, or XGBoost, is an ensemble learning technique that sequentially combines the outputs of multiple weak models, usually decision trees, to create a powerful prediction model. The prediction of an XGBoost model $F(x)$ for a sample x can be expressed as:

$$F(x) = \sum_{m=1}^{M} f_m(x) \tag{5}$$

where M is the number of weak models (trees) in the ensemble, and $f_m(x)$ is the prediction of the m-th weak model. Each weak model in XGBoost is trained sequentially to correct the errors made by the previous models. Mathematically, this can be represented as follows:

$$F_m(x) = F_{m-1}(x) + \gamma \cdot h_m(x) \tag{6}$$

where $F_{m-1}(x)$ is the prediction of the ensemble up to the $(m-1)$-th weak model, γ is the learning rate (shrinkage parameter) that controls the step size during the optimization process, and $h_m(x)$ is the prediction of the m-th weak model.

The weak models (trees) in XGBoost are typically shallow decision trees, often referred to as "stumps" or "weak learners". These trees are constructed to minimize a loss function, such as binary logistic loss for classification or mean squared error for regression.

XGBoost employs a regularization term in the objective function to control model complexity and prevent overfitting. Regularization is achieved through L1 and L2 regularization on the weights of the weak models, and a penalty term on the number of leaves in each tree.

Neural Network Approach: A multilayer perceptron (MLP) neural network is constructed using TensorFlow and Keras. Two hidden layers with 64 and 32 neurons, respectively, are part of the model design. To avoid overfitting, dropout layers with a 20% dropout rate are added after each concealed layer. The Adam optimizer and binary cross-entropy loss are used to train the model. Evaluation is based on accuracy, and the best-performing model weights are saved. Mathematically, the forward pass of the MLP can be represented as:

$$Z^{[l]} = W^{[l]} A^{[l-1]} + b^{[l]} \tag{7}$$

$$A^{[l]} = g^{[l]}(Z^{[l]}) \tag{8}$$

where $Z^{[l]}$ is the pre-activation output of layer l, $W^{[l]}$ and $b^{[l]}$ are the weights and biases of layer l, $A^{[l-1]}$ is the activation output of the previous layer, $g^{[l]}$ is the activation function of layer l, and $A^{[l]}$ is the activation output of layer l.

3.3 Proposed Hybrid Models

For hybrid models, we utilized a combination of various machine learning techniques both in their standard forms and with advanced preprocessing methods. The goal was to compare the effectiveness of each model under standard conditions with their performance after applying principal component analysis (PCA) for dimensionality reduction and GridSearchCV for optimal parameter tuning.

Principal Component Analysis (PCA) Principal component analysis (PCA) is a technique used for dimensionality reduction. Given a dataset X with n samples and p features, PCA aims to find k orthogonal vectors, called principal components, that capture the maximum variance in the data. These principal components are obtained by performing eigenvalue decomposition on the covariance matrix of X.

Let X be a matrix with its columns centered (mean subtracted). The covariance matrix of X is given by:

$$C = \frac{1}{n} X^T X. \tag{9}$$

Eigenvalue decomposition of C yields:

$$C = V D V^T \tag{10}$$

where V is a matrix whose columns are the eigenvectors of C and D is a diagonal matrix containing the corresponding eigenvalues.

The k principal components are the k eigenvectors corresponding to the k largest eigenvalues of C.

In our experiment, we set the parameter $n\_components$ in PCA to 0.95 when fitting the model, indicating that we aimed to retain 95% of the total variance in the data.

GridSearchCV GridSearchCV is a technique used for hyperparameter tuning. It exhaustively searches through a predefined grid of hyperparameters for a given estimator and selects the combination that yields the best performance based on a specified evaluation metric, typically using cross-validation.

Let E be an estimator with hyperparameters H, and G be a grid of hyperparameters to search. GridSearchCV aims to find the optimal hyperparameters h^* by evaluating the performance of E with each combination of hyperparameters in G using K-fold cross-validation.

The optimal hyperparameters h^* are determined as:

$$h^* = \arg\max_{h \in G} \frac{1}{K} \sum_{i=1}^{K} \text{Score}(E(X_{\text{train}_i}, X_{\text{test}_i}, h)) \tag{11}$$

where X_{train_i} and X_{test_i} are the training and testing sets for the i-th fold of cross-validation, and Score is the evaluation metric.

Table 1 Construction of hybrid models

Model	Standard configuration	Preprocessed configuration
Random forest	Base model	PCA + GridSearchCV
XGBoost	Base model	PCA + GridSearchCV
K-nearest neighbors (kNN)	Default settings	PCA + GridSearchCV
Logistic regression	Conventional	PCA + GridSearchCV
Neural network	Basic architecture	PCA + GridSearchCV

Construction of Hybrid Models In constructing our hybrid models, we combined various machine learning algorithms with preprocessing techniques to enhance their performance. Each model was examined in two configurations: the standard configuration, where the model was trained with default settings, and the preprocessed configuration, where we applied principal component analysis (PCA) for dimensionality reduction and GridSearchCV for hyperparameter tuning.

Table 1 summarizes the construction of these hybrid models. In the standard configuration, models such as random forest, XGBoost, k-nearest neighbors (kNN), logistic regression, and neural network were trained using their default settings. In contrast, in the preprocessed configuration, PCA was first applied to reduce the dimensionality of the feature space, followed by GridSearchCV to find the optimal hyperparameters for each model, thereby improving their accuracy and efficiency.

3.4 Evaluation Metrics

The performance of each model was analyzed based on accuracy, precision, recall (sensitivity), and $F1$-score.

Accuracy measures the proportion of correctly classified instances among all instances in the dataset. It is calculated as:

$$\text{Accuracy} = \frac{\text{True Positives} + \text{True Negatives}}{\text{Total Number of Instances}}. \tag{12}$$

Precision measures the proportion of correctly predicted positive instances out of all instances predicted as positive by the classifier. It is calculated as:

$$\text{Precision} = \frac{\text{True Positives}}{\text{True Positives} + \text{False Positives}}. \tag{13}$$

Recall (also known as sensitivity or true positive rate) measures the proportion of correctly predicted positive instances out of all actual positive instances in the dataset. It is calculated as:

$$\text{Recall} = \frac{\text{True Positives}}{\text{True Positives} + \text{False Negatives}}. \tag{14}$$

F1-score is the harmonic mean of precision and recall. It provides a balanced measure of the classifier's performance and is calculated as:

$$F1\text{-score} = 2 \times \frac{\text{Precision} \times \text{Recall}}{\text{Precision} + \text{Recall}}. \tag{15}$$

4 Results of Experiments

4.1 Initial Results

Tables 2, 3, and 4 provide a comprehensive summary of the classification metrics and test accuracy for various machine learning models in predicting loan default.

In Table 2, which focuses on non-default cases (Class 0), we observe that the models generally perform well, with precision values ranging from 0.72 to 0.78 and recall values ranging from 0.82 to 0.92. The $F1$-score, which balances precision and recall, lies between 0.78 and 0.84, indicating robust performance across the models.

Conversely, Table 3 highlights the performance of the models in predicting default cases (Class 1). Here, the models exhibit lower precision, recall, and $F1$-score compared to Class 0, indicating the difficulty in correctly identifying default instances. Notably, the Random Forest and XGBoost models show relatively higher performance compared to others in terms of recall and $F1$-score.

The overall test accuracy of the models, presented in Table 4, ranges from 0.65 to 0.7. While accuracy is an important metric, especially for balanced datasets, it might not fully represent model performance, especially in imbalanced datasets like this one.

These initial results provide insights into the strengths and weaknesses of each model in predicting loan default. Further analysis and experimentation, such as feature engineering or hyperparameter tuning, could potentially improve the models' performance.

4.2 Results with PCA and GradientSearchCV

In this subsection, we present the results obtained after applying Principal Component Analysis (PCA) and utilizing GridSearchCV for hyperparameter tuning.

Tables 5, 6, and 7 summarize the classification metrics and test accuracy for Class 0, Class 1, and overall accuracy, respectively.

In Table 5, focusing on non-default cases (Class 0), we observe slight improvements in precision, recall, and $F1$-score across all models compared to the initial

Table 2 Summary of classification metrics for class 0

Model	Precision	Recall	$F1$-score
Random Forest	0.74	0.90	0.81
XGBoost	0.75	0.84	0.79
kNN	0.75	0.82	0.78
Logistic Regression	0.72	0.92	0.81
Neural Network	0.78	0.90	0.84

Table 3 Summary of classification metrics for class 1

Model	Precision	Recall	$F1$-score
Random Forest	0.42	0.19	0.26
XGBoost	0.41	0.29	0.34
kNN	0.40	0.31	0.35
Logistic regression	0.31	0.10	0.15
Neural network	0.32	0.26	0.29

Table 4 Test accuracy of the models

Model	Accuracy
Random Forest	0.7
XGBoost	0.686
kNN	0.676
Logistic Regression	0.686
Neural network	0.65

Table 5 Summary of classification metrics for class 0

Model	Precision	Recall	$F1$-score
Random Forest	0.75	0.95	0.84
XGBoost	0.73	0.96	0.83
kNN	0.73	0.97	0.84
Logistic Regression	0.75	0.93	0.83
Neural Network	0.78	0.90	0.84

Table 6 Summary of classification metrics for class 1

Model	Precision	Recall	$F1$-score
Random Forest	0.58	0.18	0.27
XGBoost	0.53	0.11	0.18
kNN	0.57	0.10	0.16
Logistic Regression	0.50	0.19	0.28
Neural Network	0.55	0.31	0.40

Table 7 Test accuracy of the models

Model	Accuracy
Random Forest	0.73
XGBoost	0.723
kNN	0.727
Logistic regression	0.72
Neural network	0.745

results. Notably, Random Forest, XGBoost, and kNN models achieve higher recall values, indicating better performance in correctly identifying non-default instances.

However, in Table 6, which examines the performance on default cases (Class 1), we notice mixed results. While some models show improvements in precision and recall, others exhibit a slight decrease. The Neural Network model stands out with improved precision and recall compared to the initial results.

Regarding overall accuracy, Table 7 demonstrates that the models maintain relatively high accuracy scores after PCA and hyperparameter tuning. Notably, the Neural Network model achieves the highest accuracy among all models, indicating its effectiveness in capturing the underlying patterns in the data.

In our study, we use charts, see Figs. 2 and 3, to demonstrate the performance differences between two types of data models: standard and hybrid. Standard models including basic algorithms such as Random Forest, FGBoost, kNN, Logistic Regression and Neural Network are used unchanged. Hybrid - these are models with pre-processing of PCA data and finding the best parameters using GridSearchCV.

In the confusion matrix, Fig. 4, for the neural network model, out of a total of 200 instances, 17 were correctly predicted as defaults (True Positives), while 132 were correctly predicted as non-defaults (True Negatives). However, the model misclassified 37 instances as non-defaults when they were actually defaults (False Negatives), and 14 instances as defaults when they were actually non-defaults (False Positives). This indicates that while the model performed well in identifying non-default instances, it struggled with accurately predicting defaults, highlighting the need for further optimization to improve its performance in this area.

The sorted features, visually represented in Fig. 5, provide crucial insights into the factors influencing the model's predictions. These features were obtained through

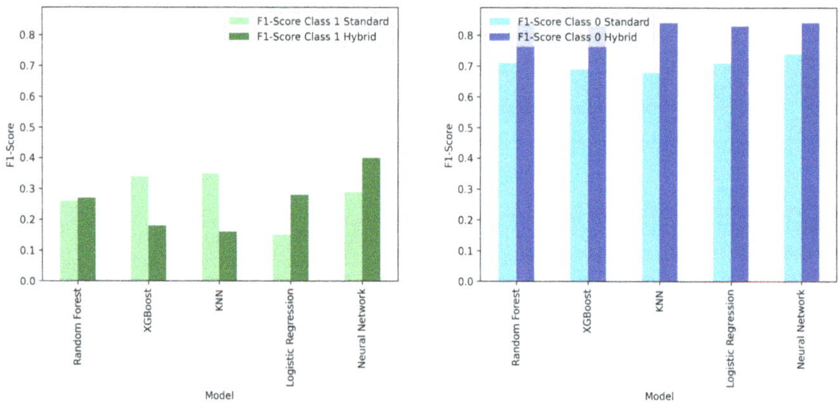

(a) F1-Score for Class 1 (b) F1-Score for Class 0

Fig. 2 $F1$-score comparison

Fig. 3 Comparison of model accuracies between standard and hybrid cases

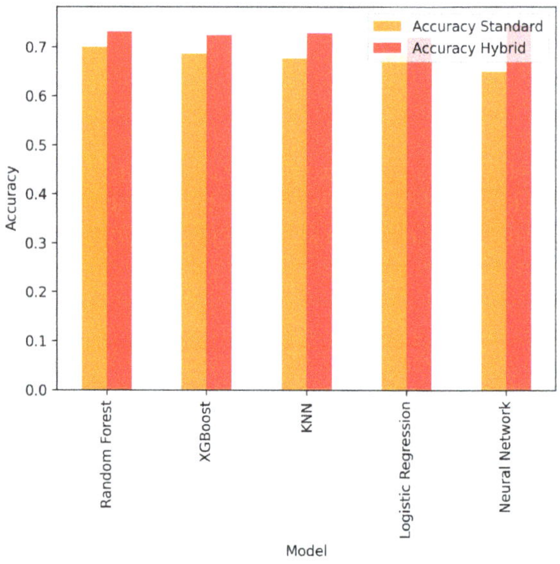

a random forest classifier, highlighting their significance in predicting outcomes. Notably, the credit score emerges as the most significant feature, indicating its substantial impact on predictions, followed closely by the ratio of debt to income and the ratio of expenditure to debt. Various expenditure and income-related ratios across different categories, including entertainment, education, and clothing, contribute significantly to the model's predictive power. Additionally, categories such as fines, health, taxes, and savings play notable roles in influencing predictions. While features related to housing, gambling, and education hold moderate importance, only the first most

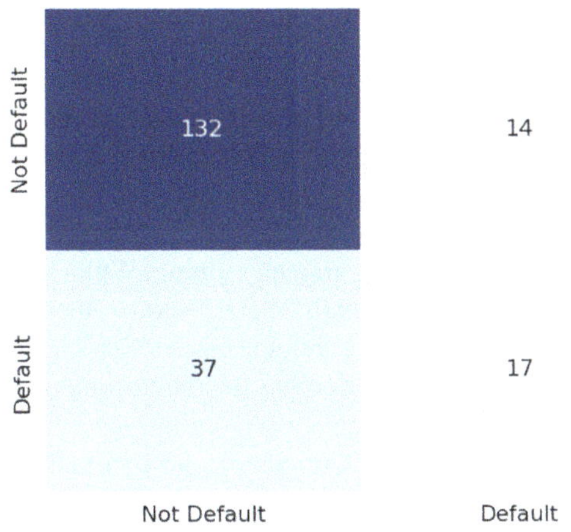

Fig. 4 Confusion matrix for neural network model

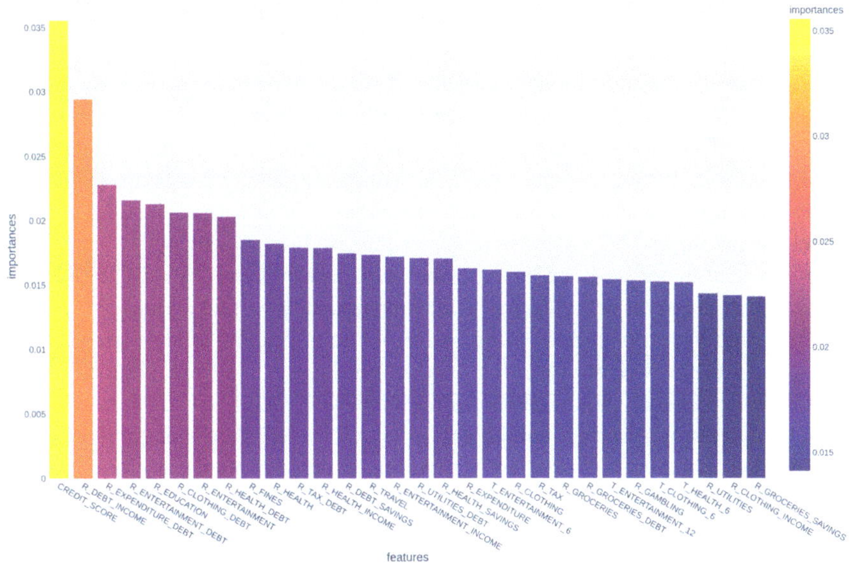

Fig. 5 Feature Importances Bar Plot

significant 30 features are depicted in the figure. Conversely, variables such as credit card ownership and mortgage show lesser significance. Variables such as savings account ownership and specific categorical attributes like debt demonstrate minimal importance in the model's predictions. Understanding the relative importance of these features offers valuable insights for decision-making in financial contexts.

5 Discussion and Conclusion

The results of our experiments reveal several important insights into the performance of various machine learning models in predicting credit default. Here, we discuss the implications of our findings and address potential areas for further investigation.

Firstly, the initial results demonstrate the baseline performance of each model without any feature engineering or hyperparameter tuning. While some models show promising results in terms of precision, recall, and $F1$-score, others exhibit deficiencies, particularly in correctly identifying default cases (Class 1). This highlights the importance of exploring different modeling techniques and optimizing model parameters to improve predictive accuracy.

Secondly, by incorporating Principal Component Analysis (PCA) and utilizing GridSearchCV for hyperparameter tuning, we observe notable improvements in the performance metrics of several models. Specifically, Random Forest, XGBoost, and kNN models demonstrate enhanced recall values for non-default cases (Class 0), indicating better detection of non-default instances. However, the performance on default cases (Class 1) varies across models, suggesting the need for further refinement to effectively identify default instances.

Furthermore, the multilayer perceptron (MLP) neural network model, with its two hidden layers comprising 64 and 32 neurons respectively, consistently performs well across different experiments, achieving high accuracy scores and demonstrating robustness in capturing complex patterns within the data. Visualizations such as confusion matrices and $F1$-score comparisons provide additional insights into the model's behavior and help identify areas for potential improvement.

In conclusion, our study underscores the importance of employing a comprehensive approach to credit default prediction, encompassing feature engineering, model selection, and hyperparameter optimization. While traditional models like Random Forest and XGBoost yield promising results, more sophisticated techniques such as Neural Networks show potential for further improvement.

Moving forward, we recommend exploring additional feature engineering techniques, such as feature selection and extraction, to enhance model performance further. Additionally, investigating ensemble methods and deep learning architectures may offer avenues for achieving even higher predictive accuracy. Ultimately, the findings of this study contribute to the ongoing efforts to develop robust credit risk assessment models, thereby facilitating more informed decision-making in financial institutions and mitigating the impact of default events.

Funding The third author acknowledges the support by a grant from the Ministry of Science and Higher Education of the Republic of Kazakhstan within the framework of the project AP19676669.

Author Contribution Ruslan Murzagaliyev conducted the experiments. Bizhigit Sagidolla, was responsible for preparing the literature review and performing the fieldwork, ensuring that the study was grounded in a solid theoretical base. Shirali Kadyrov structured the paper, wrote a significant portion of the paper, and supervised the experimental work.

Conflict of Interest The authors have no conflicts of interest to declare.

References

1. Antony, T.M., Kumar, B.S.: Predicting of credit risk using machine learning algorithms. In: International Conference on Artificial Intelligence on Textile and Apparel. pp. 99–114. Springer (2023)
2. Berg, T., Burg, V., Gombović, A., Puri, M.: On the rise of fintechs: credit scoring using digital footprints. Rev. Financ. Stud. **33**(7), 2845–2897 (2020)
3. Bluwstein, K., Buckmann, M., Joseph, A., Kapadia, S., Şimşek, Ö.: Credit growth, the yield curve and financial crisis prediction: evidence from a machine learning approach. J. Int. Econ. **145**, 103773 (2023)
4. Bracke, P., Datta, A., Jung, C., Sen, S.: Machine learning explainability in finance: an application to default risk analysis (2019)
5. Constantin, E., Constanţa, E.: Credit risk management with the purpose of optimizing the performances of the financial institutions. Ann.-Econ. Ser. **1**, 208–214 (2014), https://api.semanticscholar.org/CorpusID:154618844
6. Crouhy, M., Galai, D., Mark, R.: The essentials of risk management, vol. 1. McGraw-Hill New York (2006)
7. Dastile, X., Celik, T., Potsane, M.: Statistical and machine learning models in credit scoring: a systematic literature survey. Appl. Soft Comput. **91**, 106263 (2020)
8. Huang, Z., Chen, H., Hsu, C.J., Chen, W.H., Wu, S.: Credit rating analysis with support vector machines and neural networks: a market comparative study. Decis. Support. Syst. **37**(4), 543–558 (2004)
9. Jagtiani, J., Lemieux, C.: The roles of alternative data and machine learning in fintech lending: evidence from the lendingclub consumer platform. Financ. Manag. **48**(4), 1009–1029 (2019)
10. Khan, A.: A comparative study of credit risk management (2020). https://api.semanticscholar.org/CorpusID:218810768
11. Khandani, A.E., Kim, A.J., Lo, A.W.: Consumer credit-risk models via machine-learning algorithms. J. Bank. Financ. **34**(11), 2767–2787 (2010)
12. Lessmann, S., Baesens, B., Seow, H.V., Thomas, L.C.: Benchmarking state-of-the-art classification algorithms for credit scoring: an update of research. Eur. J. Oper. Res. **247**(1), 124–136 (2015)
13. Li, Z., Crook, J., Andreeva, G.: Dynamic prediction of financial distress using Malmquist DEA. Expert. Syst. Appl. **80**, 94–106 (2017)
14. Lin, W.Y., Hu, Y.H., Tsai, C.F.: Machine learning in financial crisis prediction: a survey. IEEE Trans. Syst. Man, Cybern. Part C (Appl. Rev.) **42**(4), 421–436 (2011)
15. Mahbobi, M., Kimiagari, S., Vasudevan, M.: Credit risk classification: an integrated predictive accuracy algorithm using artificial and deep neural networks. Ann. Oper. Res. **330**(1), 609–637 (2023)
16. McCormick, M.: Risk management. The Agile Codex (2021). https://api.semanticscholar.org/CorpusID:239733546
17. Ozbayoglu, A.M., Gudelek, M.U., Sezer, O.B.: Deep learning for financial applications: a survey. Appl. Soft Comput. **93**, 106384 (2020)
18. Qiu, X., Zhang, L., Ren, Y., Suganthan, P.N., Amaratunga, G.: Ensemble deep learning for regression and time series forecasting. In: 2014 IEEE Symposium on Computational Intelligence in Ensemble Learning (CIEL), pp. 1–6. IEEE (2014)
19. Zhao, F., Lu, Y., Li, X., Wang, L., Song, Y., Fan, D., Zhang, C., Chen, X.: Multiple imputation method of missing credit risk assessment data based on generative adversarial networks. Appl. Soft Comput. **126**, 109273 (2022)
20. Zhu, Y., Zhou, L., Xie, C., Wang, G.J., Nguyen, T.V.: Forecasting SMEs' credit risk in supply chain finance with an enhanced hybrid ensemble machine learning approach. Int. J. Prod. Econ. **211**, 22–33 (2019)

A Simulation Approach to Modeling Dengue Incidence: Evaluating Bayesian Poisson Generalized Linear Mixed Model

Nik Nur Fatin Fatihah Sapri, Wan Fairos Wan Yaacob, Siti Nur Zahrah Amin Burhanuddin, and Yap Bee Wah

Abstract Simulating weather data helps researchers predict climate patterns and analyze the transmission dynamics of vector-borne diseases. A simulation study was conducted to evaluate the model performance of Bayesian Poisson Generalized Linear Mixed Model (GLMM) using simulated data to predict the dengue incidence in Selangor. At first, the actual dengue data was collected from the Ministry of Health Malaysia, and weather data (temperature, relative humidity and rainfall) was obtained from online climate data. The simulation study began by generating daily temperature and relative humidity using ARIMA model, followed by daily rainfall using a probability distribution. In total, 16,425 data observations were simulated throughout nine districts. The study revealed that the simulated daily temperature appears to be closely matched to the actual data. Nevertheless, the simulated daily relative humidity and rainfall resulted in variability across districts. It was found that a 1% increase in relative humidity raises dengue incidence by 0.19% daily, with increases of 0.19%, 0.22%, 0.21%, and 0.17% at lags of 7, 14, 21, and 28 days, respectively. Similarly, a 1 mm increase in rainfall raises dengue incidence by 0.11%

N. N. F. F. Sapri (✉) · S. N. Z. Amin Burhanuddin
Faculty of Computer and Mathematical Sciences, Universiti Teknologi MARA, Shah Alam, Selangor, Malaysia
e-mail: nikfatinfatihah@uitm.edu.my

S. N. Z. Amin Burhanuddin
e-mail: sitinurzahrah@uitm.edu.my

W. F. Wan Yaacob
Faculty of Computer and Mathematical Sciences, Universiti Teknologi MARA, Shah Alam, Selangor, Malaysia

Institute for Big Data Analytics and Artificial Intelligence (IBDAAI), Kompleks Al-Khawarizmi, Universiti Teknologi MARA, Shah Alam, Malaysia

W. F. Wan Yaacob
e-mail: wnfairos@uitm.edu.my

Y. B. Wah
UNITAR Graduate School, UNITAR International University, Petaling Jaya, Selangor, Malaysia
e-mail: bee.wah@unitar.my

A. Shuib et al. (eds.), *Mathematical Modeling and Numerical Analysis*,
Springer Proceedings in Mathematics & Statistics 510,
https://doi.org/10.1007/978-981-96-9350-4_13

daily, with increases of 0.11%, 0.12%, 0.11%, and 0.12% at the same lags. In addition, the Bayesian model fitted in the study captured a weak spatial autocorrelation (0.0480) and a very strong temporal autocorrelation (0.9996) in the simulated dataset generated earlier. Additional weather variables and random effects can be considered in model formulation to improve the accuracy of the simulation model in future.

Keywords Bayesian · Dengue · Generalized Linear Mixed Model · Poisson · Simulation

1 Introduction

Simulating weather data plays a crucial role in epidemiological studies, as it allows researchers to systematically explore the effects of climatic variables on spread of diseases. It is particularly valuable in studying vector-borne diseases like dengue fever, where the life cycle and activity patterns of vectors such as Aedes mosquitoes are closely tied to environmental conditions. For instance, simulating varying humidity levels and temperature ranges can help predict mosquito breeding cycles and the potential for outbreaks under different climatic scenarios. Additionally, simulated weather data can be used to validate and refine epidemiological models, enhancing their predictive accuracy and reliability [1]. By analyzing the impact of different climate scenarios on disease dynamics, researchers can better understand the potential risks and consequences of climate change and inform public health strategies to mitigate these effects [2].

Identifying the best-fitted statistical distributions for a simulation study is vital because it ensures that the simulated weather data closely matches the actual data. For instance, Burhanuddin [3] employed normal, lognormal, and gamma distributions to model the pattern of daily rainfall data. The analysis showed that the daily rainfall data across all regions of Peninsular Malaysia follows a lognormal distribution. The findings were aligned with the study done by Yusuf et al. [4]. Another study done by Suhaila et al. [5] examined the mixed lognormal distribution shown to be the best fit for rainfall data from Peninsular Malaysia's rain gauge stations. Meanwhile, Malek et al. [6] and Norzaida et al. [7] found that the Weibull distribution was the best model for describing rainfall patterns. On the other hand, temperature often exhibits patterns over time. Therefore, selecting an appropriate model needs to be carefully planned to accurately capture these trends and ensure reliable predictions. In the southwestern region of Thailand, Wanishsakpong and Owusu [8] found that the ARIMA $(3,0,2)(0,1,0)_{12}$ model produced a more accurate temperature forecast than the ARIMAX model in the temperature modeling in Phuket. The empirical findings by Sankalp and Panda [9] demonstrated that the ARIMA model had superior performance in fitting the daily relative humidity data from 1981 to 2020 in five drought-prone districts (Bolangir, Bargarh, Nuapada, Kalahandi, and Phulbani) compared to the other two models (K-nearest neighbor (KNN) and Naive Bayes).

Similar findings were also revealed by Li et al. [10] that ARIMA (1,0,0) suited the best in fitting the daily average relative humidity data of a county in Gansu Province.

Previous studies have evaluated model performance through simulation work using various spatiotemporal structures within the Bayesian framework of Generalized Linear Mixed Models (GLMMs). Notably, Rushworth et al. [11], Lee et al. [12], and Napier et al. [13] employed Markov Chain Monte Carlo (MCMC) simulations for both Binomial and Gaussian distributions. Napier et al. [13] performed a simulation study to compare the performance of (MC)3 and standard MCMC algorithms in quantifying the model's ability to correctly cluster areas based on four different temporal disease trends namely constant, linearly increasing, linearly decreasing, and a change point trend. Hundred simulated datasets generated were fitted to (MC)3 and standard MCMC algorithms. In comparison, the researchers found that the MCMC algorithm less perform with overall classification just lie between 78.53 and 93.79%. Meanwhile, (MC)3 algorithm performs well with overall classification ranging between 80 and 95%. Lee et al. [12] included both spatiotemporal main effects and space–time interactions in their model formulation, with covariates following a standard normal distribution. The researchers performed simulations by exploring various combinations of spatial units ($N = 10, 20, 30, 40, 50$) and time units ($T = 10, 20, 30, 40$). The study found that when the spatial unit, $N = 20$, a large bias was produced for temporal dependence structure. Rushworth et al. [1] simulated disease counts in England from Poisson distribution as to assess the performance of four different Bayesian models for a variation of the size of expected counts ($E = 10, 50, 100$), relative risk ($A = 1, 1.5, 2$), and the time periods ($T = 1, 5, 20$). Results indicated that model 3 (model with no spatial smoothing) performed well in terms of lower root mean squared error (RMSE) and deviance information criteria (DIC). Hence, these studies provide a benchmark for this paper on implementing and evaluating Bayesian Poisson GLMMs to have a better understanding on its performance with simulated datasets, particularly focusing on scenarios relevant to epidemiological modeling.

Aedes mosquitoes have been found to transmit the dengue virus, and many studies have linked weather as a possible factor in this transmission [14–16]. The weather conditions play a pivotal role in shaping the life cycle and activities of these mosquitoes, directly impacting their ability to spread the virus. Many studies have linked weather to dengue transmission by Aedes mosquitoes. Weather affects these insects' life cycle and virus transmission. Dengue fever is positively connected with rising yearly temperatures in Bangladesh, Vietnam, Myanmar, Thailand, India, and Sri Lanka, according to Anwar et al. [17]. Bisht et al. [18] found a favorable link between temperature and entomological indices, which assess vector populations, in Delhi, India, where temperatures can exceed 47 °C. These findings supported by Tran et al. [15] that precipitation increased the Breteau Index (BI) by 0.007 and 0.023 units below and above 27.21 °C. Precipitation increased BI by 0.032 and 0.035 units in the south when the temperature was below 27.21 °C and between 27.21 and 30.17 °C. When the temperature was above 30.17 °C, relative humidity increased BI by 0.225 units. In Lahore, Pakistan, Naqvi et al. [19] reported that the dengue cases were high when rainfall reached 550 mm. These findings support the idea that weather affects mosquito hatching, survival, and dissemination.

Therefore, this study has two primary objectives: to simulate weather data (including daily temperature, relative humidity, and rainfall) and dengue incidence data for Selangor and to evaluate the performance of a Bayesian Poisson Generalized Linear Mixed Model (GLMM) using the simulated data. This paper is structured into five sections. Section 2 outlines the methodology for simulating weather and dengue incidence data. Section 3 presents the results and findings of the study. Finally, Sect. 4 concludes the paper.

2 Methodology

2.1 Data and Variables

Selangor is a major state located on the western coast of Peninsular Malaysia. It shares borders with Perak to the north, Pahang to the east, Negeri Sembilan to the south, and the Straits of Malacca to the west. The population was estimated over 6.5 million as of 2024, making it the most populous state in Malaysia [20]. It has a tropical rainforest climate characterized by clearly defined wet and dry seasons, which are influenced by two monsoon cycles: the southwest monsoon from May to September and the northeast monsoon from November to March. The simulation in this study was conducted using actual data. Hence, the dengue case data was collected from the Vector Borne and Infectious Diseases Sector (VBIDS) under the Ministry of Health (MOH) Malaysia. This dataset was collected following established protocols and records the daily incidence of dengue cases across nine districts in Selangor: Gombak, Hulu Langat, Hulu Selangor, Klang, Kuala Langat, Kuala Selangor, Petaling, Sabak Bernam, and Sepang. The data covers the period from January 1, 2013, to December 31, 2017, with a total of 16,425 observations. To add, Selangor reported the highest number of dengue incidences among other states in Malaysia [21]. This study integrates various weather factors, such as average temperature, relative humidity, and rainfall, along with lag time effects of 7, 14, 21, and 28 days as explanatory variables. The weather variables were obtained from NASA Prediction of Worldwide Energy Resources [22]. Note that the lag time effects were included in the study, indicating the duration necessary for mosquito eggs to mature into fully grown adults. Figures 1, 2, and 3 exhibit the actual weather data for each district of Selangor.

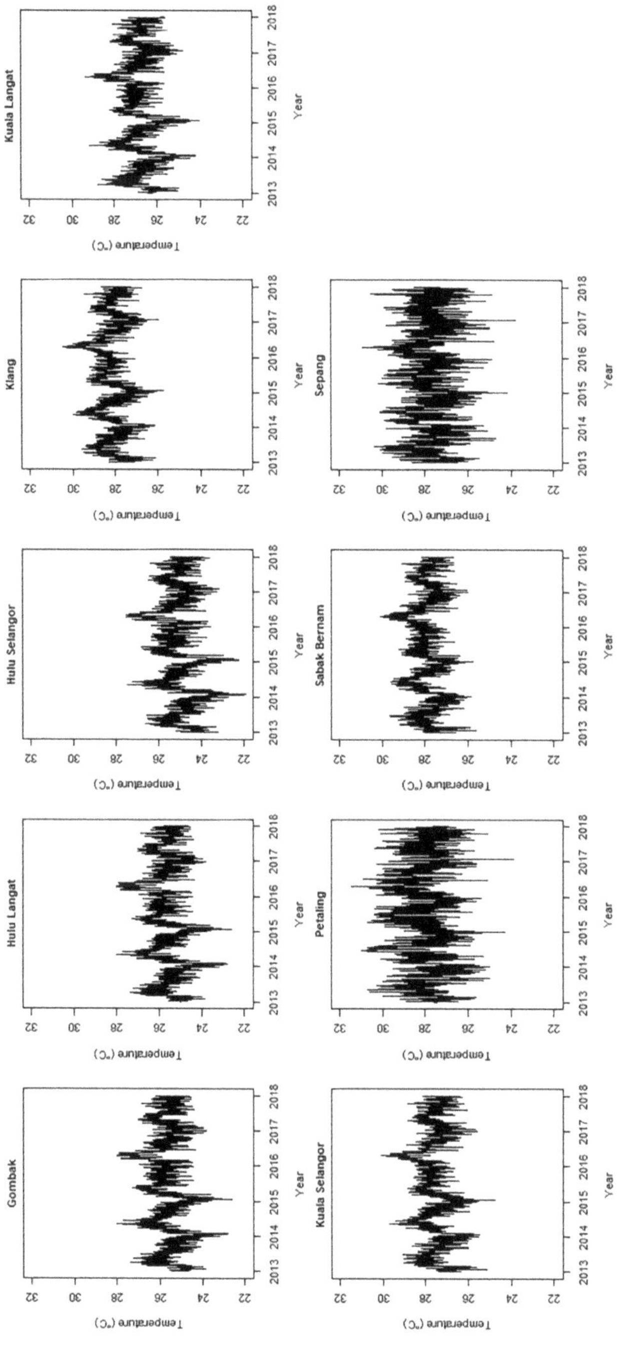

Fig. 1 Temperature (°C)

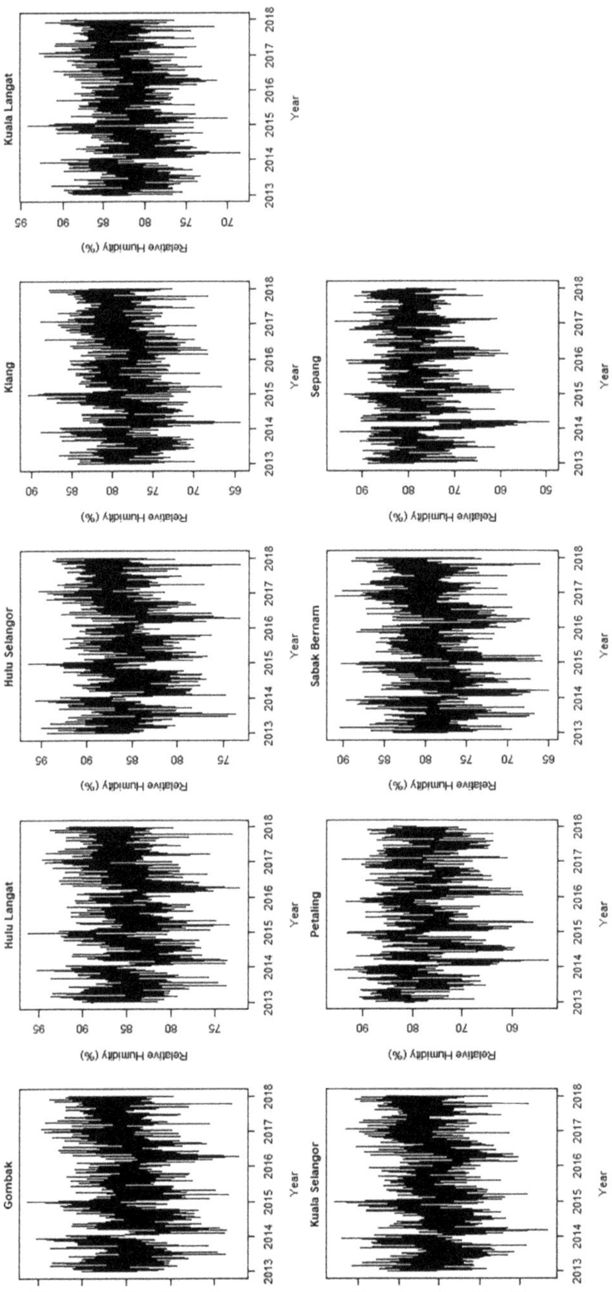

Fig. 2 Relative humidity (%)

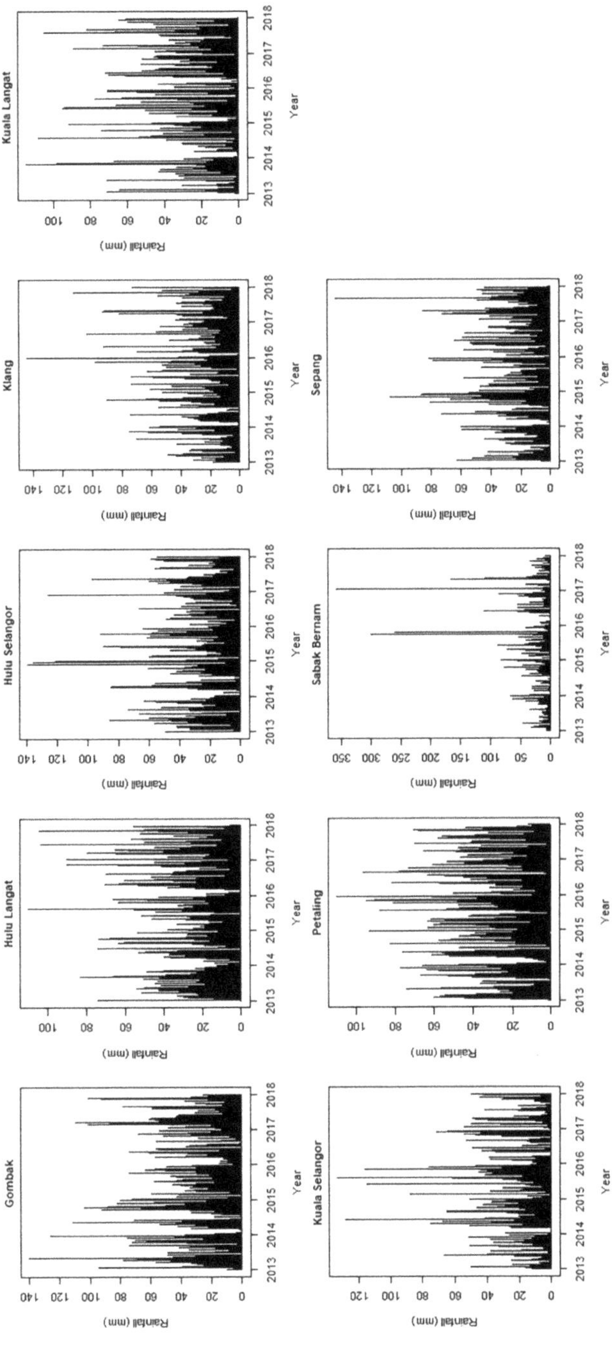

Fig. 3 Rainfall (ml)

2.2 Simulation Design and Framework

The methodology of the analysis involved two phases: Phase 1, Data Simulation, and Phase 2, Model Fitting. These phases are further explained as follows:

Phase 1: Data Simulation.

Step 1: Simulate Weather Data. The simulation study began by generating daily weather data, including temperature, relative humidity, and rainfall, for each district of Selangor. The data was simulated based on the specific distributions and characteristics of the actual weather data. To simulate daily temperature and daily relative humidity, the study used ARIMA model due to actual data of temperature and relative humidity exhibit patterns over time. Following rainfall modeling done by Burhanuddin [3], probability distributions such as lognormal distribution, gamma distribution, Weibull distribution and exponential distribution were used to simulate rainfall data. The simulation processes were carried out using R Studio. Table 1 summarizes the selected ARIMA model and selected probability distributions for each district of Selangor.

Step 2: Simulate Dengue Incidence Data.

Setting Up Model Parameters

In this study, the dengue incidence data was simulated based on Bayesian Poisson Generalized Linear Model (GLMM). To begin, the parameters of the model were fixed such that the intercept and parameter estimates, $\beta_{0...15} = 0.1$, variance parameter, $\rho_S = 0.8$, and spatial dependence parameter, $\tau^2 = 0.01$. Fixing parameters can considerably reduce the processing resources and time needed to conduct the simulation, particularly for large-scale or complex models.

Table 1 Simulation weather data

Districts	ARIMA model (temperature)	ARIMA model (relative humidity)	Probability distribution (rainfall)
Gombak	(1,1,1)	(1,1,1)	Weibull
Hulu Langat	(1,1,1)	(1,1,1)	Lognormal
Hulu Selangor	(1,1,1)	(1,1,1)	Lognormal
Klang	(2,0,1)	(1,1,1)	Lognormal
Kuala Langat	(2,0,2)	(1,1,1)	Lognormal
Kuala Selangor	(3,1,2)	(1,1,1)	Lognormal
Petaling	(2,0,1)	(1,0,3)	Weibull
Sabak Bernam	(4,0,2)	(1,1,1)	Weibull
Sepang	(3,0,1)	(3,1,1)	Weibull

Specifying Response Variable to Poisson Distribution

The dengue incidence data for each district was generated from a Poisson distribution with random effects such that $y_{\text{it}} = \text{Poisson}(\lambda_{\text{it}})$ and $\ln(\lambda_{\text{it}}) = \theta_{\text{it}}$ whereby $\theta_{\text{it}} = \beta_0 + \beta_1 x_{\text{it}} + \cdots + \beta_{15} x_{\text{it}} + \phi_{\text{it}} + \varepsilon_{\text{it}}$. Note that, β_0 is the intercept, β_1 is the regression coefficient, ρ_{S} refers to spatial dependence parameter, ρ_{T} denotes the temporal dependence parameter, τ^2 is the variance parameter, and ϕ_{it} implies the random effect component of spatiotemporal structure. This study used Bayesian inference based on Markov Chain Monte Carlo (MCMC) simulation through Gibbs sampling.

Specifying Spatial and Temporal Random Effects Parameter

The spatial and temporal random effects parameters were included in the model to capture the spatial and temporal correlations that arise due to unobserved factors. These random effects were assumed to follow conditional autoregressive (CAR) with order one which can be written as follows:

$$
\begin{aligned}
\phi_t | \phi_{t-1} &\sim N\left(\rho_{\text{T}}\phi_{t-1}, \tau^2 Q(W, \rho_{\text{S}})^{-1}\right) \quad t = 2, \ldots, T \\
\phi_1 &\sim N\left(0, \tau^2 Q(W, \rho_{\text{S}})^{-1}\right) \\
\tau^2 &\sim \text{Inverse} - \text{Gamma}(1, 0.01) \\
\rho_{\text{S}}, \rho_{\text{T}} &\sim \text{Uniform}(0, 1).
\end{aligned} \tag{1}
$$

The spatial neighborhood matrix, W, is a matrix that shows how close two areal units are to each other in space. To create matrix W, it requires spatial data at the district level which can be accessed from Global Administrative Areas (GADM) [23]. From W, the spatial precision matrix, $Q(W, \rho_{\text{S}})$, was computed using the following formula.

$$
Q(W, \rho_{\text{S}}) = \rho_{\text{S}}\left[\text{diag}(W\mathbf{1}) - W\right] + (1 - \rho_{\text{S}})I. \tag{2}
$$

Here, the spatial precision matrix was then inverted and a sample of the random effects, ϕ was generated through multivariate random normal distribution using *mvrnorm*() function. To ensure that the simulated data is as close as possible to the actual data, error terms need to be taken into account in model formulation. This is due to not all variability can be explained by the model's predictors. Thus, the error term used in the simulation study was assumed to follow a normal distribution with a mean of 0 and a standard deviation of 0.001.

Setting Up the Linear Predictor

After all the model components were computed, the simulation study was then continued by setting up a linear predictor that contains an intercept term, a set of covariates (simulated weather data and its lag), random effects (ϕ) and an error term such that;

$$
\text{LP} = \beta_0 + \beta_1 x_{\text{it}} + \beta_2 x_{\text{it}} \ldots + \beta_{15} x_{\text{it}} + \phi_{\text{it}} + \varepsilon_{\text{it}}. \tag{3}
$$

Generating Dengue Incidence, y_{it}.

Finally, the dengue incidence data can be randomly simulated from Poisson distributions with λ value which is assumed equal to a set of linear predictors generated above such `y < -rpois(16,425,LP)`.

Phase 2: Model Fitting. The simulated data was used to fit Bayesian Poisson Generalized Linear Mixed Models (GLMMs) using the `ST.CARar ()` function from the CARBayesST R package. The model specifications included a burn-in period of 20,000 iterations, a total of 120,000 samples, and thinning by 10 to minimize autocorrelation. Due to computational constraints, this process was repeated approximately 10 times. The best-fitting Bayesian Poisson GLMM was selected based on the lowest value of the deviance information criterion (DIC) obtained from the simulated data. Figure 4 summarizes the flowchart of the simulation process carried out in the study.

3 Analysis and Results

3.1 *Simulated Weather Data and Dengue Incidence Rates*

The analysis of the study began with the simulation of daily weather data, including temperature, relative humidity, and rainfall, using best-fitted statistical distributions based on actual observations of Selangor (Sect. 2.2). Figures 5, 6, and 7 provide a comprehensive view of the temporal series, depicting actual daily weather data with a black line and simulated daily weather data with a red line across each district in Selangor. Upon analyzing the simulated temperature data, it becomes apparent that there are slight deviations in the pattern from the actual temperature data (Figs. 1, 2, and 3). These discrepancies may be attributed to various factors, including differences in the ARIMA models used to generate the simulated data and the inherent variability that might exist in the actual weather patterns. However, the mean squared error (MSE) values tabulated in Table 2 successfully revealed that the MSE values for temperature are relatively low across all districts, suggesting almost accurate simulations between actual and simulated temperature data. Meanwhile, the MSE computed for simulated daily relative humidity data varies more significantly in which Petaling and Sepang have notably higher MSE values of 88.4398 and 79.8075, respectively, as compared to other districts. Figure 3 suggests that the ARIMA models used to generate the simulated data effectively captured the underlying patterns and trends in the actual relative humidity data. Unexpectedly, the MSE values for rainfall are the highest among the other weather variables, suggesting that rainfall is the most challenging variable to simulate accurately. Klang has the highest MSE value for rainfall with 1563.7770 suggesting the least accurate rainfall simulations, while Sepang has

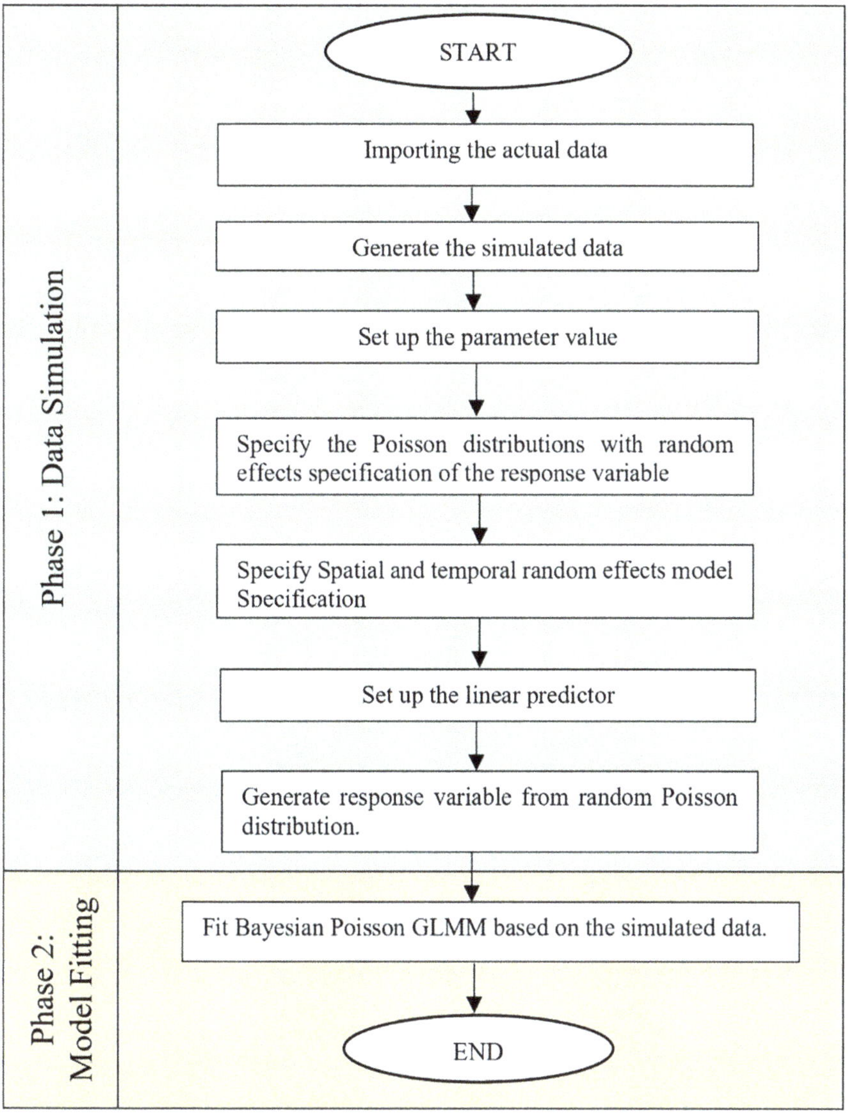

Fig. 4 Flowchart of simulation study

the lowest MSE value with 452.4179. While the simulation may not have fully replicated the complex dynamics of rainfall, it was able to generate simulated rainfall data that exhibited similarities to the actual rainfall patterns in Selangor.

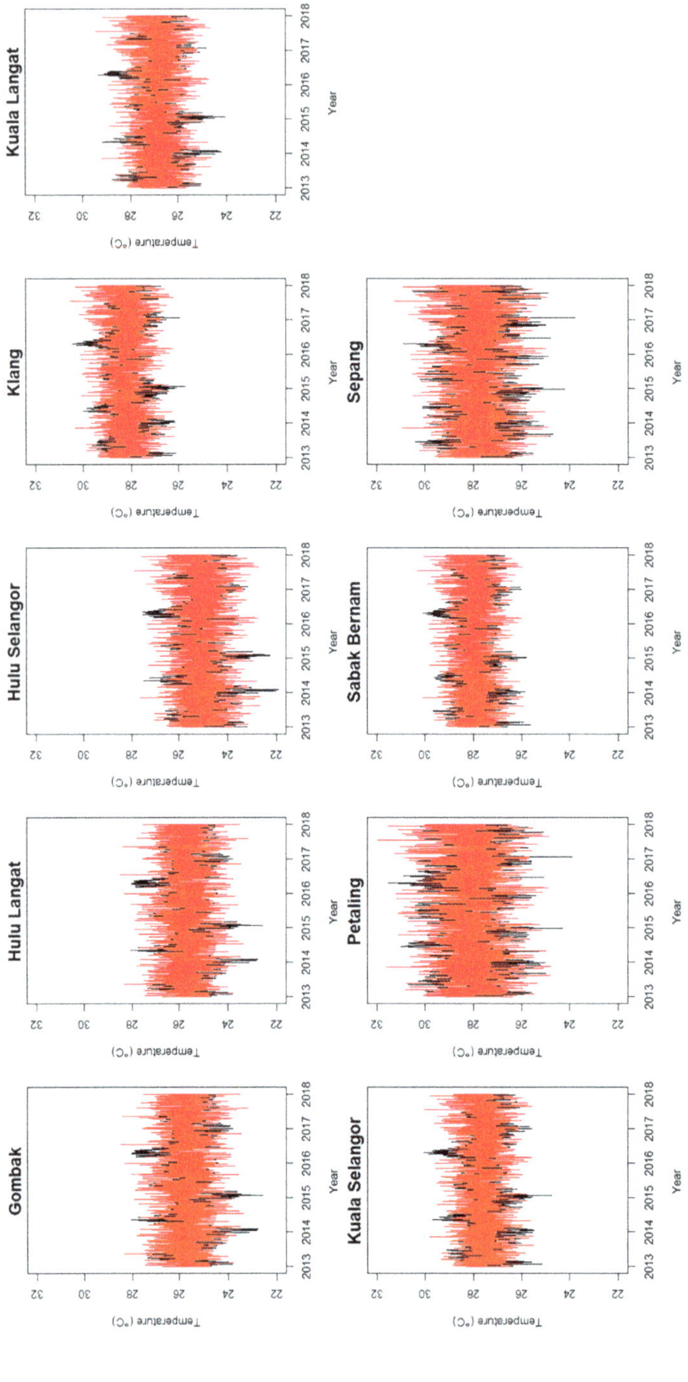

Fig. 5 Actual (black line) versus simulated (red line) temperature (°C)

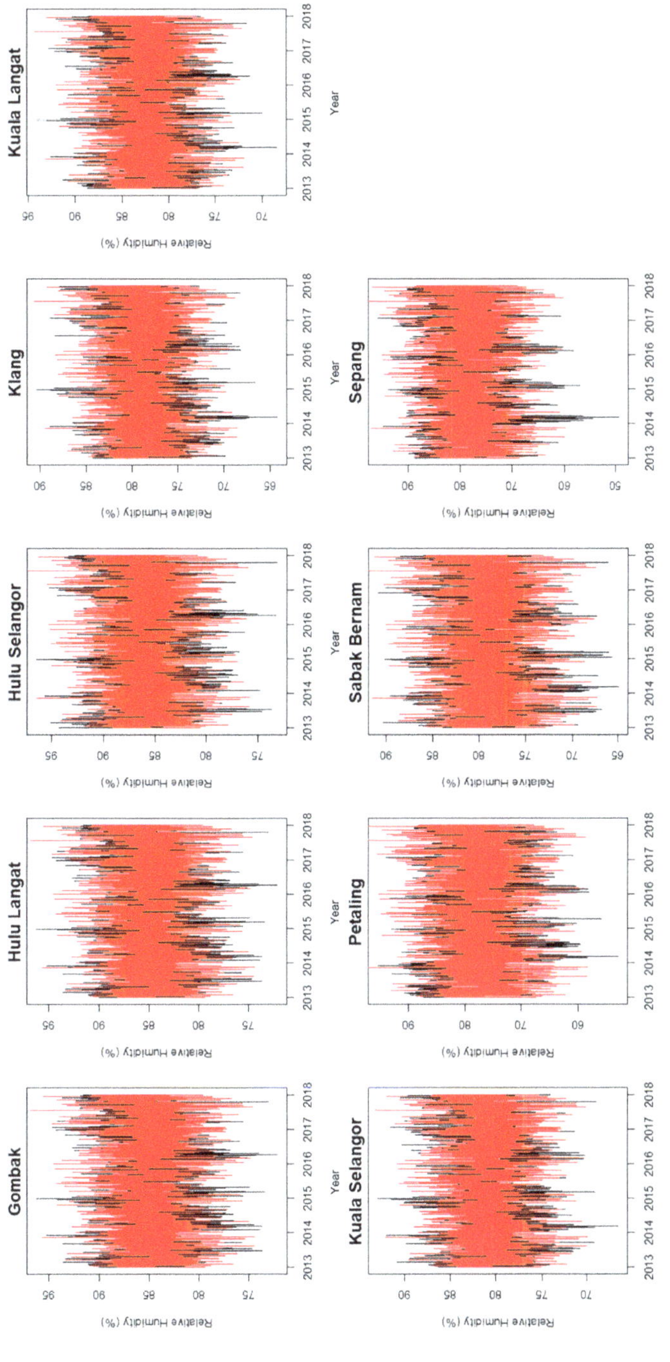

Fig. 6 Actual (black line) versus simulated (red line) relative humidity (%)

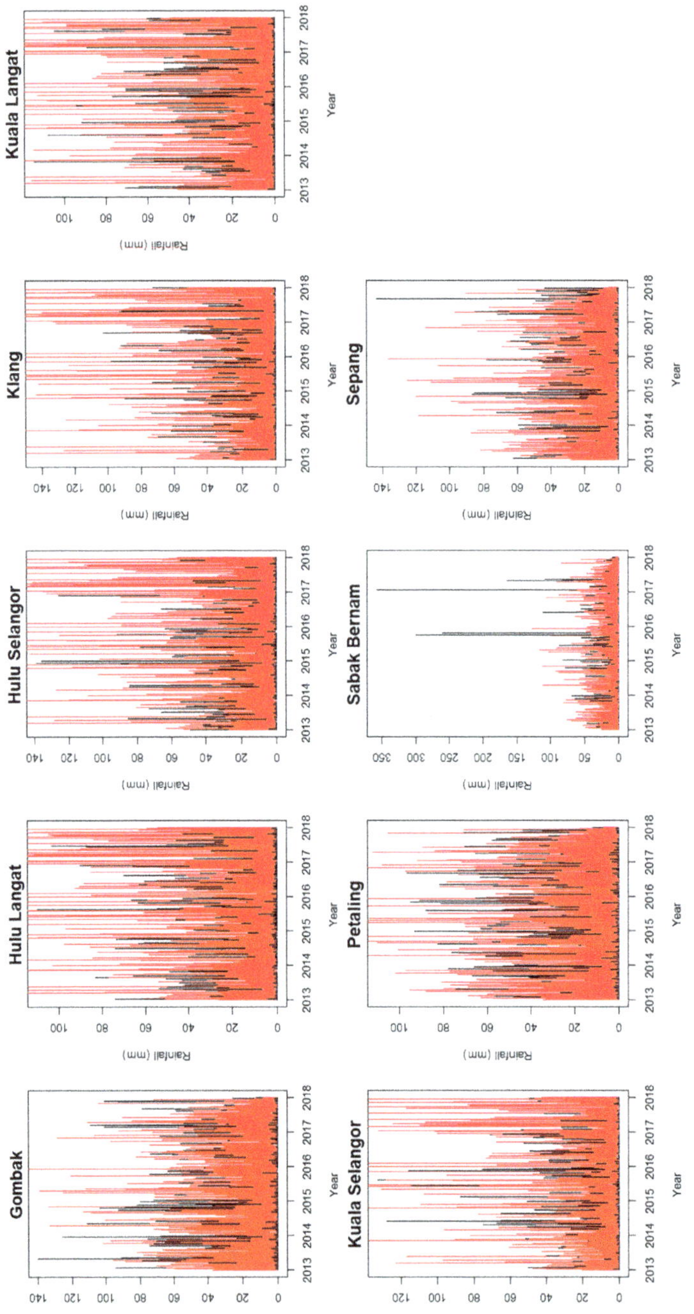

Fig. 7 Actual (black line) versus simulated (red line) rainfall (ml)

Table 2 Mean squared error between actual and simulated data

Districts	Temperature	Relative humidity	Rainfall
Gombak	1.3664	26.6840	687.3533
Hulu Langat	1.3663	26.6840	1313.3140
Hulu Selangor	1.3976	25.7926	1466.1110
Klang	1.0216	25.3570	1563.7770
Kuala Langat	1.2193	26.3317	1202.9030
Kuala Selangor	1.0824	28.3093	890.3801
Petaling	2.7278	88.4398	632.3981
Sabak Bernam	1.0110	34.0003	521.3537
Sepang	2.1603	79.8075	452.4179

3.2 *Modeling Dengue Incidence Using Simulated Dengue Data*

The application of a Bayesian Poisson GLMM using simulated data revealed significant insights into the factors influencing dengue incidence in Selangor as shown in Table 3 and Fig. 4, respectively. Checking the model convergence for the Bayesian Poisson GLMM, the Geweke diagnostics value was satisfied as the value fell within $\pm$ 2.0 as suggested by Lee [24]. This suggests that the MCMC algorithm effectively sampled from the posterior distribution, and the model's estimates were reliable. Based on Table 3, it reveals that for every 1% increase in current relative humidity and its lag, the daily risk of dengue incidence will increase by 0.19%, 0.19%, 0.22%, 0.21%, and 0.17%, respectively (Table 3). Meanwhile, for every 1 mm increase in the current day of rainfall and its lag, the daily risk of dengue incidence will increase by 0.11%, 0.11%, 0.12%, 0.11%, and 0.12%, respectively (Table 3). However, simulated temperature and its lag do not influence simulated dengue incidence rates. Next, it is also interesting to investigate the presence of spatial and temporal autocorrelation in simulated data. These random effects were modeled using conditional autoregressive (CAR) prior that can allow the variability to be captured in the simulated data. Based on Table 3, the model estimated a weak spatial autocorrelation with a value of 0.0480, suggesting little influence of neighboring spatial units on simulated dengue incidence. In contrast, the model identified a strong temporal autocorrelation with a value of 0.9996, indicating a high degree of correlation between dengue cases over time. These significant findings were then supported by Fig. 8 which Bayesian Poisson GLMM that was fitted to the simulated dataset provides evidence that the model is a good fit as the fitted values of simulated DIR (red line) are close to simulated DIR (black line). This study's significant finding supported the findings of Sapri et al. [14] and Wan Yaacob et al. [25].

Table 3 Model summary based on simulated data

Parameter	Posterior median	Expected of dengue incidence	Geweke value
Intercept	− 9.6778	–	0.5
Humid	0.0019	1.0019	0.0
Humid_lag7	0.0019	1.0019	0.4
Humid_lag14	0.0022	1.0022	− 0.2
Humid_lag21	0.0021	1.0021	− 1.7
Humid_lag28	0.0017	1.0017	− 1.9
Rain	0.0011	1.0011	1.4
Rain_lag7	0.0011	1.0011	− 0.4
Rain_lag14	0.0012	1.0012	1.4
Rain_lag21	0.0011	1.0011	0.9
Rain_lag28	0.0012	1.0012	1.1
Tau	0.0011	–	–
Rho.S (ρ_S)	0.0480	–	–
Rho.T (ρ_T)	0.9996	–	–
Mean squared error	10.0344		

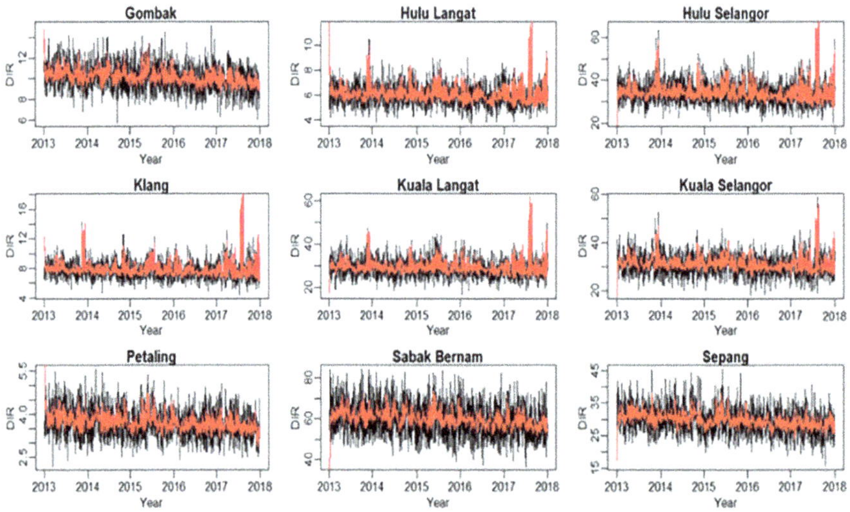

Fig. 8 Simulated DIR of Bayesian Poisson GLMM

4 Conclusions and Future Research

In this study, the researchers aim to simulate daily weather data (temperature, relative humidity, and rainfall) and daily dengue incidence rates for each district of Selangor, Malaysia based on actual data. A detailed process for simulating weather data and dengue incidence rates for Selangor was presented in the study. This process was conducted separately for each district of Selangor. This study demonstrates temperature simulations are generally accurate. There is more variability in the relative humidity simulations, with Petaling and Sepang showing significant deviations. Otherwise, rainfall simulations are the least accurate, with considerable variability across districts. Despite the higher MSE values, the simulations still capture some aspects of the actual rainfall patterns. In addition, the Bayesian Poisson GLMM provided promising results in modeling the simulated dengue incidence rates. The study found that simulated current relative humidity and its lag, as well as simulated current rainfall and its lag, were correlated with the simulated dengue incidence rates. Additionally, the model depicted a weak spatial correlation and a strong temporal correlation in the simulated dataset. Note that, the study did not include a comparison of different weather simulation models due to time and computational constraints, as well as limited data availability. However, this comparison can be conducted in future research. Future work can consider improving the accuracy of the simulation models, exploring additional weather variables, and incorporating the use of machine learning techniques for prediction.

Acknowledgements We would like to express our appreciation and record our gratitude to the School of Mathematical Sciences, College of Computing, Informatics and Mathematics, Universiti Teknologi MARA, for the financial support. We also would like to express thank you to Vector Borne and Infectious Diseases sector, Ministry of Health Malaysia, for the dengue data. The authors are also grateful to the reviewers for their insightful comments and suggestions.

References

1. Rushworth, A., Lee, D., Sarran, C.: An adaptive spatiotemporal smoothing model for estimating trends and step changes in disease risk. J. R. Stat. Soc. Appl. Stat. **66**(1), 141–157 (2017)
2. Miller, I.F., Jiranek, J., Brownell, M., et al.: Predicting the effects of climate change on the cross-scale epidemiological dynamics of a fungal plant pathogen. Sci. Rep. **12**, 14823 (2022)
3. Burhanuddin, S.N.Z.A.: The enhancement of normal ratio method through multiple imputation approach in estimating missing data with outliers for peninsular Malaysian rainfall dataset. Doctoral thesis, Universiti Teknologi MARA (2021)
4. Yusof, F., Hui-Mean, F., Suhaila, J., et al.: Rainfall characterization by application of standardised precipitation index (SPI) in Peninsular Malaysia. Theoret. Appl. Climatol. **115**, 503–516 (2014)
5. Suhaila, J., Deni, S.M., Wan, Zin, W.Z., Jemain, A.A.: Spatial patterns and trends of daily rainfall regime in Peninsular Malaysia during the southwest and Northeast Monsoons: 1975–2004. Meteorol. Atmos. Phys. **110**(1–2), 1–18 (2010)
6. Malek, A.I., Kassim, N., Kamal Ariffin, N., Abd Jalil, S., Ab Malek, H.: Modelling the distribution of rainfall amount. Math. Sci. Inform. J. 26–34 (2021)

7. Norzaida, A., Siti Musliha, M.R., Zalina, M.D., Syafrina, A.H.: Probability distributions comparative analysis in assessing rainfall process in time and space. Int. J. Civ. Eng. Technol. **8**(10), 1679–1688 (2017)
8. Wanishsakpong, W., Owusu, B.E.: Optimal time series model for forecasting monthly temperature in the southwestern region of Thailand. Model. Earth Syst. Environ. **6**, 525–532 (2020)
9. Sankalp, S., Panda, P.K.: Chapter 5—A comparative evaluation of machine learning and ARIMA models for forecasting relative humidity over Odisha districts. In: Kasiviswanathan, K.S., Soundharajan, B., Patidar, S., Jianxun H., Ojha, C.S.P. (Eds.) Development in Environmental Science, vol.14, pp. 91–105. Elsevier (2023)
10. Li, Z.Q., Zou, H.Z. and Qi, B.: Application of ARIMA and LSTM in relative humidity prediction. In: 2019 IEEE 19th International Conference on Community Technology, pp. 1544–1549. IEEE, Xi'an, China (2020)
11. Rushworth, A., Lee, D., Mitchell, R.: A spatio-temporal model for estimating the long-term effects of air pollution on respiratory hospital admissions in Greater London. Spat. Spat.-Temporal Epidemiol. **10**, 29–38 (2014)
12. Lee, D., Rushworth, A., Napier, G.: Spatio-temporal areal unit modeling in R with conditional autoregressive priors using the CARBayesST package. J. Stat. Softw. **9**(84), 1–39 (2018)
13. Napier, G., Lee, D., Robertson, C., Lawson, A.: A Bayesian space–time model for clustering areal units based on their disease trends. Biostatistics **20**(4), 681–697 (2019)
14. Sapri N.N.F.F., Wan Yaacob W.F., Yap B.W.: Quantifying the impact of climatic factors on dengue incidence using generalized linear mixed model with spatio-temporal Bayesian Poisson random effects approach. In: International Conference on Soft Computing in Data Science, vol. 1489, pp. 328–340. Springer, Singapore (2021)
15. Tran, B.L., Tseng, W.C., Chen, C.C., Liao, S.Y.: Estimating the threshold effects of climate on dengue: a case study of Taiwan. Int. J. Environ. Res. Public Health **17**(4), 1392 (2020)
16. Wan Yaacob, W.F., Sapri, N.N.F.F., Wah, Y.B.: Modeling dengue incidence in Southern Malaysia using fixed effects panel count model. Adv. Math. Sci. J. **9**(9), 7341–7351 (2020)
17. Anwar, A., Khan, N., Ayub, M., Nawaz, F., Shah, A., Flahault, A.: Modeling and predicting dengue incidence in highly vulnerable countries using panel data approach. Int. J. Environ. Res. Public Health **16**(3), 2296 (2019)
18. Bisht, B., Kumari, R., Nagpal, B.N., Singh, H., Gupta, S.K., Bansal, A., Tuli, N.R.: Influence of environmental factors on dengue fever in Delhi. Int. J. Mosq. Res. **6**(2), 11–18 (2019)
19. Naqvi, S.A.A., Jan, B., Shaikh, S., Kazmi, S.J.H., Waseem, L.A., Nasar-u-minAllah, M., Abbas, N.: Changing climatic factors favour dengue transmission in Lahore. Pakistan. Environ. **6**, 71 (2019)
20. Department of Statistics Malaysia (DoSM). https://www.dosm.gov.my/. Last accessed 23 Jan 2025
21. iDengue. https://idengue.mysa.gov.my/. Last accessed 23 Jan 2025
22. NASA Prediction of Worldwide Energy Resources. https://power.larc.nasa.gov/. Last accessed 8 Jan 2024
23. Global Administrative Areas (GADM). https://gadm.org/download_country.html. Last accessed 26 Aug 2024
24. Lee, D.: A tutorial on spatio-temporal disease risk modelling in R using Markov chain Monte Carlo simulation and the CARBayesST package. Spat. Spat. Temporal Epidemiol. **34**, 100353 (2020)
25. Wan Yaacob, W.F., Sapri, N.N.F.F., Yap, B.W.: Panel count data models: examining dengue incidence in Malaysia 2023. In: Mukherjee D. (Ed.) Applied Econometric Analysis Using Cross Section and Panel Data, pp. 571–607. Springer, Singapore (2023)

Search and Expand Hybrid Odd Lomax Fréchet Distribution Properties, Simulation, with Application

Alaa J. Nawaf, Mustafa Hassan Jumaa, Nooruldeen A. Noori, and Mundher A. Khaleel

Abstract Traditional statistical distributions often struggle with modeling asymmetric or extreme data, particularly when data is heavily skewed or concentrated in one tail. Classical distributions like the normal or exponential distributions fail to provide accurate representations in such cases, leading to poor model fitting and inadequate predictions. This issue is especially evident in applications involving survival analysis, insurance, and extreme value modeling, where the data exhibit irregular patterns that require more flexible distribution families. To address these limitations, this study proposes a new probability distribution, the Hybrid Odd Lomax Fréchet (HOLFr), according to the Odd Lomax family, which is a model for a new probability distribution consisting of four parameters, which introduces greater flexibility and accuracy in modeling real-world data with extreme characteristics. The new distribution expands upon the Odd Lomax family by incorporating additional parameters, allowing it to better handle skewness and kurtosis in complex datasets. In addition, the study will present a number of statistical properties of the hybrid distribution such as the moment generating function, survival function, risk function, Quintile function, pdf expansion, ordered statistics as well as other statistical properties that represent a few of the many mathematical and statistical features of the hybrid distribution. The parameters of the two models were also estimated using the maximum likelihood function. In order to obtain a distribution that is highly flexible to accommodate

A. J. Nawaf
General Directorate of Education, Baghdad/Al-Karkh III, Baghdad, Iraq
e-mail: alaagamilnawaf@gmai.com

M. H. Jumaa
Kirkuk Education Directorate, Kirkuk, Iraq
e-mail: mustafa.h.jumaa35515@st.tu.edu.iq

N. A. Noori (✉)
Anbar Education Directorate, Ministry of Education, Anbar, Iraq
e-mail: Nooruldeen.a.noori35508@st.tu.edu.iq

M. A. Khaleel
Departments of Mathematics, College of Computer Science and Mathemat, Tikrit, Iraq
e-mail: mun880088@tu.edu.iq

A. Shuib et al. (eds.), *Mathematical Modeling and Numerical Analysis*,
Springer Proceedings in Mathematics & Statistics 510,
https://doi.org/10.1007/978-981-96-9350-4_14

different types of real data, a Monte Carlo simulation was conducted to demonstrate the efficiency of estimating the unknown parameters of the new distribution using the maximum likelihood method. The proposed HOLFr distribution demonstrated significant flexibility and robustness in modeling asymmetric and extreme data. Through Monte Carlo simulations and real-world applications, the new distribution was compared against established models, showing superior performance in terms of flexibility and goodness-of-fit criteria such as AIC, BIC, CAIC, and HQIC. The results highlight the effectiveness of HOLFr in fitting both positively and negatively skewed data, making it a valuable tool for survival analysis and other areas requiring robust statistical models. The HOLFr distribution is a promising extension that offers enhanced accuracy in real-world data modeling, as demonstrated by its application to survival data.

Keywords OL-G family · HOLFr · Weighted moment · MLE · Monte Carlo · Real data for survival times

1 Introduction

A key gap in the existing literature is the limited flexibility of classical distributions in handling non-normal, highly skewed, or heavy-tailed data. Existing extended distributions, while attempting to address these issues, still lack the necessary adaptability to capture complex real-world data patterns effectively. For example, the conventional Frechet distribution, despite its applicability to extreme value theory, is not sufficiently flexible to fit certain asymmetric datasets with high kurtosis or varying tail behavior. A number of continuous distribution families have been developed to address this problem. Examples include: MOTL-G by [1], GG-G by [2], OF-G by [3], Weibull-G by [4], NEF-X by [5], OGR-X by [6], and OL-G by [7], which have the functions CDF and pdf family respectively:

$$G(x) = 1 - \left(1 - \frac{\mathcal{M}(x).\log(1 - \mathcal{M}(x))}{\omega}\right)^{-\delta} \tag{1}$$

$$g(x) = \frac{\delta}{\omega} m(x)\left(1 - \frac{\mathcal{M}(x).\log(1 - \mathcal{M}(x))}{\omega}\right)^{-(\delta+1)} \left[\frac{\mathcal{M}(x)}{1 - \mathcal{M}(x)} - \log(1 - \mathcal{M}(x))\right] \tag{2}$$

where $x > 0$, $\delta, \omega > 0$ are the parameters for the family, and $\mathcal{M}(x)$ is the baseline distribution.

A new distribution is needed to address the challenges posed by asymmetric and heavy-tailed data. The proposed Hybrid Odd Lomax Fréchet (HOLFr) distribution introduces additional parameters to the Odd Lomax and Frechet families, enhancing their flexibility. By incorporating more degrees of freedom, the HOLFr distribution is able to model complex data structures with better precision, providing more accurate representations of both left- and right-skewed data, as well as datasets with extreme or clustered values. This new distribution is particularly useful in applications such as survival analysis, risk management, and insurance, where precise modeling of extreme outcomes is crucial.

This study aims to propose a hybrid distribution that is characterized by high flexibility in fitting data, which is Hybrid Odd Lomax Fréchet shortly (HOLFr) according to the Odd Lomax family, which is a new probability distribution model consisting of four parameters. The cumulative distribution functions and probability density functions for the two distributions are obtained, in addition to the hazard functions and survival functions for the hybrid distribution, while plotting these functions with different values of the parameters. Conduct several simulation experiments to evaluate a small sample of MLEs to determine bias in the estimated data and mean square error. With a practical application on real data for each of the hybrid distributions, the efficiency of the distribution is determined.

The study consisted of six sections, the first is introduction, the second section of which included the new distribution with drawing the basic functions. The third section included proving some mathematical properties of the hybrid distribution. The fourth section included estimating the distribution parameters and conducting a simulation of the estimated parameters. The fifth section included the applied aspect of the distribution. The sixth and final section included it contains the most important conclusions reached during the study.

2 Hybrid Odd Lomax Fréchet (HOLFr) Distribution

Named after the French mathematician Maurice Frechet, which was first introduced in 1928, the distribution is a powerful tool for understanding and estimating rare events that have large impacts. which has been expanded and modified in various ways to enhance its flexibility and applicability in survival analysis. Examples of these expansions are: NEXF by [8], NAPF by [9], KPFD by [10], OBPF by [11], and IGoFre by [12]. There are many formulas for Frechet distribution functions, but in this study, using the CDF and PDF functions with $x > 0$, and shape parameters $\beta, a > 0$, respectively, in the form:

$$\mathcal{M}(x) = e^{-\left(\frac{\beta^a}{x^a}\right)} \tag{3}$$

$$m(x) = a\beta^a x^{-(a+1)} e^{-\left(\frac{\beta^a}{x^a}\right)}, x > 0 \tag{4}$$

Thus, by obtaining the new CDF of HOLFr distribution. By substituting Eq. (3) into Eq. (1) of shape parameters ω, δ, β, a as follows:

$$G(x) = 1 - \left(1 - \frac{e^{-\left(\frac{\beta^a}{x^a}\right)}.\log\left(1 - e^{-\left(\frac{\beta^a}{x^a}\right)}\right)}{\omega}\right)^{-\delta} \quad (5)$$

As for the new distribution pdf function, by obtain it by substituting Eq. (4) into Eq. (2) or deriving Eq. (5) as follows:

$$g(x) = \frac{\delta}{\omega} a\beta^a x^{-(a+1)} e^{-\left(\frac{\beta^a}{x^a}\right)} \left(1 - \frac{e^{-\left(\frac{\beta^a}{x^a}\right)}.\log\left(1 - e^{-\left(\frac{\beta^a}{x^a}\right)}\right)}{\omega}\right)^{-(\delta+1)}$$
$$\left[\frac{e^{-\left(\frac{\beta^a}{x^a}\right)}}{1 - e^{-\left(\frac{\beta^a}{x^a}\right)}} - \log\left(1 - e^{-\left(\frac{\beta^a}{x^a}\right)}\right)\right] \quad (6)$$

Figure 1 shows the proposed hybrid CDF function with different values of parameters, which shows the convergence of these functions between zero and one, which confirms the properties of the CDF function, which ranges within the closed interval zero and one. As for Fig. 2, which shows the PDF distribution function, which has a positive skewness and strong kurtosis for the values various parameters.

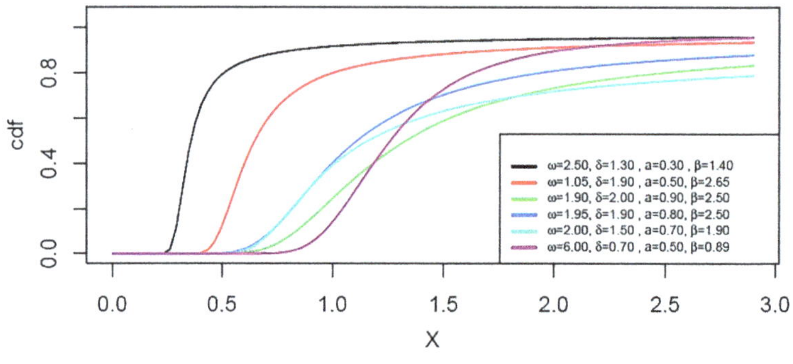

Fig. 1 Plot CDF function of HOLFr distribution with different value of shape parameters ω, δ, β, a

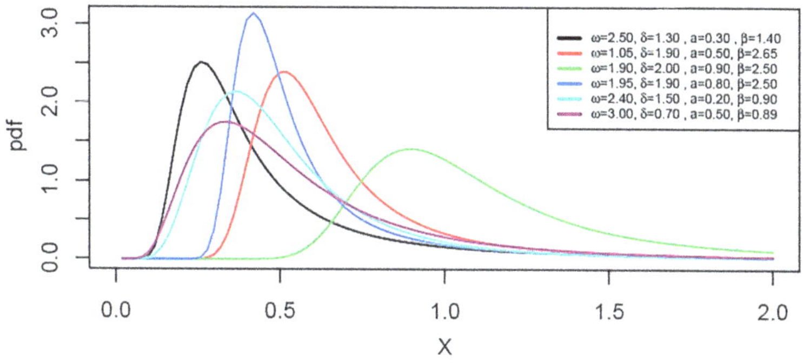

Fig. 2 Plot pdf function of HOLFr distribution with different value of shape parameters ω, δ, β, a

3 Some Mathematical Properties of HOLFr Distribution

3.1 *Reliability Analysis*

The survival function of HOLFr distribution can be inferred [13]:

$$S(x) = 1 - G(x) = \left(1 - \frac{e^{-\left(\frac{\beta^a}{x^a}\right)} . \log\left(1 - e^{-\left(\frac{\beta^a}{x^a}\right)}\right)}{\omega}\right)^{-\delta} \tag{7}$$

The hazard function of the HOLFr distribution is obtained from [14]:

$$H(x) = \frac{g(x)}{S(x)} = \frac{\frac{\delta}{\omega} a\beta^a x^{-(a+1)} e^{-\left(\frac{\beta^a}{x^a}\right)} \left[\frac{e^{-\left(\frac{\beta^a}{x^a}\right)}}{1-e^{-\left(\frac{\beta^a}{x^a}\right)}} - \log\left(1 - e^{-\left(\frac{\beta^a}{x^a}\right)}\right)\right]}{1 - \frac{e^{-\left(\frac{\beta^a}{x^a}\right)} . \log\left(1-e^{-\left(\frac{\beta^a}{x^a}\right)}\right)}{\omega}} \tag{8}$$

Figure 3 demonstrates that the hazard function of the variable can either increase or decrease, depending on the values of the parameters.

3.2 *Useful Expansions*

The CDF distribution function HOLFr from Eq. (5) will be firstly expanded using expansions of the exponential function, the logarithmic function, and the binomial expansion:

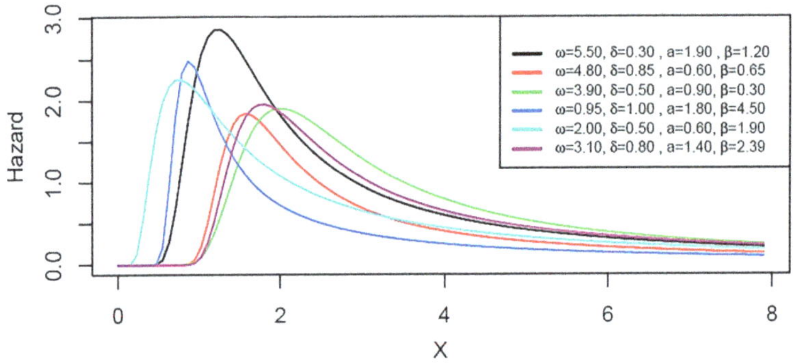

Fig. 3 Plot hazard function of HOLFr distribution with different values of ω, δ, β, a

$$\left(1-\frac{e^{-\left(\frac{\beta^a}{x^a}\right)}.\log\left(1-e^{-\left(\frac{\beta^a}{x^a}\right)}\right)}{\omega}\right)^{-\delta} = \sum_{k=0}^{\infty}\frac{\Gamma(\delta+k)\omega^{-k}}{k!\Gamma(\delta)}\left(e^{-\left(\frac{\beta^a}{x^a}\right)}\right)^k\left(\log\left(1-e^{-\left(\frac{\beta^a}{x^a}\right)}\right)\right)^k$$

From the above equation take the logarithm term and expand it in the form:

$$\left[\log\left(1-e^{-\left(\frac{\beta^a}{x^a}\right)}\right)\right]^k = \sum_{i=0}^{\infty}(-1)^i d_{k,i}\left(e^{-\left(\frac{\beta^a}{x^a}\right)}\right)^{i+k}$$

where $d_{k,i} = i^{-1}\sum_{m=1}^{i}\frac{[m(k+1)-i]}{m+1}$ for $i \geq 0$ and $d_{k,0} = 1$

From this obtain the expansion of the CDF function for the HOLFr distribution in the form:

$$G(x) = G(x) = 1 - \Psi e^{-(2k+i)\left(\frac{\beta^a}{x^a}\right)} \tag{9}$$

where $\Psi = \sum_{k=i=0}^{\infty}\frac{\Gamma(\delta+k)\omega^{-k}(-1)^i d_{k,i}}{\mathrm{K}!\Gamma(\delta)}$.

Concerning the expansion of the $(\mathrm{CDF})^z$ by the equation below:

$$G^z(x) = \left(1-\left(1-\frac{e^{-\left(\frac{\beta^a}{x^a}\right)}.\log\left(1-e^{-\left(\frac{\beta^a}{x^a}\right)}\right)}{\omega}\right)^{-\delta}\right)^z.$$

Using the binomial expansion of the above equation, an equation of the form is obtained:

$$\left(1-\left(1-\frac{e^{-\left(\frac{\beta^a}{x^a}\right)}.\log\left(1-e^{-\left(\frac{\beta^a}{x^a}\right)}\right)}{\omega}\right)^{-\delta}\right)^{z} = \sum_{m=0}^{\infty}(-1)^m\binom{z}{j}\left(1-\frac{e^{-\left(\frac{\beta^a}{x^a}\right)}.\log\left(1-e^{-\left(\frac{\beta^a}{x^a}\right)}\right)}{\omega}\right)^{-\delta m}$$

The $(\mathrm{CDF})^z$ function is derived by utilizing the expansions employed to enlarge the CDF function:

$$G^z(x) = \mathrm{Q}e^{-\frac{\beta^a}{x^a}(2k+i)} \tag{10}$$

where $\mathrm{Q} = \sum_{m=k=i=0}^{\infty}\frac{(-1)^{m+i}\Gamma(\delta m+k)\omega^{-k}d_{k,i}}{k!\Gamma(\delta m)}\binom{z}{j}$, and $d_{k,i} = i^{-1}\sum_{m=1}^{i}\frac{[m(k+1)-i]}{m+1}$ for $i \geq 0$ and $d_{k,0} = 1$.

Form Eq. (6) the pdf function becomes:

$$g(x) = \sum_{k=0}^{\infty}\frac{\delta\Gamma(\delta+1+k)\omega^{-(k+1)}}{k!\Gamma(\delta+1)}\frac{\left(e^{-\left(\frac{\beta^a}{x^a}\right)}\right)^{k+2}}{1-e^{-\left(\frac{\beta^a}{x^a}\right)}}\left(\log\left(1-e^{-\left(\frac{\beta^a}{x^a}\right)}\right)\right)^{k}a\beta^a x^{-(a+1)} - \sum_{k=0}^{\infty}\frac{\delta\Gamma(\delta+1+k)\omega^{-(k+1)}}{k!\Gamma(\delta+1)}\left(e^{-\left(\frac{\beta^a}{x^a}\right)}\right)^{k+1}\left(\log\left(1-e^{-\left(\frac{\beta^a}{x^a}\right)}\right)\right)^{k+1}a\beta^a x^{-(a+1)}$$

Using logarithm expansion of $\left(\log\left(1-e^{-\left(\frac{\beta^a}{x^a}\right)}\right)\right)^{k}$, $\left(\log\left(1-e^{-\left(\frac{\beta^a}{x^a}\right)}\right)\right)^{k+1}$, and $\frac{1}{1-e^{-\left(\frac{\beta^a}{x^a}\right)}}$ get:

$$\left[\log\left(1-e^{-\left(\frac{\beta^a}{x^a}\right)}\right)\right]^{k} = \sum_{i=0}^{\infty}(-1)^i d_{k,i}\left(e^{-\left(\frac{\beta^a}{x^a}\right)}\right)^{i+k}$$

$$\left(\log\left(1-e^{-\left(\frac{\beta^a}{x^a}\right)}\right)\right)^{k+1} = \sum_{l=0}^{\infty}(-1)^l d_{k,l}\left(e^{-\left(\frac{\beta^a}{x^a}\right)}\right)^{l+k+1}$$

$$\frac{1}{1-e^{-\left(\frac{\beta^a}{x^a}\right)}} = \sum_{z=0}^{\infty}(-1)^z\left[e^{-\left(\frac{\beta^a}{x^a}\right)}\right]^{z}$$

where $d_{k,i} = i^{-1}\sum_{m=1}^{i}\frac{[m(k+1)-i]}{m+1}$ for $i \geq 0$ and $d_{k,0} = 1$, and $d_{k,l} = l^{-1}\sum_{m=1}^{l}\frac{[m(k+1)-l]}{m+1}$ for $l \geq 0$ and $d_{k,0} = 1$

$$g(x) = \sum_{k=i=z=0}^{\infty} \frac{(-1)^{z+i}\delta\Gamma(\delta+1+k)\omega^{-(k+1)}}{k!\Gamma(\delta+1)} e^{-(2k+i+z+2)\left(\frac{\beta^a}{x^a}\right)} d_{k,i} a\beta^a x^{-(a+1)}$$
$$- \sum_{k=0}^{\infty} \frac{(-1)^{l}\delta\Gamma(\delta+1+k)\omega^{-(k+1)}}{k!\Gamma(\delta+1)} e^{-(2k+l+2)\left(\frac{\beta^a}{x^a}\right)} d_{k,l} a\beta^a x^{-(a+1)}$$

Finally, the extended pdf function of the HOLFr distribution is of the form:

$$g(x) = Hx^{-(a+1)} e^{-(2k+i+z+2)\left(\frac{\beta^a}{x^a}\right)} - Tx^{-(a+1)} e^{-(2k+l+2)\left(\frac{\beta^a}{x^a}\right)} \tag{11}$$

where $H = \sum_{k=i=z=0}^{\infty} \frac{(-1)^{z+i}\delta\Gamma(\delta+1+k)\omega^{-(k+1)}}{k!\Gamma(\delta+1)} d_{k,i} a\beta^a x^{-(a+1)}$, and $T = \sum_{k=0}^{\infty} \frac{(-1)^{l}\delta\Gamma(\delta+1+k)\omega^{-(k+1)}}{k!\Gamma(\delta+1)} d_{k,l} a\beta^a$.

3.3 Quantile Function

The quantile function $Q(u)$ is derived from the given relationship:

$$x = Q(u) = G^{-1}(u).$$

By substituting Eq. (3) into Eq. 11 of [7], obtaining the function $Q(u)$ for the HOLFr distribution in the form:

$$x = \left(-\frac{\beta^a}{\ln\left[\frac{\omega - \frac{\omega}{(1-u)^{\frac{1}{\delta}}}}{\omega - \frac{\omega}{(1-u)^{\frac{1}{\delta}}} + W_{-1}\left(\left(\omega - \frac{\omega}{(1-u)^{\frac{1}{\delta}}}\right) e^{-\left(\omega - \frac{\omega}{(1-u)^{\frac{1}{\delta}}}\right)}\right)}\right]} \right)^{\frac{1}{a}} \tag{12}$$

Table 1 represents a set of Quantile function values for different values of parameters, such that the value of the function increases with the value of u.

Table 1 Breakdown of the quantile function for specific parameter values of the HOLFr distribution

$(\omega, \delta, \beta, a)$					
u	(0.5, 1.3, 0.3, 1.4)	(1.05, 1.9, 0.5, 2.65)	(0.9, 2, 0.9, 2.5)	(0.9, 1.2, 0.4, 3.5)	(0.2, 2.5, 0.7, 0.9)
0.1	0.2111	0.4322	0.7517	0.3693	0.2645
0.2	0.2537	0.4802	0.8347	0.4044	0.3195
0.3	0.2938	0.5223	0.9065	0.4364	0.3672
0.4	0.3381	0.9792	0.4709	0.4147	
0.5	0.3925	0.6150	1.0605	0.5126	0.4664
0.6	0.4670	0.6769	1.1604	0.5696	0.5273
0.7	0.5864	0.7649	1.2980	0.6616	0.6062
0.8	0.8371	0.9182	1.5272	0.8618	0.7242
0.9	1.9100	1.3477	2.1119	1.7959	0.9644

The median of the HOLFr distribution can be determined by substituting $u = 0.5$ into Eq. (12), resulting in the following form:

$$\text{Median} = \left(-\frac{\beta^a}{\ln\left[\frac{\omega - \frac{\omega}{(0.5)^{\frac{1}{\delta}}}}{\omega - \frac{\omega}{(0.5)^{\frac{1}{\delta}}} + W_{-1}\left(\left(\omega - \frac{\omega}{(0.5)^{\frac{1}{\delta}}}\right)e^{-\left(\omega - \frac{\omega}{(0.5)^{\frac{1}{\delta}}}\right)}\right)}\right]} \right)^{\frac{1}{a}} \tag{13}$$

3.4 Skewness and Kurtosis Based on Quantile Function

The measurements of skewness (S) and kurtosis (K) were defined as follows [12]:

$$S = \frac{Q\left(\frac{6}{8}\right) - 2Q\left(\frac{4}{8}\right) + Q\left(\frac{2}{8}\right)}{Q\left(\frac{6}{8}\right) - Q\left(\frac{2}{8}\right)} \tag{14}$$

$$K = \frac{Q\left(\frac{7}{8}\right) - Q\left(\frac{5}{8}\right) + Q\left(\frac{3}{8}\right) - Q\left(\frac{1}{8}\right)}{Q\left(\frac{6}{8}\right) - Q\left(\frac{2}{8}\right)} \tag{15}$$

By replacing the value of $u = \frac{1}{8}, \frac{2}{8}, \frac{3}{8}, \frac{4}{8}, \frac{5}{8}, \frac{6}{8}, \frac{7}{8}$ in Eq. (12) to obtain the values of Eqs. (14) and (15) to get the skewness (S) and kurtosis (K) for HOLFr distribution.

3.5 Moments, Moment Generating Function, the Skewness (S) and Kurtosis (K) Based on Moments

Let x denote an arbitrary random variable with a pdf given by Eq. (11). The nth-moment of the HOLFr distribution can be expressed as [13]:

$$\mu_n = E\left(x^n\right)_{\text{HOLFr}} = H\int_0^\infty x^{n-a+1} e^{-(2k+i+z+2)\left(\frac{\beta^a}{x^a}\right)} dx \\ - T\int_0^\infty x^{n-a+1} e^{-(2k+l+2)\left(\frac{\beta^a}{x^a}\right)} dx$$

Let $y = (2k+i+z+2)\left(\frac{\beta^a}{x^a}\right) \Rightarrow x = \frac{\beta(2k+i+z+2)^{\frac{1}{a}}}{y^{\frac{1}{a}}} \Rightarrow dx = -\frac{\beta}{a}(2k+i+z+2)^{\frac{1}{a}} y^{-\frac{(1+a)}{a}} dy$.

Let $r = (2k+l+2)\left(\frac{\beta^a}{x^a}\right) \Rightarrow x = \frac{\beta(2k+l+2)^{\frac{1}{a}}}{r^{\frac{1}{a}}} \Rightarrow dx = -\frac{\beta}{a}(2k+l+2)^{\frac{1}{a}} r^{-\frac{(1+a)}{a}} dr$

$$\mu_n = \frac{-H\beta^{n-a+2}(2k+i+z+2)^{\frac{n-a+2}{a}}}{a} \int_0^\infty y^{-\frac{2+n}{a}} e^{-y} dy + \frac{T\beta^{n-a+2}(2k+i+z+2)^{\frac{n-a+2}{a}}}{a} \int_0^\infty r^{-\frac{2+n}{a}} e^{-r} dr$$

$$\mu_n = \frac{-H\beta^{n-a+2}(2k+i+z+2)^{\frac{n-a+2}{a}}}{a} \Gamma\left(-\frac{2+n}{a}+1\right) + \frac{T\beta^{n-a+2}(2k+i+z+2)^{\frac{n-a+2}{a}}}{a} \Gamma\left(-\frac{2+n}{a}+1\right)$$

$$\mu_n = \frac{\beta^{n-a+2}\Gamma\left(-\frac{2+n}{a}+1\right)}{a} \left[-H(2k+i+z+2)^{\frac{n-a+2}{a}} + T(2k+i+z+2)^{\frac{n-a+2}{a}}\right] \quad (16)$$

The variance of the HOLFr distribution can be calculated using the formula $\sigma^2 = \mu_2 - \mu_1^2$ to get the form:

$$\sigma^2 = \frac{\Gamma\left(1-\frac{4}{a}\right)}{a}\left[-H(2k+i+z+2)^{\frac{4-a}{a}} + \mathrm{T}(2k+i+z+2)^{\frac{4-a}{a}}\right] - \left(\frac{\beta^{3-a}\Gamma\left(-\frac{3}{a}+1\right)}{a}\left[-H(2k+i+z+2)^{\frac{3-a}{a}} + \mathrm{T}(2k+i+z+2)^{\frac{3-a}{a}}\right]\right)^2 \quad (17)$$

The skewness (S) and kurtosis (K) are determined by the following mathematical definitions [14]:

$$S = \frac{\frac{\beta^{5-a}\Gamma\left(1-\frac{5}{a}\right)}{a}\left[-H(2k+i+z+2)^{\frac{5-a}{a}} + \mathrm{T}(2k+i+z+2)^{\frac{5-a}{a}}\right]}{\left[\frac{\beta^{4-a}\Gamma\left(1-\frac{4}{a}\right)}{a}\left[-H(2k+i+z+2)^{\frac{4-a}{a}} + \mathrm{T}(2k+i+z+2)^{\frac{4-a}{a}}\right]\right]^{\frac{3}{2}}} \quad (18)$$

$$K = \frac{\Gamma\left(1-\frac{6}{a}\right)\left[-H(2k+i+z+2)^{\frac{6-a}{a}} + \mathrm{T}(2k+i+z+2)^{\frac{6-a}{a}}\right]}{\beta^{3a-2}\left[\frac{\Gamma\left(1-\frac{4}{a}\right)}{a}\left[-H(2k+i+z+2)^{\frac{4-a}{a}} + \mathrm{T}(2k+i+z+2)^{\frac{4-a}{a}}\right]\right]^2} - 3 \quad (19)$$

The moment generating function (mgf) of the HOLFr distribution can be determined using equation [15].

$$M_x(y)_{\text{HOLFr}} = \int_{-\infty}^{\infty} e^{yx} g(x)\mathrm{d}x$$

Used series expansion for e^{yx}

$$M_x(y)_{\text{HOLFr}} = \sum_{n=0}^{\infty} \frac{y^n}{n!} E(x^n) = \sum_{n=0}^{\infty} \frac{y^n}{n!} [\mu_n]$$

By Eq. (16) in above equation an equation of the form is obtained:

$$M_x(y)_{\text{HOLF}r} = \sum_{n=0}^{\infty} \frac{y^n \beta^{n-a+2} \Gamma\left(-\frac{2+n}{a}+1\right)}{an!}$$
$$\left[-H(2k+i+z+2)^{\frac{n-a+2}{a}} + \mathrm{T}(2k+i+z+2)^{\frac{n-a+2}{a}}\right] \quad (20)$$

3.6 Incomplete Moments

The formula [16] provides the incomplete moments of a random variable x.

$$\mu_r(y) = \int_0^y x^r f(x)\mathrm{d}x$$

Replacing the HOLFr distribution with a pdf substituting Eq. (11) into the previous equation yields:

$$\mu_r(y)_{\text{HOLFr}} = H \int_0^y x^{r-(a+1)} e^{-(2k+i+z+2)\left(\frac{\beta^a}{x^a}\right)}\mathrm{d}x - T \int_0^y x^{r-(a+1)} e^{-(2k+l+2)\left(\frac{\beta^a}{x^a}\right)}\mathrm{d}x$$

Let $s = (2k+i+z+2)\frac{\beta^a}{x^a}$.

If $x = 0 \Rightarrow s = 0$, if $x = y \Rightarrow s = (2k+i+z+2)\frac{\beta^a}{y^a} \Rightarrow \mathrm{d}y = -\frac{\beta}{a}(2k+i+z+2)^{\frac{1}{a}} s^{-\frac{(1+a)}{a}} \mathrm{d}s$

$$\Rightarrow \int_0^y x^{r-(a+1)} e^{-(2k+i+z+2)\left(\frac{\beta^a}{x^a}\right)}\mathrm{d}x$$

$$= -\int_0^{(2k+i+z+2)\frac{\beta^a}{y^a}} \left[\frac{\beta(2k+i+z+2)^{\frac{1}{a}}}{s^{\frac{1}{a}}}\right]^{r-(a+1)} \frac{\beta}{a}(2k+i+z+2)^{\frac{1}{a}} s^{-\frac{(1+a)}{a}} e^{-s}\,\mathrm{d}s$$

$$\Rightarrow \int_0^y x^{r-(a+1)} e^{-(2k+i+z+2)\left(\frac{\beta^a}{x^a}\right)} dx$$

$$= -\frac{\beta^{r-a}(2k+i+z+2)^{\frac{r-a}{a}}}{a} \int_0^{(2k+i+z+2)\frac{\beta^a}{y^a}} s^{-\frac{r}{a}} e^{-s}\, ds$$

$$\Rightarrow \int_0^y x^{r-(a+1)} e^{-(2k+i+z+2)\left(\frac{\beta^a}{x^a}\right)} dx$$

$$= -\frac{\beta^{r-a}(2k+i+z+2)^{\frac{r-a}{a}}}{a} \Gamma\left(1-\frac{r}{a}, (2k+i+z+2)\frac{\beta^a}{y^a}\right)$$

By same way can find

$$\int_0^y x^{r-(a+1)} e^{-(2k+l+2)\left(\frac{\beta^a}{x^a}\right)} dx = -\frac{\beta^{r-a}(2k+i+z+2)^{\frac{r-a}{a}}}{a} \Gamma\left(1-\frac{r}{a}, (2k+l+2)\frac{\beta^a}{y^a}\right).$$

Finally, the function $M_x(y)_{\text{HOLFr}}$ is obtained in the form:

$$\mu_r(y)_{\text{HOLFr}} = \frac{-H\beta^{r-a}(2k+i+z+2)^{\frac{r-a}{a}}}{a} \Gamma\left(1-\frac{r}{a}, E\right) + \frac{T\beta^{r-a}(2k+i+z+2)^{\frac{r-a}{a}}}{a} \Gamma\left(1-\frac{r}{a}, W\right) \tag{21}$$

where $E = (2k+i+z+2)\frac{\beta^a}{y^a}$, and $W = (2k+l+2)\frac{\beta^a}{y^a}$.

3.7 Probability Weighted Moments

The probabilistic weighted moments of the HOLFr distribution can be obtained using the method described in [16].

$$\varphi_{n,z} = E\left(x^n F^z(X)\right) = \int_{-\infty}^{\infty} x^n g(x) G^z(x) dx$$

By substituting Eqs. (10), and (11) in above equation to get:

$$\varphi_{n,z} = HQ \int_0^{\infty} x^{n-(a+1)} e^{-(4k+2i+z+2)\left(\frac{\beta^a}{x^a}\right)} dx - -TQ \int_0^{\infty} x^{n-(a+1)} e^{-(4k+l+i+2)\left(\frac{\beta^a}{x^a}\right)} dx.$$

Let $y = (4k+2i+z+2)\left(\frac{\beta^a}{x^a}\right)$, and $u = (4k+l+i+2)\left(\frac{\beta^a}{x^a}\right)$.

Finally get:

$$\varphi_{n,z} = \frac{Q\beta^{n-a}\Gamma\left(1-\frac{n}{a}\right)}{a}\left[T(4k+l+i+2)^{\frac{n-a}{a}} - H(4k+2i+z+2)^{\frac{n-a}{a}}\right] \quad (22)$$

3.8 Disparity Measures

Statistical instruments are utilized to quantify the magnitude of discrepancy within a certain distribution. They provide a method to evaluate and compare the extent of the discrepancy among different datasets.

3.8.1 Lorenz Curve

The random variable x is described as [17]:

$$L_F(y) = \frac{1}{\mu}\int_{-\infty}^{y} xf(x)\mathrm{d}x$$

Using Eq. (11) in the above equation to get:

$$L_F(y) = \frac{1}{\mu}\left[H\int_{-\infty}^{y} x^{-a}\mathrm{e}^{-(2k+i+z+2)\left(\frac{\beta^a}{x^a}\right)}\mathrm{d}x - T\int_{-\infty}^{y} x^{-a}\mathrm{e}^{-(2k+l+2)\left(\frac{\beta^a}{x^a}\right)}\mathrm{d}x\right].$$

Therefore, getting:

$$L_F(y) = \frac{\left[-(2k+i+z+2)^{\frac{1-a}{a}}\Gamma\left(-\frac{1}{a}-1, D\right) + \mathrm{T}(2k+l+2)^{\frac{1-a}{a}}\Gamma\left(-\frac{1}{a}-1, P\right)\right]}{\mu a\beta^{a-1}} \quad (23)$$

where $\mathrm{D} = (2k+i+z+2)\frac{\beta^a}{x^a}$, and $\mathrm{P} = (l2k+l+2)\frac{\beta^a}{x^a}$.

3.8.2 Bonferroni Curve

The random variable x is described as [17]:

$$\mathcal{B}_{\mathrm{F}}(y) = \frac{L_{\mathrm{F}}(y)}{G(x)}.$$

By substituting Eqs. (9) and (23) into the following equation, an equation of the form is obtained:

$$\mathcal{B}_{\mathrm{F}}(y) = \frac{\left[-(2k+i+z+2)^{\frac{1-a}{a}}\Gamma\left(-\frac{1}{a}-1, D\right) + T(2k+l+2)^{\frac{1-a}{a}}\Gamma\left(-\frac{1}{a}-1, P\right)\right]}{\mu a\beta^{a-1}\left(1-\Psi e^{-(2k+i)\left(\frac{\beta^a}{x^a}\right)}\right)} \tag{24}$$

3.9 Entropy

3.9.1 Rényi Entropy

The random variable x is described as [1]:

$$I_{\mathrm{R}}(\gamma)_{\mathrm{HOLFr}} = \frac{1}{1-\gamma}\log\int_0^\infty g^\gamma(x)\mathrm{d}x.$$

By substituting Eq. (11) into the equation above and employing binomial expansion, an equation of the form is obtained:

$$I_{\mathrm{R}}(\gamma)_{\mathrm{HOLF}r} = \frac{1}{1-\gamma}\log\int_0^\infty x^{-\gamma(a+1)}\sum_{m=0}^{n}(-1)^n\binom{n}{m}H.T\left(\mathrm{e}^{-(2k+i+z+2)\left(\frac{\beta^a}{x^a}\right)}\right)^m\left(\mathrm{e}^{-(2k+l+2)\left(\frac{\beta^a}{x^a}\right)}\right)^{\gamma-m}\mathrm{d}x$$

$$I_{\mathrm{R}}(\gamma)_{\mathrm{HOLF}r} = \frac{1}{1-\gamma}\sum_{m=0}^{n}(-1)^m\binom{n}{m}H.T\log\int_0^\infty x^{-\gamma(a+1)}\mathrm{e}^{-(im+zm+2\gamma k+\gamma l+2\gamma-ml)\frac{\beta^a}{x^a}}\mathrm{d}x$$

Let $y = (im+zm+2\gamma k+\gamma l+2\gamma-ml)\frac{\beta^a}{x^a} \Rightarrow x^a = \frac{\beta^a(im+zm+2\gamma k+\gamma l+2\gamma-ml)}{y} \Rightarrow x = \frac{\beta(im+zm+2\gamma k+\gamma l+2\gamma-ml)^{\frac{1}{a}}}{y^{\frac{1}{a}}} \Rightarrow \mathrm{d}x = -\frac{\beta}{a}(im+zm+2\gamma k+\gamma l+2\gamma-ml)^{\frac{1}{a}}y^{-\frac{1+a}{a}}\mathrm{d}y$, take $M = im+zm+2\gamma k+\gamma l+2\gamma-ml$

$$I_{\mathrm{R}}(\gamma)_{\mathrm{HOLFr}} = \frac{1}{1-\gamma}\sum_{m=0}^{n}(-1)^m\binom{n}{m}H.T\log\frac{-M^{\frac{1-\gamma}{a}-\gamma}}{a\beta^{\gamma(a+1)}}\int_0^\infty y^{-(\gamma+a)\left(\frac{1+a}{a}\right)}\mathrm{e}^{-y}\mathrm{d}y$$

$$I_{\mathrm{R}}(\gamma)_{\mathrm{HOLFr}} = \frac{1}{1-\gamma}\sum_{m=0}^{n}(-1)^m\binom{n}{m}H.T\log\left[\frac{-M^{\frac{1-\gamma}{a}-\gamma}}{a\beta^{\gamma(a+1)}}\Gamma\left(1-(\gamma+a)\left(\frac{1+a}{a}\right)\right)\right] \tag{25}$$

3.9.2 Arimoto Entropy

The AEN measure is computed using the subsequent equation [18]:

$$A(\gamma)_{\mathrm{HOLFr}} = \frac{\gamma}{1-\gamma}\left(\left[\int_0^\infty g^\gamma(x)\mathrm{d}x\right]^{\frac{1}{\gamma}} - 1\right).$$

From Eq. (25) we get:

$$A(\gamma)_{\mathrm{HOLFr}} = \frac{\gamma}{1-\gamma}\left(\left[\frac{-M^{\frac{1-\gamma}{a}-\gamma}}{a\beta^{\gamma(a+1)}}\Gamma\left(1-(\gamma+a)\left(\frac{1+a}{a}\right)\right)\right]^{\frac{1}{\gamma}} - 1\right) \tag{26}$$

3.9.3 Havrda and Charvat Entropy

$$\mathrm{HC}(\gamma)_{\mathrm{HOLFr}} = \frac{1}{2^{1-\gamma}-1}\left(\left[\int_0^\infty g^\gamma(x)\mathrm{d}x\right]^{\frac{1}{\gamma}} - 1\right)$$

From Eq. (26) we get:

$$\mathrm{HC}(\gamma)_{\mathrm{HOLFr}} = \frac{1}{2^{1-\gamma}-1}\left(\left[\frac{-M^{\frac{1-\gamma}{a}-\gamma}}{a\beta^{\gamma(a+1)}}\Gamma\left(1-(\gamma+a)\left(\frac{1+a}{a}\right)\right)\right]^{\frac{1}{\gamma}} - 1\right) \tag{27}$$

3.9.4 Tsallis Entropy

$$T(\gamma)_{\mathrm{HOLFr}} = \frac{1}{\gamma-1}\left(1-\int_0^\infty g^\gamma(x)\mathrm{d}x\right)$$

From Eq. (26) we get:

$$T(\gamma)_{\mathrm{HOLFr}} = \frac{1}{\gamma-1}\left(1-\frac{-M^{\frac{1-\gamma}{a}-\gamma}}{a\beta^{\gamma(a+1)}}\Gamma\left(1-(\gamma+a)\left(\frac{1+a}{a}\right)\right)\right) \tag{28}$$

1. **Estimation**

The parameters of the HOLFr distribution are determined using the maximum likelihood estimation method. The log-likelihood function for a random sample $x_1, x_2, \ldots, x_n$ is derived. The distribution adheres to the pdf of the HOLFr distribution [19].

$$L(\theta, x) = \prod_{i=1}^{n} g(x)$$

$$L(\theta, x_i) = \prod_{i=1}^{n} \frac{\delta}{\omega} a\beta^a x_i^{-(a+1)} \mathrm{e}^{-\left(\frac{\beta^a}{x_i^a}\right)} \left(1 - \frac{\mathrm{e}^{-\left(\frac{\beta^a}{x_i^a}\right)} . \log\left(1 - \mathrm{e}^{-\left(\frac{\beta^a}{x_i^a}\right)}\right)}{\omega}\right)^{-(\delta+1)}$$

$$\left[\frac{\mathrm{e}^{-\left(\frac{\beta^a}{x_i^a}\right)}}{1 - \mathrm{e}^{-\left(\frac{\beta^a}{x_i^a}\right)}} - \log\left(1 - \mathrm{e}^{-\left(\frac{\beta^a}{x_i^a}\right)}\right)\right]$$

The log-likelihood function L is obtained as:

$$\begin{aligned} L = {} & \mathrm{nlog}(\delta) - \mathrm{nlog}(\omega) + \mathrm{nlog}(a) + \mathrm{anlog}(\beta) - (a+1)\sum_{i=1}^{n} x_i - \sum_{i=1}^{n} \frac{\beta^a}{x_i^a} \\ & + \sum_{i=1}^{n} \log\left[\frac{\mathrm{e}^{-\frac{\beta^a}{x_i^a}}}{1 - \mathrm{e}^{-\frac{\beta^a}{x_i^a}}} - \log\left(1 - \mathrm{e}^{-\frac{\beta^a}{x_i^a}}\right)\right] \\ & - (\delta+1)\sum_{i=1}^{n} \log\left(1 - \frac{\mathrm{e}^{-\left(\frac{\beta^a}{x_i^a}\right)} . \log\left(1 - \mathrm{e}^{-\left(\frac{\beta^a}{x_i^a}\right)}\right)}{\omega}\right) \end{aligned} \tag{29}$$

By taking the partial derivative of the given equation with respect to the distribution parameters.

$$\begin{aligned} \frac{\partial L}{\partial a} = {} & \frac{n}{a} + n\log(\beta) - \sum_{i=1}^{n} x_i - \sum_{i=1}^{n} \frac{\beta^a(\log\beta - \log x_i)}{x_i^a} \\ & + \sum_{i=1}^{n} \frac{\frac{-\beta^a(\log\beta - \log x_i)}{x_i^a}\mathrm{e}^{-\frac{\beta^a}{x_i^a}}\left[2 - 3\mathrm{e}^{-\frac{\beta^a}{x_i^a}}\right]}{\left(1 - e^{-\frac{\beta^a}{x_i^a}}\right)^2\left[\frac{\mathrm{e}^{-\frac{\beta^a}{x_i^a}}}{1-\mathrm{e}^{-\frac{\beta^a}{x_i^a}}} - \log\left(1 - \mathrm{e}^{-\frac{\beta^a}{x_i^a}}\right)\right]} \\ & - (\delta+1)\sum_{i=1}^{n} \frac{-\frac{\beta^a(\log x_i - \log\beta)\mathrm{e}^{-\frac{\beta^a}{x_i^a}}}{x_i^a}\left[\log\left(1 - \mathrm{e}^{-\frac{\beta^a}{x_i^a}}\right) + \frac{1}{1-\mathrm{e}^{-\frac{\beta^a}{x_i^a}}}\right]}{\omega - \mathrm{e}^{-\left(\frac{\beta^a}{x_i^a}\right)} . \log\left(1 - \mathrm{e}^{-\left(\frac{\beta^a}{x_i^a}\right)}\right)} \end{aligned} \tag{30}$$

$$
\begin{aligned}
\frac{\partial L}{\partial \beta} &= \frac{an}{\beta} - \sum_{i=1}^{n} \frac{a\beta^{a-1}}{x_i^a} \\
&+ \sum_{i=1}^{n} \frac{\dfrac{\left(1-\mathrm{e}^{-\frac{\beta^a}{x_i^a}}\right)\frac{-a\beta^{a-1}}{x_i^a}\mathrm{e}^{-\frac{\beta^a}{x_i^a}} - \mathrm{e}^{-\frac{\beta^a}{x_i^a}}\frac{a\beta^{a-1}}{x_i^a}\mathrm{e}^{-\frac{\beta^a}{x_i^a}}}{\left(1-\mathrm{e}^{-\frac{\beta^a}{x_i^a}}\right)^2} + \dfrac{\frac{a\beta^{a-1}}{x_i^a}\mathrm{e}^{-\frac{\beta^a}{x_i^a}}}{1-\mathrm{e}^{-\frac{\beta^a}{x_i^a}}}}{\dfrac{\mathrm{e}^{-\frac{\beta^a}{x_i^a}}}{1-\mathrm{e}^{-\frac{\beta^a}{x_i^a}}} - \log\left(1-\mathrm{e}^{-\frac{\beta^a}{x_i^a}}\right)} \\
&- (\delta+1)\sum_{i=1}^{n} \frac{a\beta^{a-1}e^{-\frac{\beta^a}{x_i^a}}\left[\log\left(1-\mathrm{e}^{-\frac{\beta^a}{x_i^a}}\right) - \dfrac{\mathrm{e}^{-\frac{\beta^a}{x_i^a}}}{1-\mathrm{e}^{-\frac{\beta^a}{x_i^a}}}\right]}{x_i^a\left[\omega - \mathrm{e}^{-\left(\frac{\beta^a}{x_i^a}\right)}.\log\left(1-\mathrm{e}^{-\left(\frac{\beta^a}{x_i^a}\right)}\right)\right]}
\end{aligned} \tag{31}
$$

$$
\frac{\partial L}{\partial \delta} = \frac{n}{\delta} - \sum_{i=1}^{n} \log\left(1 - \frac{\mathrm{e}^{-\left(\frac{\beta^a}{x_i^a}\right)}.\log\left(1-\mathrm{e}^{-\left(\frac{\beta^a}{x_i^a}\right)}\right)}{\omega}\right) \tag{32}
$$

$$
\frac{\partial L}{\partial \omega} = \frac{n}{\omega} - (\delta+1)\sum_{i=1}^{n} \frac{\mathrm{e}^{-\left(\frac{\beta^a}{x_i^a}\right)}.\log\left(1-\mathrm{e}^{-\left(\frac{\beta^a}{x_i^a}\right)}\right)}{\omega\left[\omega - \mathrm{e}^{-\left(\frac{\beta^a}{x_i^a}\right)}.\log\left(1-\mathrm{e}^{-\left(\frac{\beta^a}{x_i^a}\right)}\right)\right]} \tag{33}
$$

Computing the solutions for the nonlinear equations $\frac{\partial L}{\partial a} = 0$, $\frac{\partial L}{\partial \beta} = 0$, $\frac{\partial L}{\partial \delta} = 0$ and $\frac{\partial L}{\partial \omega} = 0$. Provides the maximum likelihood estimates of the parameters a, β, δ and ω, in that order. An analysis was unable to find the solution. The only way to locate it was using numerical methods. These methods utilize software like as *R*, MAPLE, SAS, and other similar programs.

2. **Simulation**

A Monte Carlo simulation using *R* evaluates maximum likelihood estimators (MLEs) for the HOLFr distribution. The study included 30, 75, and 150 samples for the precise Table 2 parameters, by obtaining $N = 1000$ samples. The model parameter maximum likelihood estimators are averaged to determine the mean values. After that, [14, 19] determine bias and RMSEs.

In the Monte Carlo simulation conducted for the HOLFr distribution, the best parameters are determined based on the consistency and accuracy of the parameter estimates across different sample sizes. The study used various sample sizes (30, 75, 150) to estimate parameters such as ω, δ, a, and β using the maximum likelihood estimation (MLE) method. The performance of these estimations was evaluated using two main criteria: mean values and root mean square errors (RMSEs). The parameters

Table 2 Monte Carlo simulations conducted for the HOLFr distribution

Parameter	Sample size	Mean	RMSE	Abias	Mean	RMSE	Abias
($\omega = 1.0, \delta = 1.5, a = 2.0, \beta = 1.5$)					($\omega = 3.0, \delta = 1.5, a = 4.0, \beta = 1.5$)		
Parameter	Sample size	Mean	RMSE	Abias	Mean	RMSE	Abias
ω	30	1.0094407	3.454055e−01	0.00944069	3.174596	1.805048e + 00	0.1745959
	75	0.8856690	1.609322e−01	− 0.11433103	2.651127	4.845591e−01	− 0.3488731
	150	0.8584332	1.606539e−01	− 0.14156685	2.648200	4.061466e−01	− 0.3518004
δ	30	1.8020605	8.343010e−01	0.30206054	1.756131	6.795380e−01	0.2561314
	75	1.7303518	4.315301e−01	0.23035176	1.815465	5.785396e−01	0.3154647
	150	1.7365044	2.920694e−01	0.23650440	1.666817	2.165042e−01	0.1668175
a	30	2.048792	8.182274e−01	0.04879196	3.935564	1.305871e + 00	− 6.443601e−02
	75	1.782852	3.083573e−01	− 0.21714822	3.509821	6.778477e−01	− 4.901785e−01
	150	1.715428	3.376744e−01	− 0.28457165	3.429692	6.772288e−01	− 5.703081e−01
β	30	1.773302	7.207540e−01	0.27330242	1.789189	6.786415e−01	2.891890e−01
	75	1.770780	4.963193e−01	0.27078021	1.761870	4.658970e−01	2.618704e−01
	150	1.670217	2.227569e−01	0.17021699	1.672402	2.282147e−01	1.724022e−01
ω	30	2.844980	5.278083e + 00	5.408431e−01	8.472017	3.256312e + 01	4.472017e + 00
	75	2.109384	1.159632e + 00	1.093841e−01	4.329721	2.856580e + 00	3.297214e−01
	150	2.060800	9.047009e−01	6.080011e−02	4.262848	2.069127e + 00	2.628479e−01
δ	30	2.049944	1.345316e + 00	5.499440e−01	2.040865	1.411746e + 00	5.408648e−01
	75	1.873286	8.066465e−01	3.732862e−01	1.826215	7.381940e−01	3.262150e−01
	150	1.738157	5.408431e−01	2.381568e−01	1.715431	5.363451e−01	2.154309e−01
a	30	1.424049	2.377452e + 00	4.240486e−01	2.528272	6.935561e + 00	5.282717e−01
	75	1.182006	1.725028e + 00	1.820059e−01	1.954142	4.583835e−01	− 4.585760e−02

(continued)

Table 2 (continued)

Parameter	Sample size	Mean	RMSE	Abias	Mean	RMSE	Abias
	150	1.085966	5.155743e−01	8.596607e−02	1.936334	4.042211e−01	− 6.366623e−02
β	30	2.074992	1.700602e + 00	5.749915e−01	1.851155	8.647124e−01	3.511549e−1
	75	1.811184	7.529598e−01	3.111841e−01	1.774914	6.012434e−01	2.749145e−01
	150	1.688012	5.264557e−01	1.880124e−01	1.691948	4.626932e−01	1.919477e−01
ω	30	3.684910	9.848662e + 00	6.849096e−01	6.480208	1.639469e + 01	2.480208e + 00
	75	2.944058	8.340196e−01	− 5.594190e−02	4.385752	3.312132e + 00	3.857520e−01
	150	2.916578	6.155074e−01	− 8.342232e−02	4.373753	2.464323e + 00	3.737535e−01
δ	30	2.009212	1.108138e + 00	5.092115e−01	1.960454	1.260625e + 00	4.604539e−01
	75	1.800359	6.409079e−01	3.003593e−01	1.782058	6.788679e−01	2.820582e−01
	150	1.677022	4.379471e−01	1.770217e−01	1.666992	4.749889e−01	1.669917e−01
a	30	4.863445	2.257709e + 01	1.360516e + 00	4.201257	2.333574e + 01	2.201257e + 00
	75	3.972784	5.245425e−01	3.525543e−02	2.214128	2.050999e + 00	2.141278e−01
	150	3.865594	3.924930e−01	1.950659e−02	2.136894	1.032326e + 00	1.368936e−01
β	30	1.975132	1.102708e + 00	4.185202e−01	1.925328	1.300510e + 00	4.253282e−01
	75	1.811364	6.857818e−01	3.113637e−01	1.766988	6.481498e−01	2.669877e−01
	150	1.688516	4.734540e−01	1.885158e−01	1.661377	4.618859e−01	1.613774e−01

are judged based on their ability to converge to their true values and minimize the bias and RMSE as the sample size increases.

From the Monte Carlo simulations presented in above table, the parameter values that consistently show minimal bias and lower RMSE across the sample sizes are considered the best estimates. Based on the simulation results:

1. For ω (shape parameter):
   - The simulation reveals that for larger sample sizes (e.g., $n = 150$), ω values tend to converge toward the true value with lower RMSE. For instance, when $\omega = 3.0$, the mean estimate for $n = 150$ was 2.9165 with a small RMSE of 0.6155, showing good consistency.
   - The parameter $\omega = 3.0$ shows better performance in terms of convergence as the sample size increases.
2. For δ (scale parameter):
   - Similar to ω, the parameter δ converges closer to its true value with increasing sample size. For instance, when $\delta = 1.5$, the mean estimate for $n = 150$ is 1.677 with an RMSE of 0.4379, indicating good accuracy.
   - Thus, $\delta = 1.5$ can be considered a reliable parameter choice.
3. For a (scale parameter):
   - The parameter a also demonstrates consistent results with larger sample sizes. For $a = 4.0$, the mean estimate for = 150 is 3.8656, with a very low RMSE of 0.3925, indicating its reliability in the model.
   - $a = 4.0$ is identified as the best-performing parameter for this distribution.
4. For β (scale parameter):
   - The parameter $\beta = 1.5$ provides stable results. At $n = 150$, the mean estimate for β is 1.6885 with an RMSE of 0.4734, showing a good fit and consistent performance as the sample size increases.

From the Monte Carlo simulations, the parameters that perform the best in terms of consistency, bias reduction, and lower RMSE are:

$$\omega = 3.0$$
$$\delta = 1.5$$
$$a = 4.0$$
$$\beta = 1.5.$$

These parameters have been chosen as the best-performing parameters for the HOLFr distribution model due to their strong convergence and minimal errors in the simulation experiments. This combination provides a robust fit for the HOLFr model in terms of flexibility and precision when applied to real-world data.

3. Application

In this section, showing the implementation of the practical aspect using a set of real data represented by the survival time of 72 guinea pigs infected with tuberculosis nerves [19]. The effectiveness of HOLFr distribution is accurately displayed in these data. The app showcases the benefits of HOLFr and its excellent data compatibility.

This comparison uses six different distributions, the proposed distribution is compared with six other distributions that have the same baseline distribution used to indicate the flexibility of the proposed distribution, which are as follows:

- Beta Exponential Fréchet (BeFr) (new)
- Kumaraswamy Fréchet (KuFr) (new)
- Exponential Generalized Fréchet (EGFr) (new)
- Log Gamma Fréchet (LGFr) (new)
- [0, 1]Truncated Exponentiated Exponential Fréchet (TEEFr) (new)
- Fréchet (Fr).

In order to do this comparison, use eight metrics. The two statistical measures used are the Kolmogorov–Smirnov statistic (KS) and the Anderson–Darling statistic (A). The analysis additionally considers the Cramér-von Mises (W) statistic and the HQIC, BIC, AIC by [20, 21], and CAIC information criteria. Furthermore, it utilizes the p-value derived from the Kolmogorov–Smirnov test. These criteria are often used to assess the quality of the fit.

Table 3 shows the values of the parameters estimated by the maximum likelihood method for the data used, in addition to the values of the accuracy criteria, which show that the proposed distribution obtained the lowest standard values that are met in performance by the [0,1] TEEFr distribution, which is quite far from the proposed distribution. As for the basic Fr distribution, it represents an increase in the inaccuracy in performance regarding the data used, which indicates the flexibility and efficiency of the proposed hybrid distribution. In addition to the previous table, Figs. 5 and 6 show the extent to which the distribution matches the data, and can be seen that the HOLFr distribution is suitable for modeling positively and negatively skewed data.

Also, can be extending the HOLFr distribution to include himodal and multimodal model data that would also be an interesting area for further research (Table 4).

Table 3 Estimates of models for data

Dist	MLEs	-2L	AIC	CAIC	BIC	HQIC
HOLFr	$\boldsymbol{\omega = 0.1066579}$ $\boldsymbol{\delta = 25.0795141}$ $\boldsymbol{a = 40.4750368}$ $\boldsymbol{\beta = 0.3273817}$	**95.09529**	**198.1906**	**198.7876**	**207.2972**	**201.816**
BeFr	$\omega = 1.3683981$ $\delta = 4.9211484$ $a = 2.9049637$ $\beta = 0.7243188$	102.7963	213.6406	214.2376	222.7473	217.266
KuFr	$\omega = 2.1050208$ $\delta = 8.9649266$ $a = 1.9957996$ $\beta = 0.6450534$	99.57312	207.2478	207.8448	216.3545	210.8732
EGFr	$\omega = 6.7019951$ $\delta = 0.9951016$ $a = 4.7880547$ $\beta = 0.7017667$	100.8311	209.6678	210.2648	218.7745	213.2932
LGFr	$\omega = 1.1107672$ $\delta = 5.3546042$ $a = 3.5956842$ $\beta = 0.7329769$	102.11	212.2587	212.8557	221.3653	215.8841
[0,1] TEEFr	$\omega = 44.3248058$ $\delta = 0.8273607$ $a = 33.3540295$ $\beta = 0.4751488$	96.22994	200.4599	201.0569	209.5665	204.0853
Fr	$a = 1.052192$ $\beta = 1.171994$	118.1178	240.2356	240.4095	244.7889	242.0483

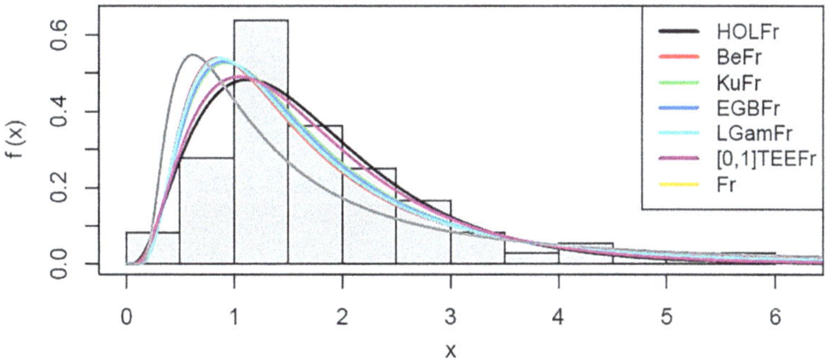

Fig. 5 Fitted densities for data

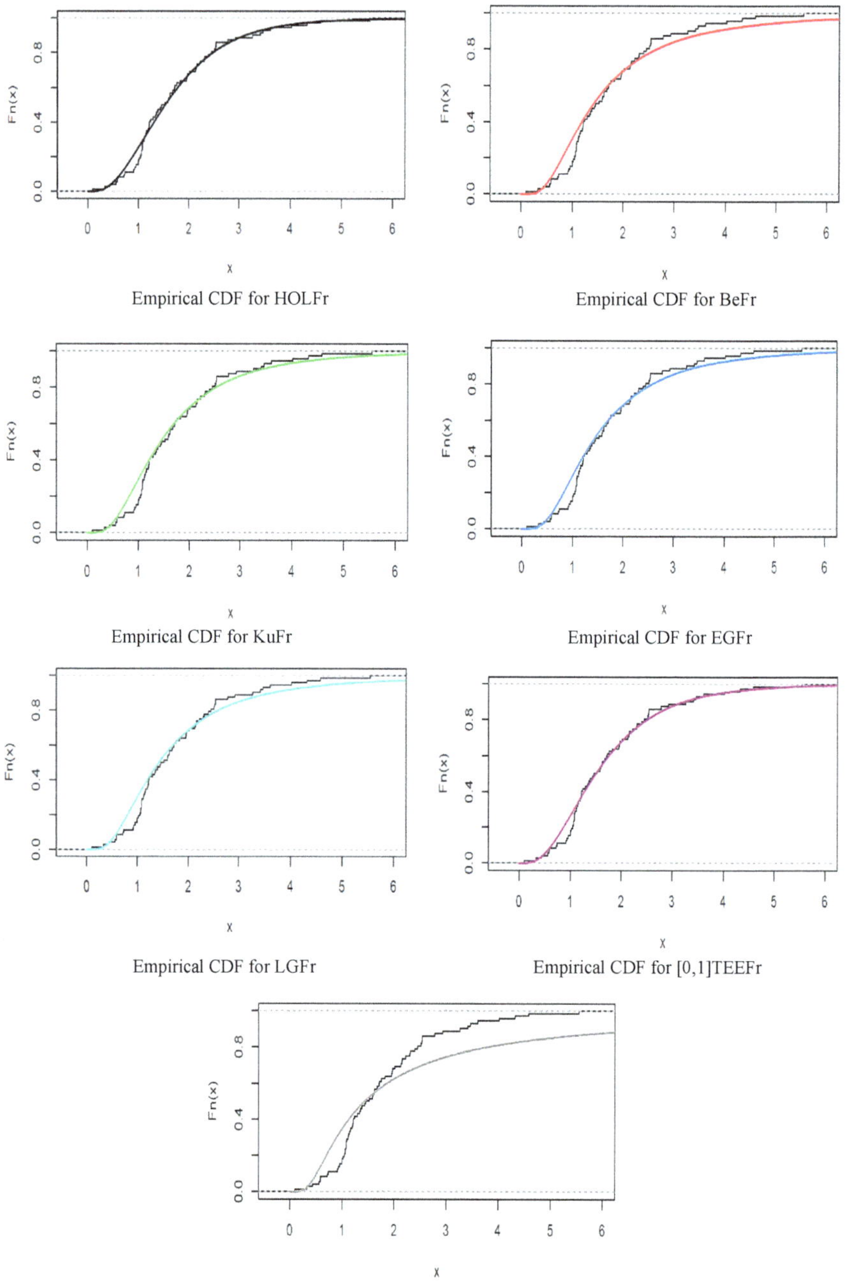

Empirical CDF for HOLFr

Empirical CDF for BeFr

Empirical CDF for KuFr

Empirical CDF for EGFr

Empirical CDF for LGFr

Empirical CDF for [0,1]TEEFr

Empirical CDF for Fr

Fig. 6 Empirical CDF for data

Table 4 Evaluate statistical metrics for the data

Dist	W	A	$K-S$	p-value
HOLFr	**0.08538912**	**0.5497654**	**0.1013971**	**0.4496848**
BeFr	0.1989575	1.355844	0.1505879	0.0763555
KuFr	0.1374487	0.9563294	0.1363578	0.137431
EGFr	0.1596745	1.104107	0.1384955	0.1262907
LGFr	21.50031	140.4778	0.9999987	0
[0,1] TEEFr	0.09117708	0.6164712	0.1136481	0.3102217
Fr	0.5284355	3.310618	0.1991305	0.00662518

4 Conclusion

The research presented a new continuous distribution with four parameters, which is the hybrid Odd Lomax Frechet (HOLFr) distribution derived from the Odd Lomax family. From graphs (2) and (4) it can be seen that the HOLFr distribution is suitable for modeling positively and negatively skewed data. Also, it is clear that the distribution can be used in studies of real data and survival data because the reliability histogram tends to decrease with increasing time, and a number of its statistical properties have been demonstrated. The hybrid distribution also demonstrated high efficiency and accuracy in describing the data used through practical application on real data, which demonstrates the efficiency and adaptability of the new distribution. The values in Table 3, as well as Figs. 5 and 6, comparing the new distribution and various other distributions show how superior is the HOLFer distribution.

Conflicts of Interest The authors assert that there are no conflicts of interest. As well as, all figures and tables included in the study are the work of researchers.

References

1. Khaleel, M.A., Oguntunde, P., Al abbasi, J.N., Ibrahim, N.A., AbuJarad, M.H.: The Marshall-Olkin Topp Leone-G family of distributions: a family for generalizing probability models. Sci. Afr. **8**, e00470 (2020)
2. Nasir, A.M., Tahir, M.H., Chesneau, C.: The odds generalized gamma-G family of distributions: properties, regressions and applications. Statistica (Bologna) **80**(1), 3–38 (2020)
3. Shrahili, M., Elbatal, I.: Truncated Cauchy power odd Fréchet-G family of distributions: theory and applications. Complexity **1**, 4256945 (2021)
4. Ibrahim, M., Ali, M.M., Yousof, H.M.: The discrete analogue of the Weibull G family: properties, different applications, Bayesian and non-Bayesian estimation methods. Ann. Data Sci. **10**, 1069–1106 (2021)
5. Muhammad, A., Khan, D.M., Aamir, M., El-Morshedy, M., Ahmad, Z., Khan, Z.: RETRACTED ARTICLE: a new flexible exponentiated-X family of distributions: characterizations and applications to lifetime data. IETE J. Res. **69**(11), 94–96 (2023)
6. Khalaf, A.A., Khaleel, M.: The new strange generalized Rayleigh family: characteristics and applications to COVID-19 data. Iraqi J. Comput. Sci. Math. **5**(3), 92–107 (2024)

7. Noori, N.A., Khalaf, A.A., Khaleel, M.A.: A new generalized family of odd Lomax-G distributions properties and applications. Adv. Theory Nonlinear Anal. Its Appl. **7**(4), 1–16 (2023)
8. Alzeley, O., Almetwally, E.M., Gemeay, A.M., Alshanbari, H.M., Hafez, E.H., Abu-Moussa, M.H.: Statistical inference under censored data for the new exponential-X Fréchet distribution: simulation and application to Leukemia data. Comput. Intell. Neurosci. **1**, 2167670 (2021)
9. Alshanbari, H.M., Gemeay, A.M., El-Bagoury, A.A.-A.H., Khosa, S.K., Hafez, E.H., Muse, A.H.: A novel extension of Fréchet distribution: application on real data and simulation. Alex. Eng. J. **61**(10), 7917–7938 (2022)
10. Alsadat, N., Aijaz, A., Jallal, M., Gemeay, A.M., Meraou, M.A., Hussam, E., Elmetwally, E.M., Hossain, M.M.: The novel Kumaraswamy power Frechet distribution with data analysis related to diverse scientific areas. Alex. Eng. J. **70**(1), 651–664 (2023)
11. Suleiman, A.A., Daud, H., Othman, M., Sawaran Singh, N.S., Ishaq, A.I., Sokkalingam, R., Husin, A.: A novel extension of the fréchet distribution: statistical properties and application to groundwater pollutant concentrations. Data Sci. Insights **1**(1), 8–24 (2023)
12. Akarawak, E.E., Adeyeye, S.J., Khaleel, M.A., Adedotun, A.F., Ogunsanya, A.S., Amalare, A.A.: The inverted Gompertz-Fréchet distribution with applications. Sci. Afr. **21**, e01769 (2023)
13. Hassan, A.S., Khaleel, M.A., Mohamed, R.E.: An extension of exponentiated Lomax distribution with application to lifetime data. Thail. Stat. **19**(3), 484–500 (2021)
14. Khalaf, A.A., Ibrahim, M.Q., Noori, N.A., Khaleel, M.A.: [0,1] Truncated exponentiated exponential Burr type X distribution with applications. Iraq J. Sci. **65**(8) (2024)
15. Noori, N.A.: Exploring the properties, simulation, and applications of the Odd Burr XII Gompertz distribution. Adv. Theory Nonlinear Anal. Its Appl. **7**(4), 60–75 (2023)
16. Teamah, A.-E.A.M., Elbanna, A.A., Gemeay, A.M.: Fréchet-Weibull distribution with applications to earthquakes data sets. Pak. J. Stat. **36**(2), 135–147 (2020)
17. Hassan, A.S., Almetwally, E.M., Khaleel, M.A., Nagy, H.F.: Weighted power Lomax distribution and its length biased version: properties and estimation based on censored samples. Pak. J. Stat. Oper. Res. **17**(2) (2021)
18. Khaleel, M.A., Oguntunde, P.E., Al Abbasi, J.N., Ibrahim, N.A., AbuJarad, M.H.: The Marshall-Olkin Topp Leone-G family of distributions: a family for generalizing probability models. Sci. AFrican **8**, e00470 (2020)
19. Jabbour, M.G., Datta, N.: A tight uniform continuity bound for the Arimoto-Rényi conditional entropy and its extension to classical-quantum states. IEEE Trans. Inf. Theory **64**(4), 2169–2181 (2022)
20. Khalaf, A.A., Noori, N.A., Khaleel, M.A.: A new expansion of the inverse Weibull distribution: properties with applications. Iraqi Stat. J. **1**(1), 52–62 (2024)
21. Abdullah, H.H., Khalaf, N.S., Noori, N.A.: Comparison of non-linear time series models (Beta-t-EGARCH and NARMAX models) with Radial Basis function neural network using real data. Iraqi J. Comput. Sci. Math. **5**(3), 26–44 (2024)

Travel Itinerary for Visiting Islamic Landmarks in Kuala Lumpur Using MTZ-TSP Model

Nurul Athirah Anuar, Amar Hanafi, Nur Athirah Azmi, Adibah Shuib, and Zati Aqmar Zaharudin

Abstract The optimal itinerary offers higher efficiency and is very crucial, especially for an individual with a restricted amount of time to visit their point of interest. The Miller-Tucker-Zemlin Traveling Salesman Problem (MTZ-TSP) is an approach that finds the shortest path that visits multiple vertices using integer linear programming technique. Utilizing the MTZ-TSP in the context of travel itinerary creation could improve travel experience in an area, especially for solo traveler. Islamic landmarks refer to places, sites, or landmarks that hold significance within the Islamic faith or culture. However, the study that used MTZ-TSP to develop travel itineraries for solo travelers with limited time, focusing on enhancing the travel experience associated with Islamic landmarks in urban area, is still lacking. Hence, this paper presents an optimized travel itinerary for visiting seven Islamic landmarks in the selected urban area, namely, Kuala Lumpur, Malaysia, using the MTZ-TSP approach. Three initial start points were selected to provide a diverse range of travel itineraries for tourists. As a result, it was found that the total travel time to traverse all of the selected Islamic landmarks would be 98 min.

Keywords MTZ-TSP · Travel itinerary · Islamic landmarks · Routing

N. A. Anuar · A. Hanafi · N. A. Azmi · Z. A. Zaharudin (✉)
Faculty of Computer and Mathematical Sciences, Universiti Teknologi MARA Cawangan Negeri Sembilan, Negeri Sembilan, Malaysia
e-mail: zati@uitm.edu.my

A. Shuib
Faculty of Computer and Mathematical Sciences, Universiti Teknologi MARA, Selangor, Malaysia

A. Shuib et al. (eds.), *Mathematical Modeling and Numerical Analysis*,
Springer Proceedings in Mathematics & Statistics 510,
https://doi.org/10.1007/978-981-96-9350-4_15

1 Introduction

The tourism industry is composed of a variety of services and businesses that are committed to fostering economic development and promoting cultural exchange. Improving the tourism sector is critical to improve tourists' travel experience especially in promoting the country to globally renowned status [6, 13]. The sector contributes millions of dollars to the country's economy, making it one of the most valuable industries.

The tourists, who are generally unfamiliar with the many points of interest (POI) in a city, find the process of creating a travel itinerary difficult and complex [15]. For tourists with time constraints, planning their travel itinerary would be important to ensure they enjoy the city with limited time [8]. The tourists can tailor their travel itinerary to visit the POIs through travel agencies, however, this approach would incur additional expenses. Moreover, this option is insignificant for tourists who have their own POIs, as the travel agencies may suggest points of interest that are incompatible with the tourist's preferences. Alternatively, the tourists can gather information from multiple tourism websites and create their own schedule and route to visit the POIs. However, creating an itinerary might be a tedious and challenging process as it involves a thorough investigation of multiple routes to reach from one POI to another. This process is time-consuming, especially when selecting optimal routes within a complex and large network. Therefore, a well-planned route to traverse the city's network must be systematically computed, ensuring that all POIs are visited.

The traveling salesman problem (TSP) is combinatorial optimization problem that ensure all the nodes within a graph G is visited exactly once, while minimizing the associated costs. TSP has been extensively used in various domains, including logistics, planning, and scheduling. Nevertheless, tackling the TSP can be very challenging as the number of potential routes increases exponentially with the number of nodes. Various methods have been developed to solve TSP, including the MTZ, i.e., a linear programming model for solving the TSP. MTZ-TSP is a formulation to solve the TSP as an LP [11], which introduces a set of that prevent sub-tour [3]. This method finds the shortest path that visits multiple locations and returns to the starting point, thus saving valuable time and resources.

The method has been utilized by Zaharudin et al. [19] for calculating traversal distance within waste collection systems. The constraint of MTZ-TSP, specifically the sub-tour elimination, has been extensively applied in addressing VRP-related issues including waste management [4, 17, 18], temporal or mobile facilities [2, 5] and healthcare services [12, 14]. Utilizing the MTZ-TSP to create the travel itinerary would be advantageous as it enables the determination of the optimal route for the travelers to follow, thereby ensuring that they can plan their journey efficiently and maximize their time at the destination.

Tourism is a global industry that directly impacts the country's economy which closely connected to various other sectors including agriculture, handicrafts, transportation, and more [16]. According to the Tourism Malaysia website [10], tourism in Malaysia had a significant growth of about 40% between 2022 and September

2023. Kuala Lumpur is Malaysia's capital city, which is a renowned tourist destination with many POIs, including notable Islamic landmarks. The city offers various places related to Islamic landmarks, such as National Palace, Islamic Arts Museum Malaysia, National Mosque of Malaysia, and Sultan Abdul Samad Jamek Mosque. These landmarks attract tourists by providing a glimpse of the diverse and rich heritage. However, the locations of these POIs are scattered across the city' network hence, required careful attention when establishing the corresponding travel itinerary. To the best of our knowledge, no study has yet addressed the systematic procedure for the establishment of a travel itinerary that encompasses the route between Islamic landmarks in Kuala Lumpur.

Therefore, this study objectively establishes the travel route for visiting the corresponding Islamic landmarks situated in Kuala Lumpur and the neighboring regions. Furthermore, three itineraries that entail varying starting points were proposed to offer tourists a wider range of options. The remainder of this paper is organized as follows: Sect. 2 focused on methodology, Sect. 3 presents the respective results, and Sect. 4 concludes the entire study.

2 Methodology

The solution for the travel itinerary of Islamic landmarks is computed in five steps, as illustrated in Fig. 1. The first step entails the identification of significant Islamic landmarks in Kuala Lumpur and its neighboring area. The second step involves the creation of a graphical network between these landmarks. The third step involves the measurement of the distance between landmarks. For the fourth step, the linear programming model, namely the MTZ-TSP, is employed to compute route sequence for these landmarks. The final step is to establish a travel itinerary to visit these landmarks.

Step 1: Identify the location of Islamic landmarks in Kuala Lumpur and its neighboring area.

There are several Islamic landmarks in Kuala Lumpur and the surrounding areas that have been identified. For this study, we chose seven landmarks based on their profound alignment with Islamic culture, heritage, and architecture. The following landmarks are the Islamic Arts Museum Malaysia, National Mosque of Malaysia,

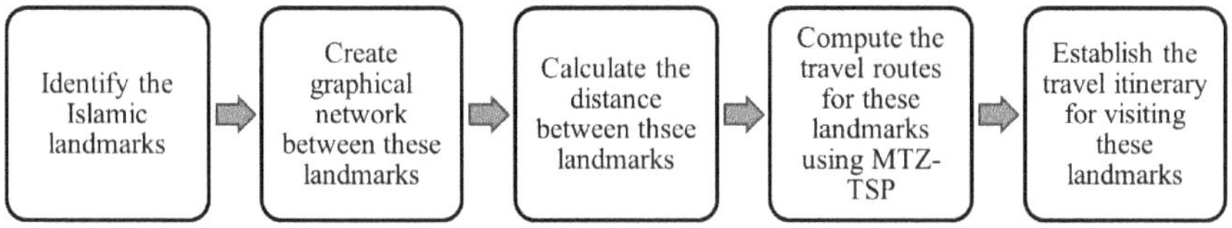

Fig. 1 Framework of computing route sequence for the Islamic landmarks in Kuala Lumpur

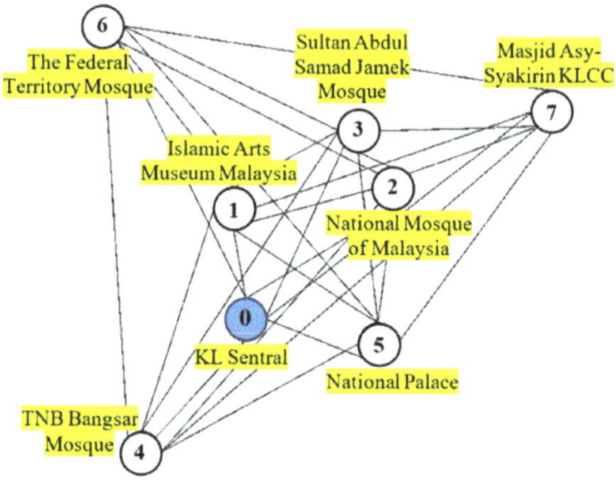

Fig. 2 Network for the Islamic landmarks in Kuala Lumpur

Sultan Abdul Samad Jamek Mosque, TNB Bangsar Mosque, the Federal Territory Mosque, Masjid Asy-Syakirin Kuala Lumpur City Center (KLCC), and the National Palace. These landmarks were chosen because they can also serve as exceptional educational resources, offering valuable knowledge about Islamic tradition and architecture. For example, the characteristics of the National Palace's architecture symbolize the distinctive features of Malay and Islamic architectural style [7, 9]. In addition, these places frequently advocate for sustainable tourism practices, which are in line with the principles of conservation and environmental care. Meanwhile, the Kuala Lumpur Sentral Station (KL Sentral) is assigned as both the starting and final depot of the travel route. The depot was selected as the largest hub that connects most of the public transport facilities in Kuala Lumpur [1].

Step 2: Create graphical network for these Islamic landmarks.

This step involves creating a graph that represents the route network between the selected Islamic landmarks. The network between landmarks and the depot is also created. The graph network as shown in Fig. 2.

The graph in Fig. 2 is based on the actual position of the landmarks. As seen from the graph, the numbering system is used to represent the selected landmarks and the depot. For instance, the depot is assigned to value "0," the Islamic Arts Museum Malaysia is assigned to value "1," the National Mosque of Malaysia is assigned to value "2," and so on. Next, the distance between these landmarks and depots is calculated.

Step 3: Establish the distance between each landmark.

The distance between each of the selected Islamic tourism destinations is gathered using Google Maps. The distance is measured in minutes and the journey is assumed to be conducted during the weekend morning only. In addition, it was assumed that the transportation for the route would be either private or through an e-hailing system. The gathered information as shown in Table 1.

Table 1 shows the distance matrix between the landmarks and depot. The symbol "*i/j*" is used to denote the set of locations represented within the graph where the

Table 1 Proximity of the chosen Islamic landmarks (in minutes) starting with KL Sentral (depot)

Area	i/j	0	1	2	3	4	5	6	7
KL Sentral	0	∞	8	7	18	14	12	12	22
Islamic Arts Museum Malaysia	1	10	∞	1	12	12	10	12	24
National Mosque of Malaysia	2	10	2	∞	14	14	12	12	24
Sultan Abdul Samad Jamek Mosque	3	12	7	6	∞	14	12	12	20
TNB Bangsar Mosque	4	16	10	9	18	∞	14	16	24
National Palace	5	12	12	10	14	18	∞	16	24
The Federal Territory Mosque	6	14	14	12	20	18	16	∞	18
Masjid Asy-Asykirin KLCC	7	20	20	18	24	22	22	20	∞

i denotes the start node that is positioned on the row of the table. Meanwhile the j denotes the end node that is positioned on the column of the table. As presented in Table 1, the symbol "∞" represents that there is no distance between the same location. Table 1 also shows that the travel distances between these locations are not symmetrical as a significant number of roads in Kuala Lumpur are one-way streets. As an example, the journey from the KL Sentral to the Islamic Arts Museum Malaysia took approximately 8 min, whereas the journey from the Islamic Arts Museum Malaysia to KL Sentral took 10 min.

Step 4: Compute the travel routes for these landmarks using MTZ-TSP.

The MTZ-TSP focused on finding the minimum routing distance so that all nodes (or points) are visited at least once. Let i the starting nodes, and j be the ending nodes, where n represents the set of these nodes. Parameter d_{ij} indicates the distance (in minutes) between node i and j. Parameter p is the maximum number of nodes that can be visited by tourists. There are three decision variables of the MTZ-TSP, namely, the x_{ij}, u_i, and u_j. The x_{ij} represents binary value of "1" if the travel from i to j is selected, and "0" otherwise. Meanwhile, decision variables of u_i and u_j are used to eliminate the sub-tour. The entire model is represented by the following equations:

$$\sum_{j=1}^{n}\sum_{i=1}^{n} d_{ij}x_{ij} \sum_{j=1}^{n}\sum_{i=1}^{n} d_{ij}x_{ij} \tag{1}$$

such that

$$\sum_{i=1, i\neq j}^{n} x_{ij} = 1; \quad \forall j = 1, 2, \ldots, n \tag{2}$$

$$\sum_{j=1, j\neq i}^{n} x_{ij} = 1; \quad \forall i = 1, 2, \ldots, n \tag{3}$$

$$u_i - u_j + px_{ij} \leq p - 1; \quad 1 \leq i \neq j \leq n \tag{4}$$

$$u_i, u_j \in \mathbb{R}; \quad \forall i, j = 1, 2, \ldots, n \tag{5}$$

$$x_{ij} \in \{0, 1\}; \quad \forall i, j = 1, 2, \ldots, n \tag{6}$$

Equation (1) represents the objective function, i.e., to minimize the total travel to all Islamic landmarks. Inequalities (2)–(6) are the constraints of MTZ-TSP. Inequality (2) and (3) strictly indicates that all the landmarks must be visited at least once. Inequality (4) is the sub-tour elimination constraints. This indicates that no landmarks will be visited more than once or in a recurring manner. Meanwhile, (5) and (6) indicate the characteristics of the decision variables, which u_i and u_j is in terms of real positive number, and the x_{ij} is a binary number.

In addition, we established the following conditions in addition to the previously mentioned assumptions: (i) the tourists have only one day to visit Kuala Lumpur; (ii) the weather and traffic conditions are consistently clear; (iii) there is reliable transportation available between the landmarks; (iv) the tourists do not encounter any difficulties in finding the landmarks; and (v) the landmarks are open and operational during the visitations. The model is then solved by using the optimization software, namely, the ILOG CPLEX 22.0, on a computer with the speed of 2.71 GHz and 12 GB of RAM.

Step 5: Establish the travel itinerary for visiting these landmarks.

The travel itinerary for visiting Islamic landmarks from and ending at KL Sentral is based on CPLEX output. Table 2 shows a summary of the output for variable x_{ij}, where the row represents the start node i and the column represents the end node j. Meanwhile, the value "1" from the table indicates that there is a traversing from i to j, whereas the value "0" indicates otherwise. For example, the first row of Table 2 indicates the traversal from node 0 to node 6. The second row shows the traversal from node 1 to node 0, while the third row portrays the traversal from node 2 to node 1. The process is then repeated for the entire rows.

The travel itinerary for visiting the selected Islamic landmarks is then determined using the actual location names provided in Table 1. The following section presents the findings in detail.

3 Results and Discussions

This section presents the results of the travel itinerary for single day that starts and ends at KL Sentral, focuses on Islamic landmarks in Kuala Lumpur. The results are presented in Table 3 and Fig. 3.

From Table 3, the minimum expected total traveling time for visiting these landmarks is 98 min. Please note that the time spent at these landmarks is excluded from

Table 2 Output for variable x_{ij}

x_{ij}	j								
		0	1	2	3	4	5	6	7
i	0	0	0	0	0	0	0	1	0
	1	1	0	0	0	0	0	0	0
	2	0	1	0	0	0	0	0	0
	3	0	0	1	0	0	0	0	0
	4	0	0	0	0	0	1	0	0
	5	0	0	0	1	0	0	0	0
	6	0	0	0	0	0	0	0	1
	7	0	0	0	0	1	0	0	0

Table 3 Route from the start node to the subsequent node and the related travel times

Start node	Subsequent node	Travel time (mins)
KL Sentral (depot)	The Federal Territory Mosque	12
The Federal Territory Mosque	Masjid Asy-Syakirin	18
Masjid Asy-Syakirin	TNB Bangsar Mosque	22
TNB Bangsar Mosque	National Palace	14
National Palace	Sultan Abdul Samad Jamek Mosque	14
Sultan Abdul Samad Jamek Mosque	National Mosque of Malaysia	6
National Mosque of Malaysia	Islamic Arts Museum Malaysia	2
Islamic Arts Museum Malaysia	KL Sentral (depot)	10
Total travel times (minutes)		98

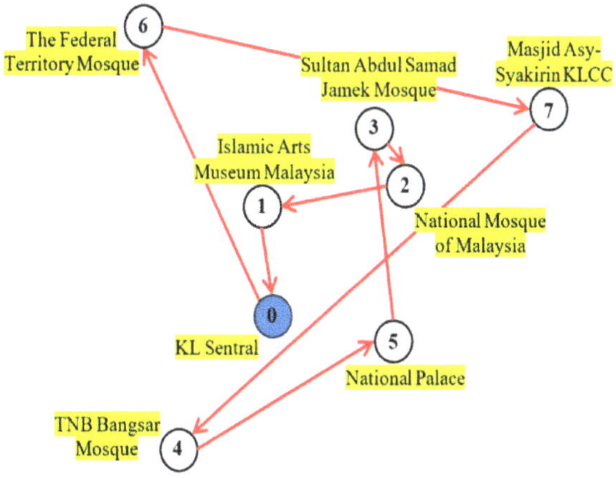

Fig. 3 Travel sequence from KL Sentral (depot) to all other Islamic landmarks

Table 4 Duration of visitation of the chosen Islamic landmarks (in mins)

Area	Visitation duration (mins)
Islamic Arts Museum Malaysia	100
National Mosque of Malaysia	100
Sultan Abdul Samad Jamek Mosque	30
TNB Bangsar Mosque	30
National Palace	100
The Federal Territory Mosque	30
Masjid Asy-Asykirin KLCC	30
Total visitation duration	420

this study. According to the information provided in the table, it is recommended for tourists to visit the first landmark, which is the Federal Territory Mosque. The estimated travel time to reach this landmark is 12 min. Although the National Mosque is the nearest landmark to KL Sentral (approximately 7 min away), the algorithm of the MTZ-TSP selects the combination of the overall traversal time for all landmarks, so it is at the minimum level.

The travel sequence for our study is depicted in Fig. 3. It was found that the travel sequence that start at KL Sentral → The Federal Territory Mosque → Masjid Asy-Syakirin → The Federal Territory Mosque → TNB Bangsar Mosque → National Palace → Sultan Abdul Samad Jamek Mosque → National Mosque → Islamic Arts Museum Malaysia → KL Sentral.

To enhance the accuracy of the results, we account for the estimated visitation duration at each landmark. Table 4 provides these assumptions. It is assumed that landmarks, excluding mosques, have a visitation duration of 120 min, while mosques are assigned a duration of 30 min.

The total time to visit all the landmarks is approximately 98 min, with the entire journey taking about 518 min, or roughly 9 h, for visiting Islamic landmarks in Kuala Lumpur. If a tourist begins their journey early in the morning, they will conclude their tour at the Masjid Asy-Asykirin KLCC, which offers a beautiful view at night.

4 Conclusion and Recommendations

The objective of this study was to establish the most effective journey sequence for visiting particular Islamic landmarks in Kuala Lumpur. The optimal travel sequence to these landmarks was determined using an ILP approach, specifically the MTZ-TSP. It has been found that the route would require 98 min to visit these landmarks. It is estimated that the tourists would need maximum of 9 h to visit these landmarks. Indeed, specific assumptions were made, such as restricting the visits to weekends and assuming consistently uncongested traffic. In addition, we do not include the duration of time spent at each of these landmarks. The study demonstrates that the use of the MTZ-TSP approach has a significant impact on the current body of research, particularly in its application within the tourism sector. The MTZ-TSP not only improves efficient routing decisions, but it also optimizes tourists' navigation and investigation of these attractions within a limited timeframe.

This study successfully accomplished its objective of enhancing travel routes for Islamic tourism attractions in Kuala Lumpur while also laying the groundwork for further research and practical implementations in the future. For future work, it is recommended to make modifications to the MTZ-TSP to incorporate a detailed itinerary that specifies the amount of time to be spent at each point. In addition, we recommend diversifying the schedule. For example, proposing a more than one-day trip to visit Islamic landmarks in Kuala Lumpur.

Acknowledgements This work was supported by Universiti Teknologi MARA (UiTM) Cawangan Negeri Sembilan for the Geran MyRA with the Project Code: 600-RMC/GPM LPHD 5/3 (165/2021).

References

1. Abdullah, F., Mahmood, W.H.W., Kamat, S.R., Othman, M.N.: Evaluation of noise level in woman train cabin at Ktm Komuter Berhad. In: Journal of Physics: Conference Series, vol. 1529, no. 4, p. 042031. IOP Publishing (2020)
2. Arias, A., Martínez, L.H., Hincapie, R.A., Granada, M.: An efficient approach to solve the combination between Battery Swap Station Location and CVRP by using the MTZ formulation. In: 2015 IEEE PES Innovative Smart Grid Technologies Latin America (ISGT LATAM), pp. 574–578. IEEE (2015)
3. Barbosa, L.H., Uchoa, E.: A branch-cut-and-price algorithm for the traveling salesperson problem with hotel selection. Comput. Oper. Res. **123**, 104986 (2020)
4. Ben-Romdhane, H., Ouertani, N., Krichen, S., Nouaouri, I.: On optimizing healthcare waste routing systems using waste separation policies: a case study. Appl. Soft Comput. **146**, 110615 (2023)
5. Buluc, E., Peker, M., Kara, B. Y., Dora, M.: Covering vehicle routing problem: application for mobile child friendly spaces for refugees. Or Spectrum, 1–24 (2022)
6. Haini, H., Wei Loon, P., Yong, S.K., Husseini, S.: Does social globalization affect the relationship between international tourism and economic growth? J. Travel Res. **63**(1), 252–269 (2024)
7. Ismail, A.S., Zhaharin, E.N.: Built form properties as sign and symbols of patron political ideology. Jurnal Kejuruteraan **29**(2), 87–96 (2017)
8. Ji, H., Zheng, W., Zhuang, X., Lin, Z.: Explore for a day? Generating personalized itineraries that fit spatial heterogeneity of tourist attractions. Inf. Manag. **58**(8), 103557 (2021)
9. Kadir, T.A.Q.R.A., Kassim, P.S.J., Latip, N.S.A.: Identity, tradition and the city: dichotomies and realities of creating an urban language of the Malay palace. Planning Malaysia **16** (2018)
10. Malaysia Tourism Statistics (2024). Retrieved August 13, 2024, from https://data.tourism.gov.my/
11. Miller, C.E., Tucker, A.W., Zemlin, R.A.: Integer programming formulation of traveling salesman problems. J. ACM (JACM) **7**(4), 326–329 (1960)
12. Ouertani, N., Ben-Romdhane, H., Nouaouri, I., Allaoui, H., Krichen, S.: A multi-compartment VRP model for the health care waste transportation problem. J. Comput. Sci. **72**, 102104 (2023)

13. Rojas-Méndez, J.I., Davies, G.: Promoting country image and tourism in new or underdeveloped markets. J. Travel Res. **63**(3), 755–768 (2024)
14. Saksuriya, P., Likasiri, C.: Vehicle routing problem with time windows to minimize total completion time in home healthcare systems. Mathematics **11**(23), 4846 (2023)
15. Taylor, K., Lim, K.H., Chan, J.: Travel itinerary recommendations with must-see points-of-interest. In: Companion Proceedings of The Web Conference 2018, pp. 1198–1205 (2018)
16. Thommandru, A., Espinoza-Maguiña, M., Ramirez-Asis, E., Ray, S., Naved, M., Guzman-Avalos, M.: Role of tourism and hospitality business in economic development. Mater. Today: Proc. **80**, 2901–2904 (2023)
17. Yuliza, E., Octarina, S., Eliyati, N., Puspita, F.M., Akmal, A.: Robust optimization model and its solution on the vehicle routing problem with time windows. In: AIP Conference Proceedings, vol. 2913, no. 1. AIP Publishing (2023)
18. Yuliza, E., Puspita, F.M., Supadi, S.S.: The robust counterpart open capacitated vehicle routing problem with time windows on waste transport problems. Bull. Electr. Eng. Inform. **9**(5), 2074–2081 (2020)
19. Zaharudin, Z.A., Shuib, A., Rodzi, Z.M., Nordin, N.A.M.: Recyclable waste collection routing in households' areas using the Mtz-Tsp model–a case study in Rasah, Negeri Sembilan. EDUCATUM J. Sci. Math. Technol. **11**(1), 8–14 (2024)

The Measurement of Water Service Provider Efficiency in Malaysia by Using Scale Directional Distance Functions Method

Nur Rasyida Mohd Rashid, Noor Asiah Ramli, Nur Syazwana Syamsul, Nur Hafizah Abdul Khalid, and Nursyahira Hidayu Robain

Abstract Water conservation is becoming a bigger challenge globally as the demand for water increases. Water losses are a concern for water supply service systems worldwide, particularly in developing nations. As a result, the problem of water losses, also identified as Non-Revenue Water (NRW), has become crucial to the administration of the water supply service. The current policy of the efficiency measurement by the service providers is impractical since NRW has been ignored as the undesirable product in the water supply system. Previously, there are many studies on measuring the efficiency of the water providers by using Data Envelopment Analysis (DEA). However, conventional DEA does not consider the undesirable output in efficiency measurement. The measurement model without the presence of undesirable outputs will have an unfair inaccurate result. Thus, Scale Directional Distance Function (SDDF) model will be applied in this study as an extension to the conventional DEA approach in measuring the efficiency of water supply providers of 14 states in Malaysia. The result displays that seven states which are Johor, Melaka, Pulau Pinang, Perak Perlis, Sabah, and Selangor achieved full efficiency score and the average efficiency score is 82.39%. In addition, this approach is able to establish target values for decreasing or extension of the outputs to make future improvements as well as to achieve full efficiency.

Keywords CCR · DEA · SBM-UO · SDDF · Water service provider efficiency

---

N. R. M. Rashid (✉) · N. S. Syamsul · N. H. A. Khalid · N. H. Robain
Faculty of Computer and Mathematical Sciences, Universiti Teknologi MARA (UiTM), Seremban, Negeri Sembilan, Malaysia
e-mail: syida@uitm.edu.my

N. A. Ramli
Faculty of Computer and Mathematical Sciences, Universiti Teknologi MARA, Shah Alam, Selangor, Malaysia

A. Shuib et al. (eds.), *Mathematical Modeling and Numerical Analysis*,
Springer Proceedings in Mathematics & Statistics 510,
https://doi.org/10.1007/978-981-96-9350-4_16

1 Introduction

Water is the fundamental natural resource and valuable economic asset not only for human beings but also for every creature in this world. This resource is crucial for ecological harmony, economic development as well as in the provision of environmental amenities and goods [7]. The demand for clean water will become more crucial in future; thus, it is crucial to have a sufficient supply of clean water [26]. In Malaysia, Suruhanjaya Perkhidmatan Air National (SPAN) is the organization with the responsibility to regulate the water services industry as well as to monitor and control the water supply [11]. Measuring water efficiency is a huge significant process and beneficial to assist the water providers to enhance and improve their services [1]. The efficient water management refers to the technical efficiency in the distribution of the water resources where the providers should define their objectives, monitor, and compare their achievements with other providers for development and improvement [20].

The water resources deficiency might occur due to climate change and other factors, thus proper planning of a nation's water supply system will avoid the current water crisis. In addition, it should be noticed that in many established nations, the issue of water loss and leakage problems can appear thus disturbing the efficiency of treated water distribution systems. Non-revenue water, often known as NRW, is a phrase frequently used in Malaysia's water delivery industry. NRW is the volume of water being charged at the consumer premises meters which diverges from the quantity of water that has been delivered [11]. In other terms, SPAN refers NRW as clean and properly maintained water that has been supplied through the water supply process but has not yet been charged to consumers [10]. Thus, the existence of NRW into the volume of the water distribution system does not bring income or revenue to the water supply providers and failure action to lower the elevated NRW volume, demonstrates a very significant operational issue [26]. The low of NRW quantity indicates that the water service has now been organized appropriately according to an outstanding distribution system [27]. Additionally, Vilanova and Balestieri also consider the unnecessary water losses to be a useful indication of the providers' efficiency, together with customer happiness, and financial stability [30]. Minimizing water waste is crucial for sustainable water management since it strengthens the financial security of the water service industry.

The absence of the undesirable output in the efficiency measurement will produce an unfair and undesirable result [10]. Thus, in this study SDDF is applied as an extension to conventional Data Envelopment Analysis (DEA)-based technique to measure the efficiency of the integrated production of desired and unwanted outputs involving the volume of NRW for 14 states in Malaysia. The SDDF technique considers unwanted outputs to enable the evaluation of water efficiency for the 14 providers in Malaysia. This innovative technique also has been used in the manufacturing industry which involves undesirable output such as in the insurance industry and food industry [29]. Finally, the water services providers will be ranked so that the inefficient state could make improvements as well as conduct benchmarking process from the efficient state.

2 Literature Review

Sufficient clean water supply is a basic human need where water also acts as an essential connection between civilization and their environment [11]. There are many mathematical methods to measure the water providers' efficiency such as Data Envelopment Analysis (DEA) [20]. Data Envelopment Analysis (DEA) technique by Charnes, Cooper, and Rhodes [4] is the efficiency measuring tool that has gained popularity in the water business industry such as in the studies by Lombardi et al. (2019) [14], Cabrera et al. [3], Romano et al. [23] and many more. This model searches to evaluate a set of decision-making units (DMUs) in terms of their relative efficiency which employ similar inputs to produce similar outputs [4]. In addition, the general picture of this model calculates the relative performance of peer units that allows the combination of a set's efficient DMUs to create an effective boundary that serves as a baseline to assess the efficiency level performance of the inefficient peer units [12].

DEA also has been widely used in many industries such as in the evaluations of the banking industry [19]; public state libraries [9]; and in sustainable supplier evaluation [22]. DEA can handle the input and output factor groups, even if the variables are in different units [25]. Numerous studies on the effectiveness of the water supply service industry have applied the DEA model from 1986, as in Romano and Guerrini [24], Byrnes et al. [2], and García-Sánchez [8]. However, the standard DEA did not consider on the undesirable output. The measurement of efficiency by conventional DEA does not consider the undesirable output such as NRW as indicator for the technical quality for water services. The NRW has been considered as important issue in Malaysia since 1986 because the incidence or NRW gives major impact on the revenue for this country. Many actions have been taken to tackle the issue with the target to achieve less than 25% of reduction for NRW in 2020 and the actions keep continues [10]. Some of past research has treated NRW as input factor in their efficiency measurement; however, this concept does not describe the real production process [17]. Thus, NRW had been considered as the undesirable output by applying the Directional Distance Approach [10]. However, the DDF model offers increasing the desirable output and simultaneously decreasing undesirable outputs [6, 21].

Thus, the extension of the DEA model which is Scale Directional Distance Function (SDDF) model will be applied in this study as the model incorporates undesirable outputs for the efficiency measurement. This model offers the expansion and contraction of the desirable output and undesirable outputs in a different proportion that also enable of the determination of the optimal direction to the frontier for each of the inefficient DMU [21].

3 Methodology

The approach for this study is as follows:

3.1 Collection of the Data

Data collection is the initial step in this research methodology. The data presented here were compiled from a report produced by the Malaysian Water Association (MWA) in collaboration with SPAN for 2020 [16]. Table 1 presents the inputs and outputs from 14 water service providers as the DMUs. The operating expenses and raw water are the inputs in this study while the revenue and water production are the desirable output and NRW as the undesirable output.

Table 1 Data of water utilities for each DMU for year 2020

DMU	OPEX (RM)	Raw water (MLD)	Water production (MLD)	Revenue (RM)	NRW (MLD)
Johor	736,167	1683	1640	853,527	426
Kedah	260,748	1388	1294	290,591	596
Kelantan	80,276	493	445	98,410	220
Labuan	27,752	71	69	16,241	20
Melaka	157,696	732	478	184,292	102
Negeri Sembilan	176,535	870	744	187,948	267
Pulau Pinang	199,037	1152	995	278,504	182
Pahang	251,605	1183	1108	153,056	588
Perak	228,555	1283	1237	350,333	379
Perlis	25,249	228	216	34,019	121
Sabah	446,482	1196	1196	224,508	618
Sarawak	230,398	1368	1192	263,605	381
Selangor	1,727,899	4648	4593	2,023,915	1545
Terengganu	126,321	664	605	123,939	188

3.2 Evaluate the Efficiency of DMUS by Taking into Consideration Inputs and Desirable Outputs Using CCR Method

In this study, the water service e provider efficiency for 14 Malaysian states is evaluated using the CCR model, which only take the desirable input and desirable output into consideration. CCR output-oriented model has been applied that emphasizes maximizing the desirable output which are water production and revenue by using Eqs. (1)-(3) [28]. Reference set is used to rank efficient DMUs as well as for benchmarking process [29]. It is possible to determine the rank of each since inefficient DMUs will refer to the efficient DMU. The ranking for the reference set is shown after the CCR model's coding had been executed in LINGO 20.0.

In 1978, Charnes, Cooper, and Rhodes introduced CCR method on the idea of efficiency measurement where DMU is recognized as the component that transforms inputs into outputs, and its efficiency was assessed [5]. The efficiency scores range from the scale from 0 to 1, with a score of less than 1 indicating inefficient units and a score of 1 indicating efficient DMUs. As a result, the effective units can be the benchmark for the inefficient units [20].

The radial model is based on the Constant Returns Scale (CRS) theory which means that a change in input quantities will result in a corresponding change in output quantities. Differently in various terms, Mahmoudi et al. [15] stated that if the input is altered positively, the output will likewise be altered favorably. In this study, output-oriented CCR will be applied which is able to increase the output while preserving the inputs. Furthermore, when optimizing output, output-oriented CCR will keep at a minimum present input rate [18]. The CCR output-oriented model as in Kao [13] is as follows:

$$\text{Min, } h_k = \sum_{r=1}^{s} u_r y_{rk}. \tag{1}$$

Subject to

$$\sum_{i=1}^{p} w_i x_{am} = 1, \tag{2}$$

$$\sum_{r=1}^{n} w_r y_{rt} - \sum_{i=1}^{m} w_i x_{it} \geq 0, t = 1, \ldots, s, \tag{3}$$

$$w_r \geq \varepsilon, (r = 1, \ldots, n),$$

$$w_r \geq \varepsilon, (i = 1, \ldots, p),$$

where

A u_r Input of weight.
w_i Output of weight.
y_{rt} Input.
x_{it} Output.
t DMU.
p Number of inputs.
n Number of outputs.
s Number of DMU.

3.3 Evaluate Efficiency of the DMUS by Considering Input and Undesirable Output by Using SDDF Model

The extended DEA-based approach which is SDDF is applied to measure the efficiency of the DMUs with the integrated formation of desired and undesirable output, which also identifies the best way to the frontier for each unit of study and offers several expansion and contraction factors [21]. SDDF technique is a model that governs the optional direction to the frontier for each unit analysis and provides dissimilar growth and reduction factors to attain the better efficiency score with slacks-based efficiency metric [24]. The SDDF model as in Ramli et al. [21] for the mth DMUs is as follows:

$$Max, h_m = \sum_{j=1}^{J} y_{yj} + \sum_{k=1}^{K} y_{uk}. \tag{4}$$

Subject to

$$\sum_{n=1}^{N} z_n y_{jn} \geq y_{jm} + \gamma_{yj} \cdot 1; \quad j = 1, 2, \ldots, J, \tag{5}$$

$$\sum_{n=1}^{N} z_n x_{in} \leq x_i m; \; i = 1, 2, \ldots, I, \tag{6}$$

$$\sum_{n=1}^{N} z_n u_{kn} = u_{km} - \gamma_{uk} \cdot 1; \quad k = 1, 2, \ldots, K, \tag{7}$$

$$z_n, \gamma_{yj}, \gamma_{yj} \geq 0; \; n = 1, 2, \ldots, N. \tag{8}$$

To obtain the efficiency score, θ_m^{SDDF} is the formulated as Eq. (9) where h_m is the inefficiency score obtained from model (4).

$$\theta_m^{\text{SDDF}} = 1 - h_m. \tag{9}$$

Then, the direction vector to the production boundary can be derived from the optimal solution to model (4). When $\sum_{j=1}^{J} \gamma_{yj} + \sum_{k=1}^{K} \gamma_{uk} > 0$, the scale direction for the jth desirable output and the kth undesirable output for each DMU can be obtained as in Eq. (10). Equation (10) refers to each entry of the direction vector from the model (4) that are being divided by the summation of the expansion rate of desirable output as well as the contraction rate of undesirable output.

However, if γ_{yj} and γ_{uk} are equal to 0, it denotes that the DMU is located on the efficient boundary, and then the direction vectors SD_i and SD_k can be chosen arbitrarily.

$$\text{SD}_j = \frac{\gamma_{yj}^*}{\sum_{j=1}^{J} \gamma_{yj} + \sum_{k=1}^{K} \gamma_{uk}}, \tag{10}$$

and

$$\text{SD}_k = \frac{\gamma_{uk}^*}{\sum_{j=1}^{J} \gamma_{yj} + \sum_{k=1}^{K} \gamma_{uk}}. \tag{11}$$

Then, the minimum and maximum directions for both of SD_j and SD_k's are between 0 and 1 where the lowest and maximum directions are $\left(0 \le \text{SD}_j, \text{SD}_k \le 1\right)$. To obtain the proper scale direction for each desired and unwanted output variable, the sum of the scale directions SD_j and SD_k must also equal 1 $\left(\text{SD}_j + \text{SD}_k = 1\right)$.

Then, from the acquired scale directions obtained, the goal or target value for each DMUs can be further measured by summing the multiplication of the intensity variable (z_n) from the model (4) with the actual value of the desirable and undesirable output $\left(y_{jn}\right)$ and (u_{kn}) respectively as Eq. (11).

$$\sum_{n=1}^{N} z_n y_{jn} \tag{12}$$

and

$$\sum_{n=1}^{N} z_n u_{kn}. \tag{13}$$

The target value will be equal to the actual value if the mth DMU having a zero value under the objective function in model (4). This is because the mth DMU is 100% completely efficient.

4 Results and Discussion

The efficiency score of CCR model can be measured based on Eq. (1) which is the CCR output-oriented model. Table 2 shows the result of efficiency score, reference set, and rank of 14 DMUs of water service providers in Malaysia. The reference set is the indicator to determine the rank when there is more than one state that shares the same value of efficiency score. Furthermore, the rank displays which state is positioned in number 1 until number 14.

Based on Table 2, there are five states that have achieved 100% efficiency score which are Johor, Perak, Perlis, Sabah, and Selangor. To determine which state rank will be in the first place, the reference set then will be used. The reference set refers to the frequency of the efficient DMUs being used as a reference set for inefficient DMUs which can evaluate the importance of efficient DMUs. The reference set for efficient DMU is itself. Since Perak has been mostly referred by the inefficient DMUs which is 8 times, then Perak will sit in the first rank. Then, the second rank is followed by Perlis, Sabah, and Johor. The average efficiency score in 14 states of Malaysia by CCR model is 95.13%. Hence, this represents that Kelantan, Melaka, Negeri Sembilan, Pulau Pinang, Sarawak and Terengganu did not achieve the average efficiency score of their 2020 performance.

As mentioned earlier by Kamarudin et al. [10], the calculation without involving undesirable output will produce unfair and unreliable results. Therefore, the efficiency score of the water service provider will be further carried out using SDDF model. Calculation of efficiency score for each DMU using SDDF model has applied

Table 2 Efficiency score using CCR model for each DMU

DMU	States	Efficiency score (%)	Reference set	Reference frequency	Rank
1	Johor	100	1	0	5
2	Kedah	96.52	9, 11	0	7
3	Kelantan	94	9, 10	0	10
4	Labuan	97.4	11, 13	0	6
5	Melaka	82.73	9, 13	0	14
6	Negeri Sembilan	88.28	9, 11	0	13
7	Pulau Pinang	91.47	9, 10	0	11
8	Pahang	96.51	9, 11	0	8
9	Perak	100	9	8	1
10	Perlis	100	10	3	3
11	Sabah	100	11	5	2
12	Sarawak	90.6	9, 10	0	12
13	Selangor	100	13	2	4
14	Terengganu	94.29	9, 11	0	9
	Average score	95.13			

Eq. (4). The results of efficiency score and the reference set have been executed by using LINGO 20.0 software and shown in Table 3. Since Pulau Pinang has been mostly referred by the inefficient DMUs which is 7 times, then Pulau Pinang will be in the first rank. The states that have efficient score of less than 100% are Kedah, Kelantan, Labuan, Negeri Sembilan, Pahang, and Sarawak. It shows that those states are inefficient in water supply service with six states have more than 50% of efficiency while Pahang with the least efficiency score of 50.27%. The overall average of efficiency score by SDDF is 82.39%. It is also can be seen that some states have better score by DEA model but obtain lower score by SDDF model such as Kedah, Kelantan and Negeri Sembilan. This is because the SDDF model has incorporate NRW in the efficiency measurement.

Table 4 shows the target value column that indicates the levels of water production and revenue, as well as non-revenue water, that each state should produce to achieve full efficiency. The change (%) column indicates in percentage terms how much the target values of water production and revenue, as well as non-revenue water, need to be increased and decreased from the actual values, respectively. The states can increase and contract the desirable and undesirable outputs by different proportions. For instance, Johor could expand its desirable output water production by 7.65% and revenue by 16.77% while concurrently persisting its undesirable output which is the non-revenue water to achieve the efficiency of water service provider since the change is 0. Furthermore, Kedah could expand its desirable output water production by 6.03% and revenue by 5.71% while concurrently contracting its undesirable output which is the non-revenue water by 0.20%. In other case, Pahang should reduce the

Table 3 Efficiency score using SDDF model for each DMU

DMU	States	Efficiency score (%)	Reference set	Reference frequency	Rank
1	Johor	100	1	4	3
2	Kedah	58.06	1, 7, 9	0	11
3	Kelantan	57.8	7, 9, 10	0	12
4	Labuan	92.17	1, 7, 9	0	8
5	Melaka	100	5	0	5
6	Negeri Sembilan	50.48	7	0	13
7	Pulau Pinang	100	7	7	1
8	Pahang	50.27	1, 7, 9	0	14
9	Perak	100	9	6	2
10	Perlis	100	10	2	4
11	Sabah	100	11	0	5
12	Sarawak	67.12	7, 9, 10	0	10
13	Selangor	100	13	0	5
14	Terengganu	77.57	1, 7, 9	0	9
	Average score	82.39			

Table 4 Target value and change from the SDDF model

States	Scales direction			Target value			Change (%)		
	Water production	Revenue	NRW	Water production	Revenue	NRW	Water production	Revenue	NRW
Johor	0.076	0.168	0.000	1765	996,694	426.00	7.65	16.77	0
Kedah	0.060	0.057	0.002	1372	307,186	594.79	6.03	5.71	– 0.20
Kelantan	0.021	0.019	0.002	454	100,313	219.51	2.08	1.93	– 0.22
Labuan	0.003	0.003	0.001	69	16,293	19.98	0.32	0.32	– 0.12
Melaka	0.022	0.036	0.001	489	190,967	101.94	2.23	3.62	– 0.06
Negeri Sembilan	0.035	0.037	0.001	770	194,890	266.62	3.47	3.69	– 0.14
Pulau Pinang	0.046	0.055	0.001	1041	293,747	181.88	4.64	5.47	– 0.07
Pahang	0.052	0.030	0.004	1165	157,660	585.77	5.17	3.01	– 0.38
Perak	0.058	0.069	0.001	1308	374,453	378.59	5.77	6.88	– 0.11
Perlis	0.010	0.007	0.004	218	34,246	120.57	1.01	0.67	– 0.35
Sabah	0.056	0.044	0.003	1263	234,413	616.31	5.58	4.41	– 0.27
Sarawak	0.056	0.052	0.001	1258	277,261	380.45	5.56	5.18	– 0.14
Selangor	0.214	0.398	0.001	5577	2,828,910	1543.82	21.42	39.77	– 0.08
Terengganu	0.028	0.024	0.002	622	126,958	187.72	2.82	2.44	– 0.15

quantity of non-revenue water from its current value of 588 Million Liters Per Day (MLD) to its desired value of 586 Million Liters Per Day (MLD) which 0.38% of reduction.

5 Conclusions

Based on the findings, it can be concluded that SDDF model is applicable to be the relevant efficiency measurement for water service providers in Malaysia that are facing with water losses. NRW should not be abandoned when the efficiency measurement is being conducted by the water service providers as this can affect the score performance. This study has applied the SDDF method for measuring efficiency of water service providers in Malaysia that provides an accurate score which also considers desirable inputs and outputs, and undesirable outputs. The approach also established target values for reduction or expansion of the outputs for the inefficient DMUs to achieve full efficiency. The potential reduction of NRW suggested by the SDDF model will be very beneficial to the government in order to reduce wastage in water industry. The findings also can be used as direction for future improvement and benchmarking process among the water service providers can be conducted further. Future studies are advised to use Spearman's rank correlation test to compare the full ranking result with the SPAN performance indicators to assess the appropriateness of the DEA model that is applied in this study as a substitute performance indicator for water supply services in Malaysia. In addition, integration of this model with another model such as Goal Programming model and Analytical Network Process (ANP) should also be conducted in the future research.

Acknowledgements The authors would like to show appreciation to the Malaysian Water Association (MWA) for providing dataset in this study.

Author Contribution Nursyahira Hidayu Robain analyzed the data for the study. Nur Rasyida Mohd Rashid, Nur Syazwana Syamsul, and Nur Hafizah Abdul Khalid drafted the manuscript; Noor Asiah Ramli has participated in revision of the intellectual content. All authors have read and approved the final manuscript.

Funding The authors received no specific funding for this work.

Conflict of Interest The authors have no conflicts of interest to declare.

References

1. Abu-Bakar, H., Williams, L., Hallett, S.H.: A review of household water demand management and consumption measurement. J. Clean. Prod. **292**, 125872 (2021)
2. Byrnes, P., Grosskopf, S., Hayes, K.: Efficiency and ownership: further evidence. Rev. Econ. Stat. 337–341 (1986)

3. Cabrera, E., Estruch-Juan, E., Molinos-Senante, M.: Adequacy of DEA as a regulatory tool in the water sector. The impact of data uncertainty. Environ. Sci. Policy **85**, 155–162 (2018)
4. Charnes, A., Cooper, W.W., Rhodes, E.: Measuring the efficiency of decision-making units. Eur. J. Oper. Res. **2**(6), 429–444 (1978)
5. Cooper, W.W., Seiford, L.M., Tone, K., Zhu, J.: Some models and measures for evaluating performances with DEA: past accomplishments and future prospects. J. Prod. Anal. **28**, 151–163 (2007)
6. Färe, R., Grosskopf, S.: Directional distance functions and slacks-based measures of efficiency: some clarifications. Eur. J. Oper. Res. **206**(3), 702 (2010)
7. García-Ávila, F., Avilés-Añazco, A., Sánchez-Cordero, E., Valdiviezo-Gonzáles, L., Ordoñez, M.D.T.: The challenge of improving the efficiency of drinking water treatment systems in rural areas facing changes in the raw water quality. S. Afr. J. Chem. Eng. **37**, 141–149 (2021)
8. García-Sánchez, I.M.: Efficiency measurement in Spanish local government: the case of municipal water services. Rev. Policy Res. **23**(2), 355–372 (2006)
9. Guccio, C., Mignosa, A., Rizzo, I.: Are public state libraries efficient? An empirical assessment using network data envelopment analysis. Socioecon. Plann. Sci. **64**, 78–91 (2018)
10. Kamarudin, N., Ismail, W.R., Ramli, N.A.: Malaysian water utilities performance with the presence of undesirable output: a directional distance function approach. Jurnal Teknologi **78**(4–3), 17–22 (2016)
11. Kamarudin, N., Rosmanira Ismail, W., Azri Mohd, M.: Network data envelopment analysis as instrument for evaluating water utilities' performance (Analisis Pengumpulan Data Rangkaian sebagai Instrumen untuk Mengukur Prestasi Utiliti Air). J. Qual. Meas. Anal. JQMA **14**(2) (2018)
12. Kamarudin, N., Ismail, W.R.: Establishment of performance indicators for Malaysian water utilities with the presence of undesirable output. Jurnal Teknologi **78**(4–4), 99–105 (2016)
13. Kao, C. Kao, C.: Slacks-based efficiency measures. In: Network Data Envelopment Analysis: Foundations and Extensions, pp. 65–87. Springer, Heidelberg (2017)
14. Lombardi, G.V., Stefani, G., Paci, A., Becagli, C., Miliacca, M., Gastaldi, M., Almeida, C.M.V.B.: The sustainability of the Italian water sector: an empirical analysis by DEA. J. Clean. Prod. **227**, 1035–1043 (2019)
15. Mahmoudi, M.J., Fathi, B., Sajadifar, H., Shahsavari, A.: Measuring efficiency of water and wastewater company: a DEA approach. Res. J. Appl. Sci. Eng. Technol. **4**(12), 1642–1648 (2012)
16. Malaysian Water Association (MWA): Malaysia Industri Guide 2020 (2020)
17. Munisamy, S.: Efficiency and ownership in water supply: evidence from Malaysia. Int. Rev. Bus. Res. Pap. **5**(6), 148–260 (2009)
18. Ng, K.H., Wong, S.C., Yap, P.K., Khezrimotlagh, D.: A survey on Malaysia's banks efficiency: using data envelopment analysis. Sch. J. Econ. Bus. Manag. **1**(11), 586–592 (2014)
19. Othman, F.M., Mohd-Zamil, N.A., Rasid, S.Z.A., Vakilbashi, A., Mokhber, M.: Data envelopment analysis: a tool of measuring efficiency in banking sector. Int. J. Econ. Financ. Issues **6**(3), 911–916 (2016)
20. Pan, Z., Wang, Y., Zhou, Y., Wang, Y.: Analysis of the water use efficiency using super-efficiency data envelopment analysis. Appl Water Sci **10**(6), 1–11 (2020)
21. Ramli, N.A., Munisamy, S., Arabi, B.: Scale directional distance function and its application to the measurement of eco-efficiency in the manufacturing sector. Ann. Oper. Res. **211**, 381–398 (2013)
22. Rashid, N.R.M., Halim, B.A., Zawawi, I.S.M., Abdullah, H.N., Hassan, N.I.A.A., Ismail, N.A.: Integration of data envelopment analysis with analytical hierarchy process and entropy measurement in the optimization of supplier selection. MATEMATIKA: Malays. J. Ind. Appl. Math. 1–14 (2023)
23. Romano, G., Guerrini, A., Molinos-Senante, M.: Water utility efficiency assessment in Italy by accounting for service quality: an empirical investigation. Util. Policy **45**, 97–108 (2017)
24. Romano, G., Guerrini, A.: Measuring and comparing the efficiency of water utility companies: a data envelopment analysis approach. Util. Policy **19**(3), 202–209 (2011)

25. Rostamzadeh, R., Akbarian, O., Banaitis, A., Soltani, Z.: Application of DEA in benchmarking: a systematic literature review from 2003–2020. Technol. Econ. Dev. Econ. **27**(1), 175–222 (2021)
26. Rosmanira Ismail, W., Kamarudin, N.: Modified Russell network model for determining Malaysia water supply services performance. J. Qual. Meas. Anal. JQMA **16**(2), 231–241 (2020)
27. Salleh, H.B.M.: Evaluation of non-revenue water and leakage in public and private water supply systems. (Doctoral dissertation). Heriot-Watt University (2000)
28. Teknologi, J., Kamarudin, N., Rosmanira Ismail, W., Ramli, A., Mara, T., Alam, S.: Malaysian water utilities performance with the presence of undesirable output: A directional distance function approach. Jurnal Teknologi **78** (2016)
29. Tone, K.: A slacks-based measure of efficiency in data envelopment analysis. Eur. J. Oper. Res. **130**(3), 498–509 (2001)
30. Vilanova, M.R.N., Balestieri, J.A.P.: Modeling of hydraulic and energy efficiency indicators for water supply systems. Renew. Sustain. Energy Rev. **48**, 540–557 (2015)

Factors Influencing Vegetables' Freshness Score for Optimization Model of the Postharvest Fresh Vegetables Supply Chain

Adibah Shuib, Najihan Awang Ali, Nor Aslily Sarkam, and Ismail Rakibe

Abstract Managing the freshness of postharvest vegetables in the supply chain poses significant challenges. Freshness, a critical quality indicator, directly impacts the fitness for consumption and consumer purchasing decisions. Previous studies have identified various factors influencing vegetables' freshness, including consumer perspectives, physiological characteristics, temperature management, and delivery time. However, limited research compares the factors across different stakeholders in the supply chain. This study aims to identify influential factors affecting vegetables' freshness from the perspectives of producers, wholesalers, retailers, and transport providers. A stratified sampling survey was conducted with 151 respondents to identify key freshness determinants. Multiple linear regression analyses revealed that mode of transportation, delivery time, biological control, and temperature control significantly influence postharvest vegetable freshness. These factors were utilized to formulate a freshness score, an objective function for optimizing vehicle routing in vegetable distribution. The findings underscore the importance of integrating these criteria into supply chain practices to enhance freshness, reduce losses, and improve efficiency. Strategies proposed include best practices for postharvest storage and handling to preserve quality, extend shelf life, and minimize waste. To ensure efficient supply chain operations and higher consumer satisfaction, stakeholders must prioritize these critical factors. This research provides valuable insights into optimizing the fresh vegetable supply chain, contributing to improved sustainability and resilience.

A. Shuib · N. A. Ali
Faculty of Computer and Mathematical Sciences, Universiti Teknologi MARA (UiTM), Shah Alam, Selangor, Malaysia

N. A. Ali (✉) · N. A. Sarkam
Faculty of Computer and Mathematical Sciences, Universiti Teknologi MARA (UiTM) Perak Branch, Perak, Malaysia
e-mail: najihanali@uitm.edu.my

I. Rakibe
Faculty of Plantation and Agrotechnology, Universiti Teknologi MARA (UiTM), Merlimau, Melaka, Malaysia

A. Shuib et al. (eds.), *Mathematical Modeling and Numerical Analysis*,
Springer Proceedings in Mathematics & Statistics 510,
https://doi.org/10.1007/978-981-96-9350-4_17

Keywords Postharvest fresh vegetables supply chain · Freshness criteria · Freshness influential factors · Multiple linear regression · Freshness score

1 Introduction

Fresh vegetables are a collection of plants enriched with nutrients, playing a crucial role in balancing a healthy diet [1–3]. The United Nations (UN) announced the year 2021 as the International Year of Fruits and Vegetables to increase awareness of consuming these products in alignment with Sustainable Development Goals (SDGs) of SDG 2 (Zero Hunger) and SDG 3(Good Health and Well-being) [4]. Several scholars support promoting the benefits of fresh vegetables for an enhancing lifestyle, boosting the immune system and preventing non-communicable diseases [5–7].

The vegetables supply chain has significance for the economic sustainability of the agricultural sector [8]. Efficient supply chain management, especially in the postharvest stage, is crucial for maintaining freshness while optimizing timely delivery, quality control, transparency in origin and handling practices, and promoting sustainability through waste reduction and eco-friendly packaging for the entire circular economic process [8–10]. Figure 1 represents an overview of the stakeholders involved in the fresh vegetables supply chain.

The postharvest fresh vegetables supply chain starts with farmers who grow and harvest them. Postharvest vegetables then passed into the hands of transport providers, which move the vegetables from farms to distribution centers where they are stored, sorted out, and prepared for further distribution. These vegetables are then sent to wholesalers and retailers that either sell in bulk or directly to consumers. Consumers, as the end-users buy fresh vegetables for personal consumption. The chain represents fresh vegetables distribution from farm to fork.

However, managing the freshness of vegetables through various stakeholders in postharvest vegetables supply chain, in Malaysia particularly, poses significant challenges. Figure 2 illustrates the detailed flowchart of Malaysia's postharvest fresh vegetables supply chain from farmers (both contract and independent) to the end-users. It starts with raw input by the farmers in the form of pesticides, fertilizers, and seedlings. From there, the farmers' produce is directed to either the collectors, wholesalers, importers, Federal Agricultural Marketing Authority (FAMA), or private traders. These intermediaries then distribute the vegetables to various

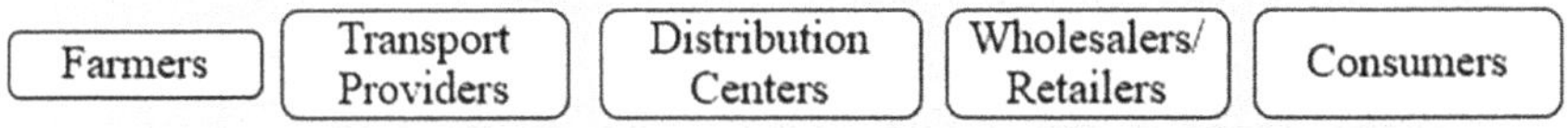

Fig. 1 Stakeholders of the postharvest fresh vegetables supply chain

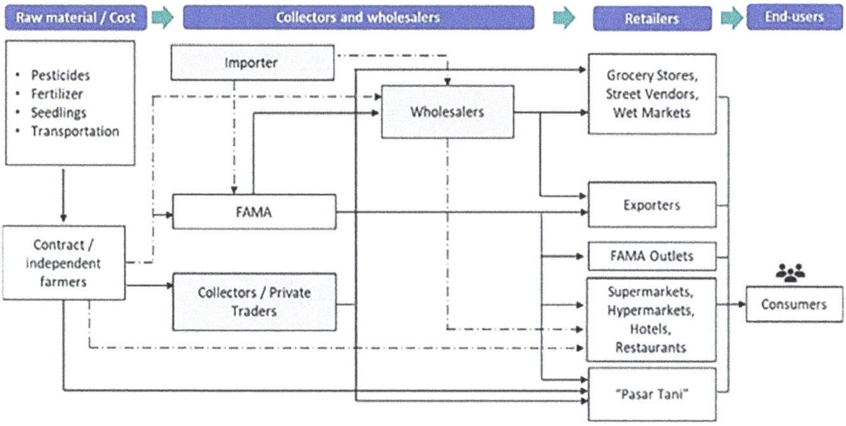

Fig. 2 Malaysia postharvest fresh vegetables supply chain. *Source* [11]

retailers including grocery stores, wet markets, supermarkets, hotels, and restaurants before finally reaching the consumers. Some produce is also exported or sold through specific outlets like FAMA outlets or "Pasar Tani"—a local farmer's market. The dotted lines in Fig. 2 represent alternative routes that farmers can take within the supply chain, illustrating the flexibility of the distribution process. The key stages of the supply chain, indicated by blue boxes, involve raw material input, collectors and wholesalers, retailers, and final end-users, providing a structured overview of the entire process.

The complexity of the process is further compounded by Malaysia's heavy reliance on imports to satisfy the need for vegetables, especially leafy greens. Most vegetables imported come from China, Indonesia, Australia, Thailand, Vietnam, United Kingdom, Netherlands, Spain and Bangladesh. This situation leads to a high dependency on external sources and impacts food security and agricultural sustainability. Consequently, this situation has contributed to increase food waste as high as 14% globally, totaling 931 million metric tons owing to the inability of the companies to make sure that the end user gets fresh leafy green produce [12–14]. Therefore, preserving freshness is essential to minimize food waste and meet consumer demands.

Freshness criteria and freshness factors are key indicators of vegetables quality that directly impact fitness for human consumption and consumer purchasing decisions [15]. In the context of evaluating fresh vegetables' quality, freshness criteria are defined as specific standards or benchmarks to meet predefined quality standards. Meanwhile, freshness factors concern elements or variables that influence the ability to meet those criteria [16, 17].

Figure 3 presents the freshness quality which encompasses freshness criteria and factors influencing the freshness criteria throughout vegetables supply chain. The mapping of freshness quality is synthesized based on analysis from standard Food Agriculture Organization (FAO), FAMA with agencies and several authors [17–22].

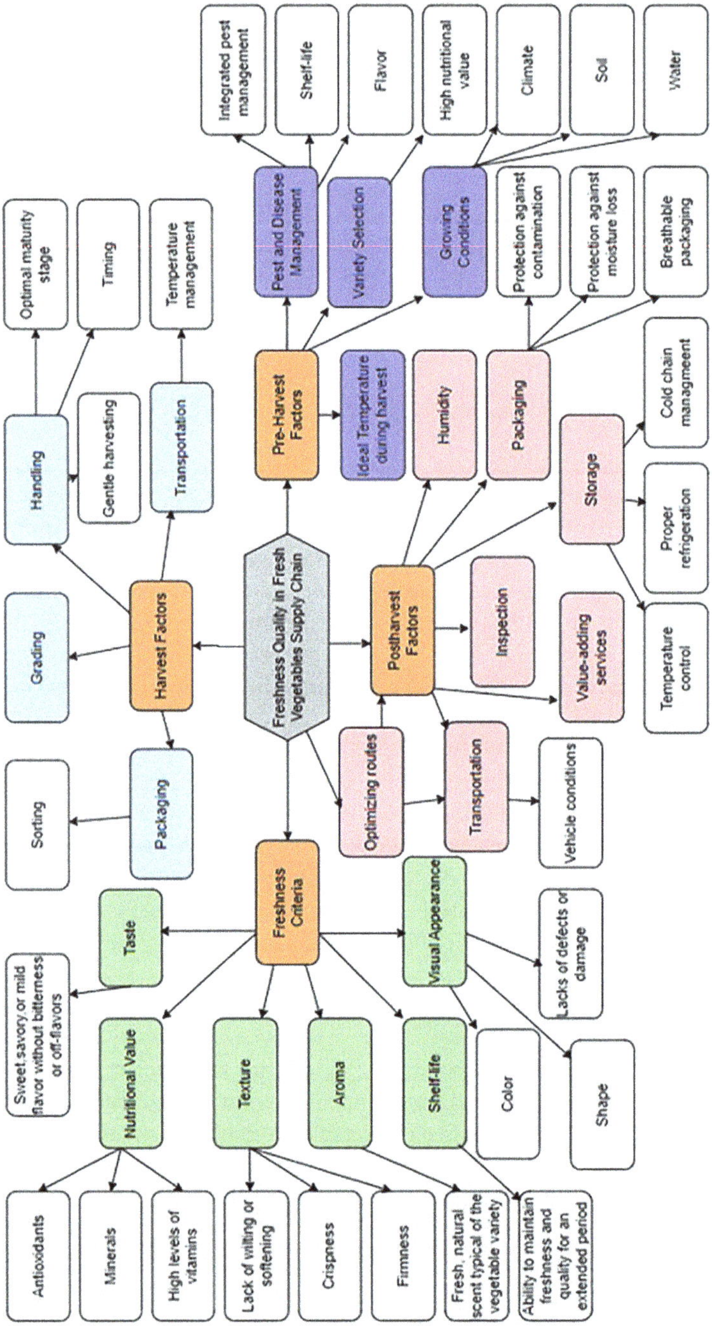

Fig. 3 Freshness quality mapping in fresh vegetables supply chain

Figure 3 highlights that freshness factors in preharvest, harvest and postharvest activities require detailed handling for each session. Additionally, the handling process during the postharvest stage is critical compared to others, as deterioration occurs rapidly, especially due to microbial spoilage, nutrient oxidation, and physical damage [23–26].

Our study aims at developing a freshness score of fresh vegetables to ensure vegetables' freshness is maintained at optimum level across the supply chain. To establish this freshness score, this study needs to determine the most significant factors that influence the freshness of the fresh vegetables supplied among three stakeholders, suppliers (farmers), distribution (logistics transport providers) and marketing (FAMA), wholesalers, and retailers during the postharvest stage. The freshness score is to be used as one of the objective functions of the green mixed integer goal programming model (G-MIGP) for vehicle routing problems with the soft time windows of the fresh vegetables' distribution.

Past studies have classified the factors affecting vegetables' freshness under certain criteria, which include consumer perspective, physiological and biological features, temperature management control, and postharvest delivery time. However, no studies compared the similarities and differences among these factors under different stakeholders. Thus, our study aims to identify the influential factors of vegetables' freshness based on three main stakeholders of the fresh vegetables supply chain: producers, wholesalers, retailers, and transport providers. This paper discusses the crucial part in developing freshness score for the fresh vegetables, which is determining the most influential factors that affect the freshness criteria of the postharvest fresh vegetables supply chain. The remainder of the paper is organized as follows: Sect. 2 materials and methods discusses the methodology, Sect. 3 results and discussion present the results and discussions concerned, and, finally, Sect. 4 conclusion summarizes the output and recommendation.

2 Materials and Methods

This study employed quantitative methods to collect data from targeted respondents using a survey questionnaire. The stratified sampling technique has been employed. In the first stage, the questionnaire was designed to align with the specific research context by incorporating findings from previous literature [27–30] and through collaborative discussion with local authorities. Five academic experts reviewed the questionnaire to ensure the questions were well-structured and understandable, and the metrics appropriately measured the freshness factors in the postharvest fresh vegetables supply chain. The face-to-face survey technique entails direct interviews and interaction with the respondents to ensure comprehensive responses from all participants. A total of 151 samples were collected, consisting of stakeholders involved in the postharvest fresh vegetables supply chains, including producers, wholesalers, retailers, and transport providers. Leafy green fresh vegetables such as spinach, leaf mustard, and poh choy were selected for their relevance, as indicated by the import dependency ratio analysis [31].

The vegetables group was selected due to their cultivation in both areas of the study scope (highland and lowland) as well as their high supply and demand which cater to the preferences and needs of consumers. This study focuses on lowland farms located in Kinta, Muallim and Batang Padang, Perak farms, as well as highland farms in Cameron Highlands. The contact lists of the respondents were obtained from the directory of the Department of Agriculture (DOA) and Municipal Council, based on their respective locations.

The structured questionnaire in this study divided into four parts: A, B, C and D, as follows:

- Part A: Demographic Information
- Part B: Transport Details
- Part C: Postharvest Activities (Time of harvest, Sorting and Grading Technique, Packaging Quality, Storage Conditions, Mode of Transportation, Length of Delivery Time, Biological Control and Temperature Control)
- Part D: Freshness Satisfaction.

Part A contains questions concerning the respondent's profile, types of crops, location and background. Part B consists of questions asking for details on transportation which include type of vehicle(s), delivery origin and destination, distance travel, and whether the vehicle is equipped with freezer or cold storage. Part C is composed of statements regarding aspects of each of the independent variables of the study. The respondents were required to rate the statement using the Likert scale from 1 to 5, where 1 refers to "strongly disagree" and 5 to "strongly agree." Meanwhile, Part D consists of questions related to stakeholder satisfaction with freshness of their products and their company.

Figure 4 shows the factors influencing freshness criteria in the postharvest fresh vegetables supply chain.

The consumer perspective is excluded from the study, as it focuses solely on stakeholders ranging from producers to retailers and wholesalers. Based on the framework shown in Fig. 4, eight (8) hypotheses are proposed in this study. These hypotheses are as the following:

H1: Time of harvest has a significant effect on the postharvest of fresh vegetables freshness.

H2: Sorting and grading technique has a significant effect on the postharvest fresh vegetables freshness.

H3: Packaging quality has a significant effect on the postharvest of fresh vegetables freshness.

H4: Storage conditions have a significant effect on the postharvest of fresh vegetables freshness.

H5: Mode of transportation has a significant effect on the postharvest of fresh vegetables freshness.

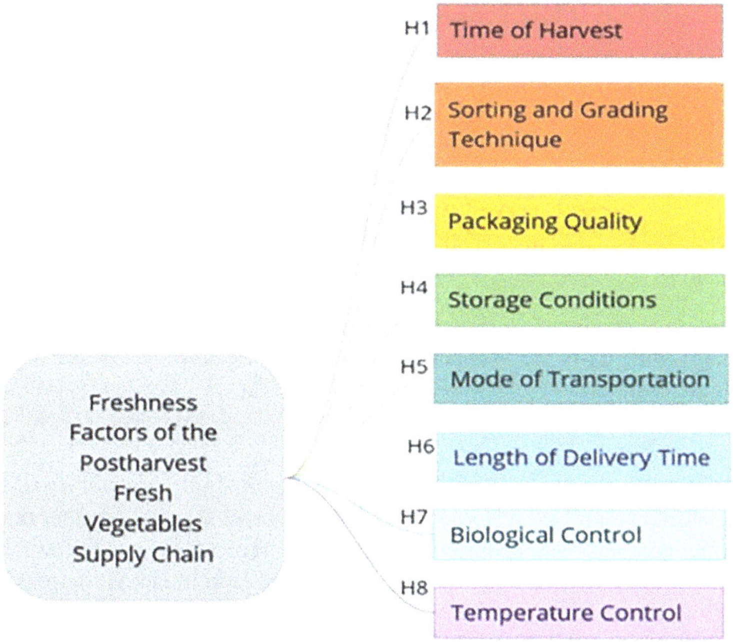

Fig. 4 Freshness factors of the postharvest fresh vegetables supply chain (modified and adapted from: [28, 32, 33]

H6: Length of delivery time has a significant effect on the postharvest of fresh vegetables freshness.
H7: Biological control has a significant effect on the postharvest of fresh vegetables freshness.
H8: Temperature control has a significant effect on the postharvest of fresh vegetables freshness.

The hypothesis will be rejected if the *p*-value is less than the significance level of 0.05 [34]. The analysis in our study starts with the correlation analysis. According to [35], the Pearson correlation coefficient (r) represents the degree (strength) of linear relationship and direction between two sets of measurements or variables. Table 1 shows the general guidelines for interpreting the correlation coefficient.

Multiple linear regression is an extension from simple linear regression where the statistical analysis technique provides the relationship between a dependent variable known as criterion and several independent variables known as predictor [37]. In general, the multiple linear regression model can be written as in Eq. (1).

$$Y = \beta_0 + \beta_1 X_1 + \beta_2 X_2 + \cdots + \beta_n X_n + \varepsilon \quad (1)$$

where Y is the dependent variable, X_1, X_2, …, X_n, are the independent variables, ε known as the residual term while β_0, β_1, β_2, …, β_n, are the list of the regression

Table 1 Interpretation of Pearson correlation coefficient (r)

Pearson's correlation coefficient (r)	Interpretation
0.9–1.0 (− 0.9 to − 1.0)	Very high positive (negative) correlation
0.7–0.9 (− 0.7 to − 0.9)	High positive (negative) correlation
0.5–0.7 (− 0.5 to − 0.7)	Moderate positive (negative) correlation
0.3–0.5 (+− 0.3 to − 0.3)	Low positive (negative) correlation
0.0–0.3 (0.0 to − 0.3)	Negligible correlation

Source [36]

coefficients with β_0 is the constant term. Four (4) assumptions of the regression analysis are required to be followed: (1) The residuals or errors are approximately normally distributed; (2) The value of the residuals is independent; (3) No issue of multicollinearity; (4) No issue of outlier [38].

For normality, the analysis involved is normal probability-probability (P-P) plot of regression standardized residuals. This graphical tool assesses whether the residuals (differences between the observed values and predicted values) follow a normal distribution [39]. Another issue for the linear regression model is the presence of multicollinearity, which might severely decrease the model's efficiency. If the variance inflation factor (VIF) equals 1, it indicates that the predictor is non-collinear; values between 1 and 5 indicate moderately collinear, while values equal to or greater than 5 and up to 10 indicate high collinearity. VIF values greater than 10 indicate a serious collinearity problem [40].

3 Results and Discussion

This section presents the results of the statistical analyses conducted to determine the most influential factors in postharvest vegetables' freshness. These analyses include the correlation analysis, the normality test, and the multiple regression analysis.

3.1 Correlation Analysis

Table 2 displays the correlation analysis between postharvest handling and factors influencing freshness criteria in the fresh vegetables supply chain and the p-values for significance.

Table 2 Result of correlation analysis

Var	Mean	Std. Dev	TH	SG	PQ	SC	MT	DT	BC	TC	F
TH	4.1407	0.42188	1								
SG	4.2036	0.41674	0.741**	1							
PQ	4.1545	0.45991	0.446**	0.612**	1						
SC	4.1404	0.43622	0.292**	0.474**	0.653**	1					
MT	4.1642	0.4567	0.075	0.145	0.293**	0.483**	1				
DT	4.1722	0.4112	− 0.085	0.031	0.225**	0.493**	0.574**	1			
BC	4.1755	0.53447	0.097	0.207*	0.359**	0.57**	0.658**	0.744**	1		
TC	4.181	0.59471	0.026	0.162*	0.368**	0.482**	0.571**	0.666**	0.721**	1	
F	4.0993	0.51322	0.094	0.211**	0.343**	0.521**	0.582**	0.677**	0.726**	0.688**	1

Sample Size: 151; $^{*}p < 0.05$, $^{**}p < 0.001$

Note Var variable, *TH* time of harvest, *SG* sorting and grading technique, *PQ* packaging quality, *SC* storage condition, *MT* mode of transportation, *DT* length of delivery time, *BC* biological control, *TC* temperature control, *F* freshness factor

3.2 Assumptions of Multiple Linear Regression

Several fundamental assumptions must be satisfied and verified to ensure the validity and reliability of results obtained and essential for accurate model interpretation and prediction. These assumptions include the existence of a linear relationship between the dependent (outcome) variable and the independent variables, multivariate normality (the residuals between observed and predicted values are normally distributed), no multicollinearity (independent variables are not too highly correlated with each other), variance inflation factor (VIF) below 10 which signifies no problematic multicollinearity, homoscedasticity (variance of error terms or residuals should be consistent across all levels of the independent variables) and having at least two independent variables [38, 41].

Normality Test of the Residuals

The normal probability plot is a graphical method for evaluating whether or not a dataset is approximately normally distributed [42]. Figure 5 presents the normal P-P plot of regression standardized residuals for the freshness (F), the dependent variable, to test the normality assumption. The plotted points are mostly close to the normal distribution line indicating a normal distribution of residual. There are minor deviations at some intervals, but not substantial enough to affect the overall judgment of normality. The normal distribution of residuals therefore supports the assumption of normality, and thus the validity of the regression model is justified.

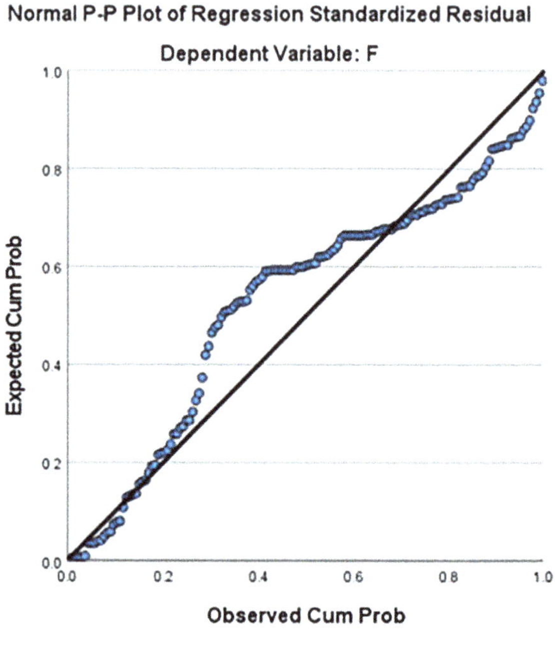

Fig. 5 Normal P-P plot of regression standardized residuals

Autocorrelation Test

The residual independence assumption was verified using the Durbin-Watson test statistics, as shown in Table 3. The Durbin-Watson test statistics in this study is 1.394. The value of 1.394 has shown that the residuals were likely independent since it falls within the recommended range 1.0–3.0.

Multicollinearity Test

Another issue for the linear regression model is the presence of multicollinearity, which might severely decrease the model's efficiency. From Table 4, the results show that each predictor has a tolerance value greater than 0.1, while variance inflation factor (VIF) values are more than one but less than five. Thus, in this study, the regression model does not present a multicollinearity issue.

Existence of the Outlier

Table 5 presents Cook's distance values to measure the issue of outliers and to inform about influential data points that may disproportionately affect the regression model's parameters. The maximum value of 0.170 is significantly below the commonly used threshold of 1, indicating that there are no influential data points in this model. Such results suggest the robustness of regression results, entirely unaffected by any single observation. This increases the validity and reliability of regression analysis in predicting freshness factors in vegetables at the postharvest stage.

Table 3 Durbin-Watson test statistics

Model	R	R^2	Adjusted R^2	Std. error of the estimate	Durbin-Watson
1	0.787	0.620	0.598	0.32530	1.394

Table 4 Multicollinearity test statistics

Variable(s)	Collinearity statistics	
	Tolerance	VIF
Time of Harvest (TH)	0.432	2.312
Sorting and Grading Technique (SG)	0.338	2.957
Packaging Quality (PQ)	0.438	2.281
Storage Conditions (SC)	0.401	2.496
Mode of Transportation (MT)	0.526	1.902
Length of Delivery Time (DT)	0.365	2.740
Biological Control (BC)	0.295	3.384
Temperature Control (TC)	0.414	2.417

Table 5 Cook's distance analysis

Cook's distance	Statistics
Minimum	0.000
Maximum	0.170
Mean	0.010
Std. Deviation	0.024
N	151

3.3 *Regression Model*

The coefficient of determination, known as R-squared (R^2), represents the proportion of the variance for the dependent variable that is predictable by the independent variables in the model. R^2 is a goodness-of-fit measure which value can be between 0 and 1 where a higher R^2 indicates a better fit for the model. Meanwhile, adjusted R^2 compares models with different numbers of variables. Since R^2 always increases upon adding more independent (predictor) variables to a model, the adjusted R^2 is a modified version of R^2 that adjusts the number of predictors in a regression model. Adjusted R^2 is a corrected goodness-of-fit (model accuracy) and its value is always less than or equal to R^2. Adding a predictor that is useful in predicting the response (dependent) variable (Y) will result in a value in adjusted R^2 that is greater than R^2. However, adding a predictor that is not useful in predicting Y, the adjusted R^2 will decrease to a value lower than R^2.

Based on Table 6, the value of R^2 is 0.620 which means 62% of the total variation of the freshness of postharvest fresh vegetables can be explained by the regression line using the significant freshness influential factors, another 38% will be explained to other factors. However, based on the value of the coefficient of correlation, R value exceeding 0.6 is considered favorable since it indicates a strong positive linear relationship between the factors that influence the freshness of the postharvest vegetables ($r = 0.787$). The adjusted R^2 is 0.598, which means additional freshness influential factors do not add value to this model. The value of R^2 represents a good fit in agricultural contexts since a lot of complexity is involved in perishable goods logistics [43, 44]. Therefore, logistics and distribution networks become crucial to maintaining the freshness of the products and are directly related to influencing the purchasing decision of the consumers [45, 46].

Subsequently, the right-tailed F test is used to check if the overall regression model is statistically significant. The null hypothesis (H_0) states that the model has no independent (predictor) variable(s) or $H_0 : Y = \beta_0$, while the alternative hypothesis is $H_a : Y = \beta_0 + \beta_1 X_1 + \beta_2 X_2 + \cdots + \beta_n X_n$. The F-statistics, calculated by dividing

Table 6 Model summary

Model	R	R^2	Adjusted R^2	Std. error of the estimate
1	0.787	0.620	0.598	0.32530

mean square of the regression (MSR) by the mean square of the residuals (MSE), is used to assess whether the regression model with independent variables provides a better fit to the data than a model without them [47]. The null hypothesis is rejected if F test statistics are larger than F critical value at a certain level of significance (or if p-value is less than the level of significance). Analysis of variance (ANOVA) in Table 7 shows the F-statistic is 28.920, which is greater 1.94, and the p-value is less than 0.05, thus, null hypothesis is rejected. This indicates that regression model $Y = \beta_0 + \beta_1 X_1 + \beta_2 X_2 + \cdots + \beta_n X_n$ is statistically significant and provides better fit than the model without independent variables, $Y = \beta_0$.

Multiple linear regression (MLR) is used to determine the relationship between two or more independent variables and one dependent variable. The MLR analysis is employed to explore the relationship between some freshness factors (independent variables) and the freshness (dependent variable) of postharvest fresh vegetables. The t test is used to check the significance of individual regression coefficients (β_i) where $i = 1, 2, \ldots, 8$, in the MLR model.

Table 8 presents the estimated coefficients of independent variables (β_i), standard errors, t statistic values, and p-values for the MLR analysis. At a significance level of 0.05, the analysis shows that mode of transportation (MT) with $\beta_5 = 0.166$ and length of delivery time (DT), where $\beta_6 = 0.346$, are significant factors affecting freshness (dependent variable). This means appropriate transportation and minimum delivery times of postharvest fresh vegetables are crucial to freshness. In addition, biological control (BC) ($\beta_7 = 0.380$) and temperature control (TC) with $\beta_8 = 0.201$ are also statistically significant factors that influence freshness in postharvest fresh vegetables supply chain. This finding signifies the BC and TC measures in maintaining the quality and shelf life of fresh produce.

On the other hand, the independent variables which are not significant determinants of freshness are time of harvest (TH), sorting and grading technique (SG), packaging quality (PQ), and storage conditions (SC) where the p-values are greater than 0.05.

Table 7 ANOVA test

Model	Sum of squares (SS)	df	Mean square	F	p-value
Regression	24.483	9 (df1)	3.060 (MSR)	28.920	< 0.001
Residual	15.027	142 (df2)	0.106 (MSE)		
Total	39.510	150			

Note *** p-value≤ 0.05

Table 8 MLR analysis

Model	Unstandardized Coefficients		Standardized Coefficients Beta	t	Significance
	β_i	Std. Error			
1 (Constant)	– 0.403	0.445		– 0.904	0.367
Time of Harvest (TH)	0.022	0.096	0.018	0.232	0.829
Sorting and Grading Technique (SG)	0.070	0.110	0.057	0.641	0.523
Packaging Quality (PQ)	– 0.011	0.087	– 0.008	– 0.125	0.901
Storage Conditions (SC)	0.075	0.096	0.057	0.783	0.435
Mode of Transportation (MT)	0.166	0.080	0.148	2.074	0.040***
Length of Delivery Time (DT)	0.346	0.100	0.277	3.452	0.001***
Biological Control (BC)	0.380	0.085	0.396	4.473	0.000***
Temperature Control (TC)	0.201	0.097	0.113	2.068	0.040***

Note *** p-value ≤ 0.05, β_i, $i = 1, 2, \ldots, 9$: coefficients of the independent variables

Based on the regression coefficients (β_i), as displayed in Table 8, the MLR model is formulated as follows:

$$Y = -0.403 + 0.166X_5 + 0.346X_6 + 0.380X_7 + 0.201X_8 \quad (2)$$

where the most influential factors of the postharvest fresh vegetables freshness are:

X_5 Mode of transportation
X_6 Length of delivery time
X_7 Biological control
X_8 Temperature control.

Determinants of postharvest vegetables freshness, as described by the MLR model (2), can easily be justified. Effective transportation management is a critical factor highlighted among scholars due to significant roles in handling, logistics, networking, and connectivity as well as distribution issues caused by poor road of the infrastructure, mechanical damage from improper loading and unloading practice, and lack of refrigerated transport. Consequently, these issues impact food security, sustainability, and economic development [48–52].

As fresh vegetables are highly perishable and have limited shelf lime, they require the shortest delivery time to avoid spoilage and reduce postharvest losses and waste [53]. Often, freshness deterioration decreases customer satisfaction and profitability for stakeholders in the supply chain [54, 55]. Also, handling biological control agents is crucial to control the postharvest pathogens and avoid spoilage [56]. Furthermore, optimal temperature control guarantees the quality and extends shelf life of vegetables [57].

4 Conclusion

This study identifies the critical factors influencing the freshness of postharvest fresh vegetables supply chain involving producers, wholesalers, retailers, and transport providers to establish the freshness score. The findings reveal that mode of transportation, length of delivery time, biological control, and temperature control are the most significant factors or predictors that influence the freshness of the postharvest vegetables. Although time of harvest, sorting and grading technique, packaging quality, and storage conditions are practically important, they did not show significant statistical impacts as freshness criteria in this case study. MLR analysis and models were employed to capture the complex relationship which multiple independent variables have on the single dependent variable. Specifically, the t test in MLR was used to determine if a particular independent variable is statistically significant in the MLR model. This study has identified the most critical factors influencing fresh vegetables' freshness which need to be prioritized by fresh vegetables supply chain stakeholders to enhance efficiency in supply chain management, maximize fresh vegetables quality, extend shelf life, reduce postharvest losses and waste, and fulfill customer satisfaction. Future studies should explore further optimization techniques and support by technological innovations such as Internet of Things (IoT), data mining, and artificial intelligence (AI) for the sustainability.

Acknowledgements The authors are grateful to the Universiti Teknologi MARA (UiTM) for all the support given to this study.

Author Contribution All authors, Najihan Awang Ali, Adibah Shuib, Nor Aslily Sarkam, and Ismail Rakibe, have contributed to coming up with the concept and objectives of this paper. Adibah Shuib has been responsible in overseeing the overall organization of the paper and the write-up of the introduction, literature review, and the conclusion. The review processes were done after going through a series of specific discussions to determine more comprehensive research methods and anticipated results. Najihan Awang Ali identified the methodology that needed to be implemented, reviewed the keywords and processes involved to achieve the intended objectives. Najihan also carried out all data analysis involved. Meanwhile, Nor Aslily Sarkam checked the statistical analysis and interpretation of the results including the write-up on results and discussion. Ismail Rakibe worked on the review of articles concerning the themes and sub-themes considered particularly on those related to agriculture and fresh vegetables. All authors provided critical feedback on analysis and results which shaped this paper and contributed toward the completion of this paper.

Funding The author(s) received no specific funding for this work.

Conflict of Interest The authors have no conflicts of interest to declare.

References

1. Liu, X., Le Bourvellec, C., Yu, J., Zhao, L., Wang, K., Tao, Y., et al.: Trends and challenges on fruit and vegetable processing: Insights into sustainable, traceable, precise, healthy, intelligent, personalized, and local innovative food products. Trends Food Sci. Technol. **125**, 12–25 (2022)
2. Tort, Ö.Ö., Vayvay, Ö., Çobanoğlu, E.: A systematic review of sustainable fresh fruit and vegetable supply chains. Sustainability **14**, 1573 (2022)
3. Wallace, T.C., Bailey, R.L., Blumberg, J.B., Burton-Freeman, B., Chen, C. O., Crowe-White, K.M., et al.: Fruits, vegetables, and health: a comprehensive narrative, umbrella review of the science and recommendations for enhanced public policy to improve intake. Crit.Al Rev. Food Sci. Nutr. **60**(13), 2174–2211 (2020)
4. FAO (Food and Agriculture Organization of the United Nations): Fruit and Vegetables—Your Dietary Essentials. The International Year of Fruits and Vegetables, Background Paper. Food and Agriculture Organization, Rome (2021). Available online: https://www.fao.org/publications/card/en/c/CB2395EN. Last accessed 18 Oct 2024
5. Harris, J., de Steenhuijsen Piters, B., McMullin, S., Bajwa, B., de Jager, I., Brouwer, I.D.: Fruits and vegetables for healthy diets: Priorities for food system research and action. In: von Braun, J., Afsana, K., Fresco, L.O., Hassan, M.H.A. (eds.) Science and Innovations for Food Systems Transformation, pp. 1–2. Springer, Cham (2023)
6. Khalid, W., Arshad, M.S., Ranjha, M.M.A.N., Różańska, M.B., Irfan, S., Shafique, B., Rahim, M.A., Khalid, M.Z., Abdi, G., Kowalczewski, P.Ł: Functional constituents of plant-based foods boost immunity against acute and chronic disorders. Open Life Sci. **17**(1), 1075–1093 (2022)
7. Baidya, B.K., Sethy, P.: Importance of fruits and vegetables in boosting our immune system amid the COVID-19. Food Sci. Rep. **1**, 50–55 (2020)
8. Martínez-Moreno, M.M., Buitrago, E.M., Yniguez, R., Puig-Cabrera, M.: Circular economy and agriculture: Mapping circular practices, drivers, and barriers for traditional table-olive groves. Sustain. Prod. Consum. **46**, 430–441 (2024)
9. Kumar, A., Agrawal, S.: Challenges and opportunities for agri-fresh food supply chain management in India. Comput. Electron. Agric. **212**, 108161 (2023)
10. Perdana, T., Kusnandar, K., Perdana, H.H., Hermiatin, F.R.: Circular supply chain governance for sustainable fresh agricultural products: minimizing food loss and utilizing agricultural waste. Sustain. Prod. Consum. **41**, 391–403 (2023)
11. MyCC Homepage, Supply Chain of Vegetables Industry. Final Report: Market Review on Food Sector under Competition Act 2010. Malaysia Competition Commission (MyCC). https://www.mycc.gov.my/market-review. Last accessed 20 July 2024
12. Diaz-Ruiz, R., Costa-Font, M., López-i-Gelats, F., Gil, J.M.: Food waste prevention along the food supply chain: a multi-actor approach to identify effective solutions. Resour. Conserv. Recycl. **149**, 249–260 (2019)
13. FAO.: The State of Food and Agriculture 2019. Moving forward on food loss and waste reduction. Rome (2019)
14. UNEP.: Food Waste Index Report 2021. Nairobi. United Nations Environment Programme (UNEP). Available at https://www.unep.org/resources/report/unep-food-waste-index-report-2021. Last accessed 20 July 2024
15. Akter, T., Mahamud, S., Iqbal, M., Swarna, T.A., Nabi, N., Ahamed, M.S.: Local vegetable freshness classification using transfer learning approaches. In: 2022 32nd International Conference on Computer Theory and Applications (ICCTA), pp. 90–95. IEEE, Alexandria, Egypt (2022)
16. Cocetta, G.: Quality or freshness? How to evaluate fruits and vegetables during postharvest. Adv. Crop. Sci. Technol. **2**(3), 1–2 (2014)
17. Acedo, A.L., Rahman, M.A., Buntong, B., Gautam, D.M.: Vegetable postharvest training manual. World Vegetable Center, Taiwan **71**(2016)
18. Kader, A.A., Rolle, R.S.: The role of post-harvest management in assuring the quality and safety of horticultural produce. Food Agric. Org. **152** (2004)

19. Faye, D.: Safety and Quality of Fresh Fruit and Vegetables: A Training Manual for Trainers. United Nations, New York and Geneva, pp. 10–30, 71–72 (2007)
20. Esteso, A., Alemany, M.M., Ortiz, Á., Panetto, H.: Increasing the sustainability of a fresh vegetables supply chain through the optimization of funding programs: a multi-objective mathematical programming approach. J. Ind. Eng. Manag. (2022)
21. Zhong, X., Zhang, M., Tang, T., Adhikari, B., Ma, Y.: Advances in intelligent detection, monitoring and control for preserving the quality of fresh fruits and vegetables in the supply chain. Food Biosci. (2023)
22. Ran, W., Chen, Y.: Fresh produce supply chain coordination based on freshness preservation strategy. Sustainability **15**(10), 8184 (2023)
23. Kasampalis, D.S., Tsouvaltzis, P., Ntouros, K.D., Gertsis, A., Moshou, D., Siomos, A.S.: Rapid nondestructive postharvest potato freshness and cultivar discrimination assessment. Appl. Sci. (2021)
24. Shah, H.M.S., Singh, Z., Kaur, J., Hasan, M.U., Woodward, A., Afrifa-Yamoah, E.: Trends in maintaining postharvest freshness and quality of Rubus berries. Compr. Rev. Food Sci. Food Saf. **22**(6), 4600–4643 (2023)
25. Mubarok, S., Suminar, E., Abidat, A.H., Setyawati, C.A., Setiawan, E., Buswar, A.S.: Overview of melatonin's impact on postharvest physiology and quality of fruits. Horticulturae (2023)
26. Putnik, P., Kovačević, D.B.: Fresh-cut apples spoilage and predictive microbial growth under modified atmosphere packaging. In: Ravishankar Rai, V., Bai, J.A. (eds.) Food Safety and Protection, pp. 29–46. CRC Press, Florida (2017)
27. Péneau, S., Linke, A., Escher, F., Nuessli, J.: Freshness of fruits and vegetables: consumer language and perception. Br. Food J. **111**(3), 243–256 (2009)
28. Ramjan, M.D., Ansari, M.T.: Factors affecting the quality of fruits and vegetables. J. Med. Plants **6**(6), 16–18 (2018)
29. Ruslan, N.A., Man, N., Mohd Nawi, N., Ding, P.: Factors that influence the implementation of postharvest handling practices among fresh vegetable producers in selected states in Malaysia. J. Int. Food Agribus. Mark. **25**(sup1), 87–97 (2013)
30. Sabuddin, A.T.: Postharvest handling practices on vegetable crops at different stages of the distribution chain: A case study in Selangor, Malaysia. Master's Thesis, Wageningen University & Research (2017)
31. DOSM: Supply and Utilization Accounts Selected Agricultural Commodities Malaysia 2016–2020. Department of Statistics Malaysia, Putrajaya. https://v2.dosm.gov.my. Last accessed 22 July 2024 (2021)
32. Amanullah, A., Nawi, N., Kamarulzaman, N., Shamsudin, M.: Factors influencing post-harvest losses of apples among growers in Paktia, Afghanistan. Food Res. **4**(6), 2313–2321 (2020)
33. Aulakh, J., Regmi, A.: Post-Harvest Food Losses Estimation: Development of Consistent Methodology. Food and Agriculture Organization of the United Nations, Rome (2013)
34. Archdeacon, T.J.: Correlation and Regression Analysis: A Historian's Guide. University of Wisconsin Press, Wisconsin (1994)
35. Cohen, J.: Statistical Power Analysis for the Behavioral Sciences. Routledge, New York (2013)
36. Jaadi, Z.: Everything you need to know about interpreting correlations. Towards Data Science, Oct 15, 2019. https://towardsdatascience.com/everything-you-need-to-know-about-interpreting-correlations-2c485841c0b8. Last accessed 20 July 2024
37. Hair, J.F., Jr., Black, W.C., Babin, B.J., Anderson, R.E.: Multivariate Data Analysis, 8th edn. Cengage Learning, Boston (2018)
38. Nopiah, Z.M., Zulkipli, F., Basri, N.E.A., Kie, C.J.: A study of waste generation prediction using correlation and multiple regression analyses. Int. J. Ind. Eng. Technol. Oper. Manag. **1**(2), 54–57 (2023)
39. Cox, D.R., Snell, E.J.: A general definition of residuals. J. Roy. Stat. Soc. Ser. B (Methodol.) **30**(2), 248–265 (1968)
40. Shrestha, N.: Detecting multicollinearity in regression analysis. Am. J. Appl. Math. Stat. **8**(2), 39–42 (2020)

41. Uyanık, G.K., Güler, N.: A study on multiple linear regression analysis. Procedia Soc. Behav. Sci. **106**, 234–240 (2013)
42. Chambers, J. M., Cleveland, W. S., Kleiner, B., Tukey, P.A.: Graphical methods for data analysis. Wadsworth Brooks/Cole, Pacific Grove (1983)
43. Saha, M., Giri, R.N.: Freshness-keeping effort vs. sustainability: an efficient approach for perishable supply chain system. Int. J. Syst. Sci. Oper. Logist. **11**(1), 2297835 (2024)
44. Esteso, A., Alemany, M.M.E., Ortiz, A., Iannacone, R.: Integrating freshness and profitability in horticultural supply chain design. Cent.L Eur. J. Oper. Res. 1–44 (2024).
45. Zhang, Y.: A review of research on the reliability assessment of fresh product distribution networks considering product perishability constraint. Front. Bus. Econ. Manag. **15**(1), 373–377 (2024)
46. Pauzi, A.S.B.M., Sapry, H.R.M., Ahmad, A.R., Simarmata, J.: The consumer perspective on factors that influence the freshness of cold storage goods to their intention to purchase from grocery stores. In: Analyzing Education Sustainability and Innovation: Multidisciplinary Research Perspectives, pp. 67–71. Springer, Cham (2024)
47. Sureiman, O., Mangera, C.M.: F-test of overall significance in regression analysis simplified. J. Pract. Cardiovasc. Sci. **6**(2), 116–122 (2020)
48. Alulu, J., Makyao, M., Huyskens-Keil, S., Lenz, B., Muendo, K.M., Mganilwa, Z., Mithöfer, D.: Distribution, transportation, and coordination in African indigenous vegetables value chains: a scoping review. Front. Environ. Econ. **2**, 1113826 (2023)
49. Al-Dairi, M., Pathare, P.B., Al-Yahyai, R., Opara, U.L.: Mechanical damage of fresh produce in postharvest transportation: current status and future prospects. Trends Food Sci. Technol. **124**, 195–207 (2022)
50. Udovč, A., Bonča, S.: Sustainable transport in the fruit and vegetable supply chain. In: Sustainable Logistics, pp. 179–199. Productivity Press, Florida (2022)
51. Issa, I.M., Munishi, E.J., Mubarack, K.: Challenges facing transportation of urban fruits and vegetables supply chain continuum in Dares Salaam, Tanzania: Supply chain, transportation, challenges, fruits and vegetables, urban areas. Bus. Educ. J. **8**(1) (2022)
52. Iordăchescu, G., Ploscutanu, G., Pricop, E.M., Baston, O., Barna, O.: Postharvest losses in transportation and storage for fresh fruits and vegetables sector. J. Int. Sci. Publ. **7**, 244–251 (2019)
53. Guritno, A.D., Khuriyati, N.: An application of RAPAGrisk (Rapid Agricultural Supply Chain Risk Assessment) method on fresh vegetables for identifying and reducing damage during delivery to consumers. KnE Life Sciences **4**(2), 1–8 (2018)
54. Zhang, Y., Zhang, W., Wu, W., Farag, M.A., Wang, L., Xiao, S., Jiang, W.: Critical assessment of the delivery methods of chemical and natural postharvest preservatives for fruits and vegetables: a review. Crit. Rev. Food Sci. Nutr. 1–23 (2023)
55. Cheraghalipour, A., Roghanian, E.: A review of optimization researches in the field of agricultural supply chain. Mod. Concepts Dev. Agron. **5**(4), 556–560 (2020)
56. Moradinezhad, F., Ranjbar, A.: Advances in postharvest diseases management of fruits and vegetables: a review. Horticulturae **9**(10), 1099 (2023)
57. Leneveu-Jenvrin, C., Charles, F., Barba, F.J., Remize, F.: Role of biological control agents and physical treatments in maintaining the quality of fresh and minimally-processed fruit and vegetables. Crit. Rev. Food Sci. Nutr. **60**(17), 2837–2855 (2020)

Spatial and Association Analyses of Road Accident Impact Severity in Malaysia During COVID-19

Muhammad Naeim Mohd Aris, Muhammad Fadhirul Anuar Mohd Azami, Siti Rahimah Batcha, Shalini Nagaratnam, and Zi Chen Cheng

Abstract The study examines the relationship between the impact severity of road accidents and the spatial grouping of Malaysian states during the COVID-19 pandemic in 2020 and 2021, addressing the lack of analysis on these trends in Malaysia. The road accidents impact severity is divided into two subgroups, which are injury severity level (i.e., death, bad injury, and light injury) and accident severity category (i.e., fatal, serious, and minor accidents). Nine combinations of these severity levels and categories for each observed year are analyzed using linear regression analysis. The significance and correlation for each combination is observed and studied. Further analysis is performed using quadrant analysis, grouping states based on the mean-centered positions. Subsequently, chi-square analysis is applied to these quadrant-based groups to explore associations between the spatial distribution of states and the combinations of severity levels and categories. The study aims to explore the spatial trends of road accidents in Malaysia, highlighting how accident and injury severities are distributed across different states during the pandemic. The findings showed that only the accident severity categories directly linked to the injury severity level have a strong linear relationship, regardless of the year. Pulau Pinang demonstrated a worsening trend, shifting from Quadrant III

M. N. Mohd Aris (✉)
Universiti Kebangsaan Malaysia, Bangi, Selangor, Malaysia
e-mail: muhammadnaeim@ukm.edu.my

M. F. A. Mohd Azami · Z. C. Cheng
Sunway University, Bandar Sunway, Selangor, Malaysia
e-mail: fadhirulma@sunway.edu.my

Z. C. Cheng
e-mail: zichen151204@gmail.com

S. R. Batcha
Faculty of Computer and Mathematical Sciences, Universiti Teknologi MARA, Shah Alam, Selangor, Malaysia
e-mail: rahimahbatcha@uitm.edu.my

S. Nagaratnam
Monash University Malaysia, Bandar Sunway, Selangor, Malaysia
e-mail: shalini.nagaratnam@monash.edu

A. Shuib et al. (eds.), *Mathematical Modeling and Numerical Analysis*,
Springer Proceedings in Mathematics & Statistics 510,
https://doi.org/10.1007/978-981-96-9350-4_18

in 2020 to Quadrant I in 2021 across all combinations, while Selangor remained in Quadrant I for all cases throughout the pandemic. The impact severity for death and serious accident, as well as for bad injury and fatal accident, showed no significant relationships between the years 2020 and 2021.

Keywords Chi-square analysis · COVID-19 · Impact severity · Quadrant analysis · Regression analysis · Road accidents

1 Introduction

Road accidents in Malaysia remain one of the critical concerns when dealing with the country's road traffic safety. As one of Asia's developing countries, Malaysia suffers from high rates of road accidents. According to researchers in [1], approximately 30,000 accidents recorded annually, and approximately 1% of these cases involve fatal accidents. As mentioned by researchers in [2], human, vehicle, and environment are the main factors that contribute to a traffic accident. In 2020, transport accidents were reported as one of Malaysia's top five causes of death, with more than 3.5% of total deaths resulting from road traffic accidents [3]. Studies have recently forecasted that the traffic accidents will become the fifth leading cause of death globally in 2030 [4].

The novel coronavirus 2019 (COVID-19) pandemic, which rapidly spread to over 60 countries by early 2020, was declared by the World Health Organization (WHO) as a global pandemic in March 2020. Approximately, 118,000 confirmed cases and over 4000 deaths across 114 countries were reported in the beginning [5]. To mitigate the spread of the COVID-19 pandemic, most governments worldwide, including Malaysia, imposed various preventive measures, such as lockdowns, travel restrictions, and social distancing. Malaysia's government implemented the first lockdown, called Movement Control Order (MCO), in March 2020 in responses to the drastic increase of COVID-19 cases. It was then followed by the Conditional Movement Control Order (CMCO), Recovery Movement Control Order (RMCO), and various state-specific MCOs, under the National Recovery Plan (NRP) [6, 7]. These restrictions primarily include domestic and international travel bans, the closure of educational institutions, and the shutdown of all non-essential private and government premises [8]. In consequence, these restrictions have caused a significant change in travel behavior, which directly gave impacts on the road traffic volumes and safety performance, and substantially reduce the proportion of road traffic deaths. According to researchers in [9], the heavy traffic on the roads is likely to result in collisions involving a variety of vehicles.

Based on the literature, various studies have been conducted in different areas to explore the impact of imposing preventive measures during COVID-19 pandemic on traffic volume and road safety. Researchers in [10] expected to enhance the understanding of the association between traffic volume and rate of accidents at intersections in Adelaide, Australia. The findings showed a positive association between those two variables at lower traffic volumes. Researchers in [11] explored the impact of traffic volume on accidents involving various types of vehicles. The findings inferred that the number of passenger cars has a significant impact on no injury accidents. Besides, the study also concluded that there is a positive correlation between the volume of light non-passenger vehicles and the probability of the occurrence of serious accidents. Here, it is worth mentioning that the reduction in mobility has increased the severity level of the accidents [12–14].

Referring to the literature, these analyses do not take into account exploring the spatial trends of road accidents in Malaysia, by relying on the combination between the injury severity level and accident severity category during the COVID-19 pandemic using linear regression, quadrant, and chi-square analyses. The findings of this study are expected to inform the correlation, spatial distribution across Malaysian states, and association between each combination of impact severity of road accidents for the years 2020 and 2021. The significance of this study is to discover patterns in accident severity across different Malaysian states, revealing states with higher or lower cases across all the combinations. Additionally, this study also provides valuable knowledge for developing and implementing adaptable road safety measures by understanding how the pandemic impacted accident severity levels. The implementation of road safety measures, regulations and enforcement strategies in Malaysia during the pandemic is expected to hugely contribute to the severity levels and categories across different states. Travel restrictions during the pandemic generally impact the road usage, which directly could influence the frequency and severity of accidents.

2 Data

The data for this study, accessed in December 2023, are taken from the official website of Malaysian Institute of Road Safety Research (MIROS) for the road accidents occurring from 2020 to 2021 across Malaysian states. The website provides various attributes and details, allowing for an in-depth analysis of road safety trends. There are 14 states involved in this study, which include Selangor, Wilayah Persekutuan Kuala Lumpur, Perak, Melaka, Johor, Pulau Pinang, Kedah, Kelantan, Sabah, Sarawak, Terengganu, Perlis, Negeri Sembilan, and Pahang.

It should be noted that the impact severity of road accident is categorized as two groups, which are injury severity level, consisting of death, bad injury, and light injury, and accident severity category, consisting of fatal, serious, and minor accidents. The focus of this study is to research the spatial patterns of road accidents in Malaysia, highlighting the distribution of the combination of injury severity level and accident severity category across different states in Malaysia during the COVID-19 pandemic. In this context, the data are independent of number of COVID-19 cases in Malaysia.

3 Methodology

In the context of this study, three different statistical analyses are involved, which are linear regression, quadrant, and chi-square analyses. Each analysis is applied to assess the data from various perspectives. Thorough explanations of these analyses are provided in Sects. 3.1, 3.2, and 3.3, respectively.

3.1 Linear Regression Analysis

This study develops a linear regression for each combination of injury severity level-accident severity category to model the relationship between the independent and dependent variables. In this analysis, the accident severity category count is treated as the predictor, while the injury severity level count is treated as the response. The linear regression model is defined in Eq. (1).

$$\hat{y} = a + bx, \tag{1}$$

where $\hat{y}$ denotes the injury severity level count computed using the developed linear regression, x denotes the accident severity category count, a denotes the y-intercept, and b denotes the linear coefficient.

Additionally, Pearson coefficient of correlation, R, and coefficient of determination, R^2, are also computed in this study to determine the correlation and goodness of fit for every combination of road accidents impact severity for the years 2020 and 2021. The formula of R and R^2 are expressed in Eqs. (2) and (3), respectively.

$$R = \frac{\sum_{i=1}^{n}(x_i - \overline{x}_i)(y_i - \overline{y}_i)}{\sqrt{\sum_{i=1}^{n}(x_i - \overline{x}_i)^2 \sum_{i=1}^{n}(y_i - \overline{y}_i)^2}}, \tag{2}$$

$$R^2 = 1 - \frac{\sum_{i=1}^{n}(y_i - \hat{y}_i)^2}{\sum_{i=1}^{n}(y_i - \overline{y}_i)^2}, \tag{3}$$

where x denotes the accident severity count, $\overline{x}$ denotes the average of the accident severity count, y denotes the injury severity count for every combination, $\overline{y}$ denotes the average of the injury severity count, and n denotes the total number of Malaysian states.

3.2 Quadrant Analysis

Next, this study employs quadrant analysis to categorize each combination of injury severity level and accident severity category by segmenting the data into four distinct quadrants based on the mean values. This analysis helps in identifying patterns of impact severity of road accidents across Malaysian states during the COVID-19 period. The following is the steps involved in the quadrant analysis, as adopted by [15].

Step 1: Calculate the means

The mean values for each variable of the combination are computed using Eq. (4).

$$\mu = \frac{\sum_{i=1}^{n} w_i}{n}, \tag{4}$$

where μ denotes the mean value for each variable, w denotes the variable count, and $n = 14$.

Step 2: Determine the thresholds

To divide the data into quadrants, thresholds are determined based on the mean values (i.e., reference points), which is computed in Step 1. Each quadrant is defined as follows.

- Quadrant I: Both variables exceed the respective means.
- Quadrant II: The first variable (in y-axis) exceeds its mean, but the second variable (in x-axis) is below its mean.
- Quadrant III: Both variables are below each of the means.
- Quadrant IV: The first variable (in y-axis) is below its mean, but the second variable (in x-axis) exceeds its mean.

Step 3: Plot the data

Each combination of injury severity level and accident severity category is plotted with the respective threshold values.

Step 4: Analyze the patterns

The spatial trends of impact severity of road accidents across Malaysian states during the COVID-19 period within each quadrant are further studied.

3.3 Chi-Square Analysis

Further analysis using chi-square is conducted on the outcomes from the quadrant analysis to determine whether there is a significant spatial association between each combination of injury severity level and accident severity category across different states for the years 2020 and 2021. This analysis is based on the following hypotheses.

- Null Hypothesis, H_0: The spatial distribution of the combination of impact severity of road accidents is independent of the year.
- Alternative Hypothesis, H_1: The spatial distribution of the combination of impact severity of road accidents is not independent of the year.

A contingency table for each combination for the years 2020 and 2021 is constructed. Note that each cell in the contingency table for the year 2020 represents the observed frequency. The expected frequency (for the year 2021) is then computed using Eq. (5).

$$f_{\exp} = \frac{\text{Total of Row} \times \text{Total of Column}}{\text{Grand Total}}, \tag{5}$$

where $f_{\exp}$ denotes the expected frequency. Hence, the chi-square statistic is computed using Eq. (6).

$$\chi^2 = \sum \frac{\left(f_{\text{obs}} - f_{\exp}\right)^2}{f_{\exp}}, \tag{6}$$

where χ^2 denotes the chi-square statistic, and f_{obs} represents the observed frequency. It is important to note that there are eight classes, with four quadrants for each year. Next, the degree of freedom, df, for the chi-square analysis is evaluated using Eq. (7),

$$\text{df} = (\text{Number of Rows} - 1) \times (\text{Number of Columns} - 1). \tag{7}$$

The p-value for each class is then evaluated based on its respective degree of freedom. In the context of this study, 5% level of significance is exercised for the decision making. All the formula is referenced from [16].

4 Results and Discussion

This study used Microsoft Excel and SPSS software to process the road accident data. Note that linear regression analysis was conducted for all combinations. However, for demonstration purposes, only the linear regression models of road accidents impact severity that are highly correlated are demonstrated in this study. Figures 1 and 2 show the regression analysis outputs for the years 2020 and 2021, respectively.

Referring to Figs. 1a and 2a, fatal accident variable is statistically significant to the injury severity level of death with the respective p-values of lesser than 0.05, with the Pearson correlation coefficients of 0.9976 and 0.9980, respectively. The resulting p-values for severity category of serious accident for the years 2020 and

(a) Model	Coefficients	*p*-values
Constant	1.873	0.812
Fatal Accident	1.072	< 0.001
Dependent Variable: Death		

(b) Model	Coefficients	*p*-values
Constant	2.373	0.641
Serious Accident	1.037	< 0.001
Dependent Variable: Bad Injury		

(c) Model	Coefficients	*p*-values
Constant	7.138	0.076
Minor Accident	1.001	< 0.001
Dependent Variable: Light Injury		

Fig. 1 Regression analysis for highly correlated linear model for the year 2020; **a** death versus fatal accident, **b** bad injury versus serious accident, and **c** light injury versus minor accident

(a) Model	Coefficients	*p*-values
Constant	2.370	0.737
Fatal Accident	1.061	< 0.001
Dependent Variable: Death		

(b) Model	Coefficients	*p*-values
Constant	8.051	0.451
Serious Accident	1.139	< 0.001
Dependent Variable: Bad Injury		

(c) Model	Coefficients	*p*-values
Constant	58.612	0.081
Minor Accident	1.035	< 0.001
Dependent Variable: Light Injury		

Fig. 2 Regression analysis for highly correlated linear model for the year 2021; **a** death versus fatal accident, **b** bad injury versus serious accident, and **c** light injury versus minor accident

2021 also reveal a significant relationship with the severity level of bad injury, as shown in Figs. 1b and 2b. The Pearson correlation coefficients obtained from this combination are 0.9975, for the year 2020, and 0.9919, for the year 2021. For the combination between light injury and minor accident, shown in Figs. 1c and 2c, the predictor also significantly contributes to the linear regression model to explain the severity category of minor accident with the respective *p*-values of lesser than 0.05, with the Pearson correlation coefficients of 0.9999 and 0.9973, respectively. The computed coefficient of determination (R^2) for each pair for the years 2020 and 2021 are tabulated in Table 1.

Based on Table 1, it can be observed that each highly positive correlated linear regression model for both years, as shown in Figs. 1 and 2, produced a high R^2, indicating a good fit of the model in representing the road accidents impact severity. In other words, the high R^2 indicates that most of the changes in the injury severity data can be explained by the accident severity data, reflected in the high Pearson coefficient of correlation.

These findings show that a high correlation naturally occurs when the accident severity category is directly linked to the injury severity level, regardless of the year. Fatal, serious, and minor accidents are likely to cause deaths, bad injuries, and light injuries, respectively. From the regression analysis, this study is able to pre-determine the linearity trend of the highly positive correlated road accidents impact severity demonstrated in the quadrant analysis. The plots from the quadrant analysis are illustrated in Figs. 3, 4, and 5, for the years 2020 and 2021.

Referring to Fig. 3a, b, the death and fatal accident counts for Selangor, Johor, Perak, Kedah remained in Quadrant I for both years, while Pahang shifted from Quadrant I in 2020 to Quadrant III in 2021, indicating an improvement in death and

Table 1 Coefficient of determination for linear regression models of road accidents impact severity in Malaysia during COVID-19

Year	Injury severity level	Accident severity category	R^2
2020	Death	Fatal	0.9952
	Death	Serious	0.4588
	Death	Minor	0.4098
	Bad	Fatal	0.4170
	Bad	Serious	0.9951
	Bad	Minor	0.5700
	Light	Fatal	0.4194
	Light	Serious	0.6191
	Light	Minor	0.9998
2021	Death	Fatal	0.9959
	Death	Serious	0.5252
	Death	Minor	0.0998
	Bad	Fatal	0.5599
	Bad	Serious	0.9838
	Bad	Minor	0.0023
	Light	Fatal	0.1238
	Light	Serious	0.0048
	Light	Minor	0.9945

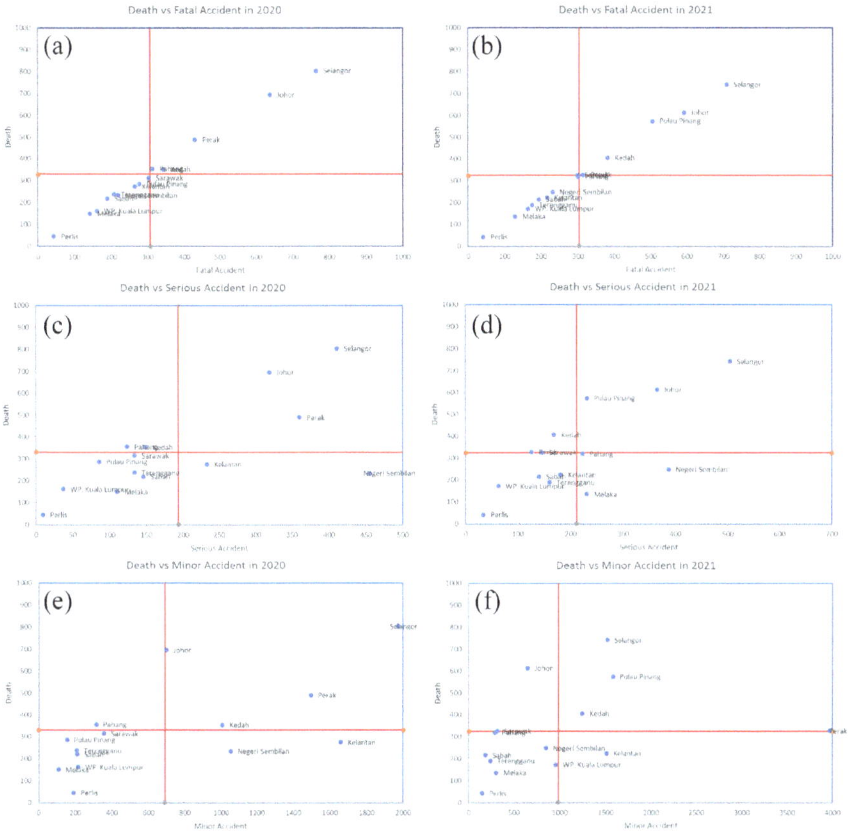

Fig. 3 Quadrant analysis for death and each accident severity category; **a** death versus fatal accident for 2020, **b** death versus fatal accident for 2021, **c** death versus serious accident for 2020, **d** death versus serious accident for 2021, **e** death versus minor accident for 2020, and **f** death versus minor accident for 2021

fatal accident counts in the second year of the COVID-19 pandemic (i.e., decrement in the number of cases of death and fatal accident). However, Pulau Pinang indicates an increment in the number of cases of death and fatal accident from the year 2020 to 2021, and Sarawak also shifted from Quadrant III in 2020 to Quadrant II in 2021. For Fig. 3c, d, the death and serious accident counts for Selangor and Johor remained in Quadrant I along the COVID-19 period. However, Perak shifted from Quadrant I in 2020 to Quadrant II in 2021, indicating a decrement in the number of serious accident cases throughout the COVID-19 period in that state. Again, Pulau Pinang demonstrated an increment trend in the number of cases of death and serious accident. The state shifted from Quadrant III in 2020 to Quadrant I in 2021. Next, for the death and minor accident counts, Selangor, Perak, and Kedah remained in Quadrant I for both years, and Wilayah Persekutuan Kuala Lumpur, Sabah, Terengganu, Melaka, and Perlis remained in Quadrant III for both years, as depicted in Fig. 3e, f. Pulau

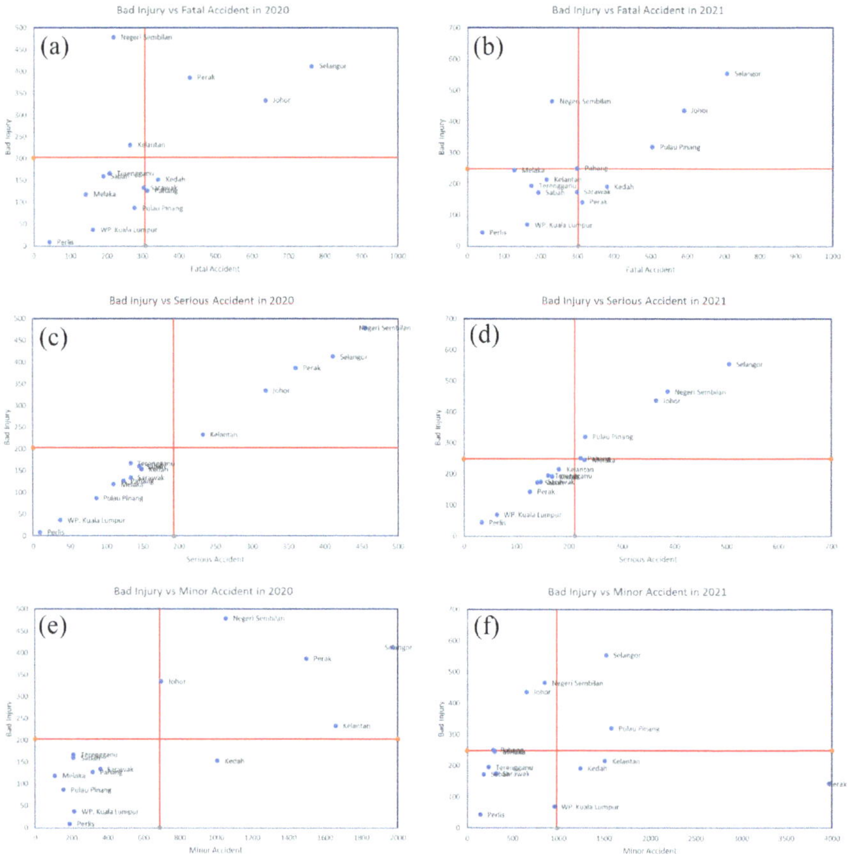

Fig. 4 Quadrant analysis for bad injury and each accident severity category; **a** bad injury versus fatal accident for 2020, **b** bad injury versus fatal accident for 2021, **c** bad injury versus serious accident for 2020, **d** bad injury versus serious accident for 2021, **e** bad injury versus minor accident for 2020, and **f** bad injury versus minor accident for 2021

Pinang showed the same trend for death and minor accident as the state shifted from Quadrant III in 2020 to Quadrant I in 2021.

Based on Fig. 4a, b, there were no changes in Selangor and Johor for both years, as both states remained in Quadrant I. Meanwhile, Perak shifted from Quadrant I, in the year 2020, to Quadrant IV, in the year 2021, indicating an improvement in the severity level of bad injury count (i.e., decrement in the number of cases of bad injury). There was no improvement in terms of the number of states in Quadrant III, which remained at seven for both years. Pulau Pinang demonstrated an increment trend in the number of cases of bad injury and fatal accident, as it shifted from Quadrant III in 2020 to Quadrant I in 2021. Next, for severity level of bad injury and severity category of serious accident, the number of states in Quadrant I remained at five for both years, with Selangor, Negeri Sembilan, and Johor staying in the

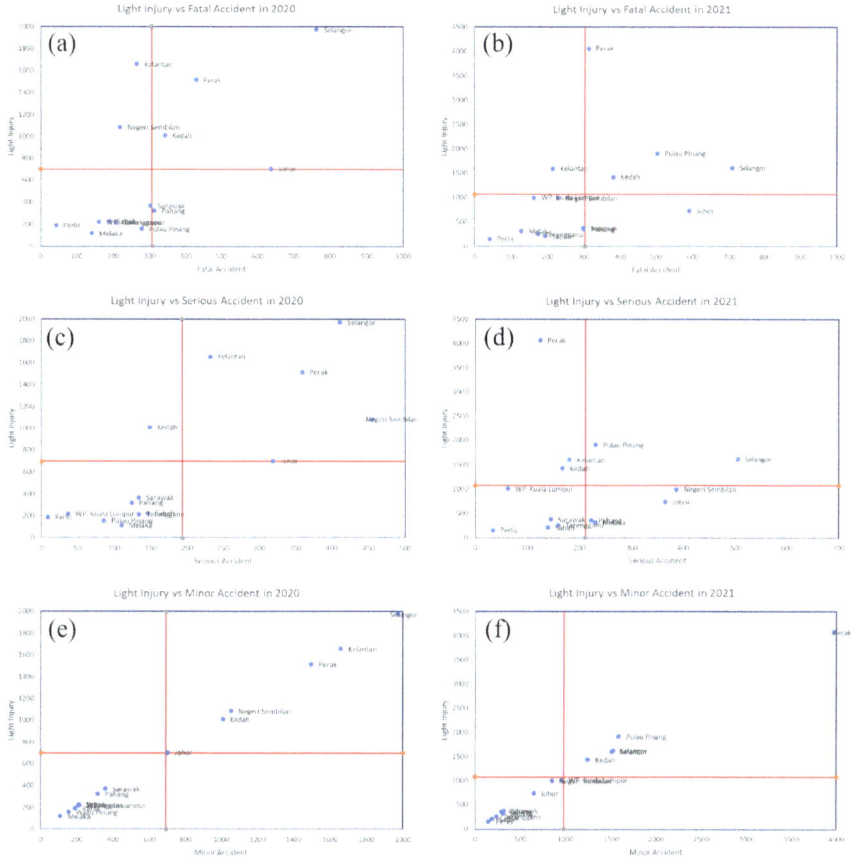

Fig. 5 Quadrant analysis for light injury and each accident severity category; **a** light injury versus fatal accident for 2020, **b** light injury versus fatal accident for 2021, **c** light injury versus serious accident for 2020, **d** light injury versus serious accident for 2021, **e** light injury versus minor accident for 2020, and **f** light injury versus minor accident for 2021

quadrant, as depicted in Fig. 4c, d. However, Melaka showcased a worsening trend in terms of serious accident cases as it shifted from Quadrant III, in 2020, to Quadrant IV, in 2021. Additionally, Perak and Kelantan showed an improvement in the bad injury and serious accident counts, as both states shifted from Quadrant I in 2020 to Quadrant III in 2021. Pulau Pinang and Pahang demonstrated a worsening trend as both states shifted from Quadrant III in 2020 to Quadrant I in 2021. Referring to Fig. 4e, f, there was a significant improvement in Quadrant I as the number of states in the quadrant reduced from five states, in 2020, to two states, in 2021, with Selangor staying in the quadrant for both years. Pulau Pinang, again, showcased the trend where the number of cases of bad injury and minor accident in the state increase from the year 2020 to 2021.

Referring to Fig. 5a, b, Selangor, Perak, and Kedah remained in Quadrant I for both years. Johor showed an improvement in terms of light injury counts in 2021. However, Pulau Pinang showed a significant shifting trend for light injury and fatal accident cases from Quadrant III in 2020 to Quadrant I in 2021. Meanwhile, Johor demonstrated an improvement in light injury counts, indicating a decrement in the number of the case throughout the COVID-19 period. For light injury and serious accident counts, there was a significant improvement in terms of the number of states remain in Quadrant I, reduced from five states, in 2020, to only two states, in 2021, with Selangor staying in the quadrant for both years. Additionally, Wilayah Persekutuan Kuala Lumpur, Sarawak, Sabah, Terengganu, and Perlis remained in Quadrant III for both years, and Pulau Pinang shifted from Quadrant III in 2020 to Quadrant I in 2021. Based on Fig. 5e, f, Negeri Sembilan and Johor showcased an improving trend from 2020 to 2021, as both states shifted from Quadrant I to Quadrant III. Selangor, Kedah, Kelantan, and Perak remained in Quadrant I for both years. However, Pulau Pinang shifted significantly from Quadrant III, in 2020, to Quadrant I, in 2021, indicating an increment in the number of light injury and minor accident cases during the COVID-19 period.

The findings were further studied by performing chi-square analysis for independence to test whether the distribution of road accidents across the impact severity significantly differs between 2020 and 2021. Table 2 shows the chi-square analysis for death and fatal accident.

It should be noted that the data comprises two dimensions, which are death and fatal accident. These impact severities were classified into high death–high fatal, high death–low fatal, low death–high fatal, and low death–low fatal, relying on the quadrant analysis. Table 2 presents the observed frequencies for each combination along with the chi-square values and standardized residuals. The critical value at 5% level of significance with 3 degrees of freedom is 7.815 (taken from the chi-square distribution table). Since the calculated chi-square statistic (24.39) is greater than the critical value (with p-value $= 2.07 \times 10^{-5} < 0.05$), the null hypothesis is rejected. This implies that there is a significant relationship between the road accident distributions

Table 2 Chi-square analysis for death and fatal accident

Category (impact severity)	Year	Contribution to χ^2	Standardized residual
High death–high fatal	2020	5.79	2.40
High death–low fatal	2020	3.21	− 1.79
Low death–high fatal	2020	3.21	− 1.79
Low death–low fatal	2020	1.78	1.33
High death–high fatal	2021	3.83	1.96
High death–low fatal	2021	2.12	− 1.46
Low death–high fatal	2021	2.86	− 1.69
Low death–low fatal	2021	1.59	1.26
		24.39	

across the impact severity of death and fatal accident in the years 2020 and 2021. The high death–high fatal for 2020 has the largest standardized residual, indicating that this combination significantly contributes to the trend.

The dimensions involved in Table 3 are death and serious accident. The impact severities were classified into high death–high serious, high death–low serious, low death–high serious, and low death–low serious. Referring to Table 3, the calculated chi-square statistic (2.30) is smaller than the critical value (with p-value = 0.5130), the null hypothesis fails to be rejected. This implies that there is no significant relationship between the road accident distributions across the impact severity of death and serious accident in the years 2020 and 2021. All the standardized residuals are below the threshold for significance.

Next, the dimensions involved in Table 4 are death and minor accident. The impact severities were classified into high death–high minor, high death–low minor, low death–high minor, and low death–low minor. Referring to Table 4, the calculated chi-square statistic (8.87) is greater than the critical value mentioned above (with p-value = 0.0311), the null hypothesis is rejected. This implies that there is a significant relationship between the road accident distributions in the years 2020 and 2021 across the impact severity of death and minor accident. However, all the standardized residuals are not exceeding the threshold value, indicating that the observed and expected values are not significantly different.

The data comprises dimensions of bad injury and fatal accident. The impact severities for Table 5 were classified into high bad–high fatal, high bad–low fatal, low bad–high fatal, and low bad–low fatal. The contingency table shows that the calculated chi-square statistic (4.00) is smaller than the critical value (with p-value = 0.2619), the null hypothesis fails to be rejected. This implies that there is no significant relationship between the road accident distributions in the years 2020 and 2021 across the impact severity of bad injury and fatal accident. All the standardized residuals are below the threshold for significance.

For this case, the impact severity of bad injury and serious accident was classified into high bad–high serious, high bad–low serious, low bad–high serious, and low

Table 3 Chi-square analysis for death and serious accident

Category (impact severity)	Year	Contribution to χ^2	Standardized residual
High death–high serious	2020	0.55	0.74
High death–low serious	2020	0.35	– 0.59
Low death–high serious	2020	0.67	– 0.82
Low death–low serious	2020	0.43	0.66
High death–high serious	2021	0.18	0.42
High death–low serious	2021	0.11	– 0.34
Low death–high serious	2021	0.01	– 0.08
Low death–low serious	2021	0.00	0.06
		2.30	

Table 4 Chi-square analysis for death and minor accident

Category (impact severity)	Year	Contribution to χ^2	Standardized residual
High death–high minor	2020	2.11	1.45
High death–low minor	2020	1.37	− 1.17
Low death–high minor	2020	0.67	− 0.82
Low death–low minor	2020	0.43	0.66
High death–high minor	2021	1.15	1.07
High death–low minor	2021	0.74	− 0.86
Low death–high minor	2021	1.46	− 1.21
Low death–low minor	2021	0.94	0.97
		8.87	

Table 5 Chi-square analysis for bad injury and fatal accident

Category (impact severity)	Year	Contribution to χ^2	Standardized residual
High bad–high fatal	2020	0.83	0.91
High bad–low fatal	2020	0.46	− 0.68
Low bad–high fatal	2020	0.46	− 0.68
Low bad–low fatal	2020	0.25	0.50
High bad–high fatal	2021	0.83	0.91
High bad–low fatal	2021	0.46	− 0.68
Low bad–high fatal	2021	0.46	− 0.68
Low bad–low fatal	2021	0.25	0.50
		4.00	

bad–low serious. Based on Table 6, the calculated chi-square statistic (24.27) is greater than the critical value (with p-value $= 2.19 \times 10^{-5} < 0.05$); hence, the null hypothesis is rejected. This implies that there is a significant relationship between the road accident distributions in the years 2020 and 2021 for the road accidents impact severity of bad injury and serious accident. Since the standardized residuals for high bad–high serious (2020) and high bad–high serious (2021) are more than 2 (i.e., threshold for significance), it implies that these pairs make the most significant contributions to the analysis.

Next, for Table 7, the impact severity of bad injury and minor accident was classified into high bad–high minor, high bad–low minor, low bad–high minor, and low bad–low minor. Referring to Table 7, the calculated chi-square statistic (10.86) is greater than the critical value mentioned above (with p-value $= 0.0125$), the null hypothesis is rejected. This implies that there is a significant relationship between the road accident distributions in the years 2020 and 2021 for the road accidents impact severity of bad injury and minor accident. The high bad–high minor for 2020

Table 6 Chi-square analysis for bad injury and serious accident

Category (impact severity)	Year	Contribution to χ^2	Standardized residual
High bad–high serious	2020	4.69	2.17
High bad–low serious	2020	3.03	– 1.74
Low bad–high serious	2020	3.54	– 1.88
Low bad–low serious	2020	2.29	1.51
High bad–high serious	2021	4.69	2.17
High bad–low serious	2021	3.03	– 1.74
Low bad–high serious	2021	1.82	– 1.35
Low bad–low serious	2021	1.18	1.08
		24.27	

Table 7 Chi-square analysis for bad injury and minor accident

Category (impact severity)	Year	Contribution to χ^2	Standardized residual
High bad–high minor	2020	4.69	2.17
High bad–low minor	2020	3.04	– 1.74
Low bad–high minor	2020	1.82	– 1.35
Low bad–low minor	2020	1.18	1.08
High bad–high minor	2021	0.00	0.03
High bad–low minor	2021	0.00	– 0.02
Low bad–high minor	2021	0.08	– 0.28
Low bad–low minor	2021	0.05	0.23
		10.86	

has the largest standardized residual, indicating that this pair significantly contributes to the analysis.

The combinations of impact severity of light injury and fatal accident for Table 8 are high light–high fatal, high light–low fatal, low light–high fatal, and low light–low fatal. The contingency table, as shown in Table 8, shows that the calculated chi-square statistic (11.03) is greater than the critical value (with p-value = 0.0116), the null hypothesis is rejected. This implies that there is a significant relationship between the road accident distributions in the years 2020 and 2021 for the impact severity of light injury and fatal accident. All the standardized residuals are below the threshold for significance.

For this case, the dimensions are light injury and serious accident. Here, the impact severities were classified into high light–high serious, high light–low serious, low light–high serious, and low light–low serious. Referring to Table 9, the calculated chi-square statistic (10.15) is greater than the critical value (with p-value = 0.0173); hence, the null hypothesis is rejected. This implies that there is a significant relationship between the road accident distributions in the years 2020 and 2021 for the

Table 8 Chi-square analysis for light injury and fatal accident

Category (impact severity)	Year	Contribution to χ^2	Standardized residual
High light–high fatal	2020	1.61	1.27
High light–low fatal	2020	0.89	− 0.95
Low light–high fatal	2020	1.21	− 1.10
Low light–low fatal	2020	0.67	0.82
High light–high fatal	2021	2.74	1.66
High light–low fatal	2021	1.53	− 1.24
Low light–high fatal	2021	1.53	− 1.24
Low light–low fatal	2021	0.85	0.92
		11.03	

Table 9 Chi-square analysis for light injury and serious accident

Category (impact severity)	Year	Contribution to χ^2	Standardized residual
High light–high serious	2020	2.96	1.72
High light–low serious	2020	1.92	− 1.38
Low light–high serious	2020	3.14	− 1.77
Low light–low serious	2020	2.03	1.43
High light–high serious	2021	0.00	0.03
High light–low serious	2021	0.00	− 0.02
Low light–high serious	2021	0.06	0.25
Low light–low serious	2021	0.04	− 0.20
		10.15	

impact severity of light injury and serious accident. Since all the standardized residuals are not exceeding the threshold value, it shows that the observed and expected values are not significantly different.

Lastly, referring to Table 10, the dimensions involved are light injury and minor accident. The combinations of the impact severity are high light–high minor, high light–low minor, low light–high minor, and low light–low minor. Based on Table 10, the calculated chi-square statistic (28.00) is significantly greater than the critical value mentioned above (with p-value $= 3.63 \times 10^{-6} < 0.05$); hence, the null hypothesis is rejected. This proves that there is a significant relationship between the road accident distributions across the impact severity of light injury and minor accident in the years 2020 and 2021. The high light–high minor for 2020 has the largest standardized residual, indicating that this pair significantly contributes to the analysis. It should be noted that the main focus of this study is on data from the two years during the COVID-19 pandemic, the chi-square analysis only provides valuable preliminary insights.

Table 10 Chi-square analysis for light injury and minor accident

Category (impact severity)	Year	Contribution to χ^2	Standardized residual
High light–high minor	2020	5.63	2.37
High light–low minor	2020	3.64	− 1.91
Low light–high minor	2020	3.14	− 1.77
Low light–low minor	2020	2.03	1.43
High light–high minor	2021	4.69	2.17
High light–low minor	2021	3.04	− 1.74
Low light–high minor	2021	3.54	− 1.88
Low light–low minor	2021	2.29	1.51
		28.00	

5 Conclusion

The main goal of this study is to explore the associations of the spatial distribution of 14 Malaysian states across the severity levels and categories of road accidents during the COVID-19 pandemic. The linear regression analysis showed that six models were well-fitted for the years 2020 and 2021 with high coefficient of determination, where the combinations are death–fatal accident, bad injury–serious accident, and light injury–minor accident, for each year. It can be inferred that the accident severity category across all combinations is likely to cause the particular injury level, leading to a high positive correlation between the variables.

Then, quadrant analysis was performed for each combination for both years to investigate the spatial distribution across to the impact severity of road accidents. Generally, the results from the quadrant analysis imply inconsistent patterns across different states in Malaysia, including both improving and worsening trends, in the impact severity of road accidents throughout the COVID-19 pandemic. Several states showcased persistent pattern in certain quadrant, and various factors may contribute to this, such as demographic, traffic volume, socioeconomic factor, and pandemic impact. States with higher populations and traffic volumes (e.g., urban area) experience consistently higher impact severity of road accidents. Besides, economic development in several states may also influence the frequency and severity of road accidents. During the COVID-19 pandemic, certain states have consistent progression in the number of infections due to the lockdown implementation. However, some states experience minimal progression, resulting in a persistent pattern of impact severity of road accidents. For example, Selangor remained in Quadrant I for all cases throughout the COVID-19 pandemic, while Pulau Pinang continued to demonstrate a worsening trend, shifting from Quadrant III in the year 2020 to Quadrant I in the year 2021 across all combinations. The regulations implemented by the government of Malaysia during the COVID-19 pandemic did not significantly affect the road accident impact severity in these states.

Referring to the chi-square analysis findings, seven combinations were successfully proven to have a significant relationship between the impact severity of road accident distributions in the years 2020 and 2021, which are combinations of death and fatal accident, death and minor accident, bad injury and serious accident, bad injury and minor accident, light injury and fatal accident, light injury and serious accident, and light injury and minor accident. Researchers in [17] also observed an improvement in severe and fatal accident cases during the lockdown period, implying a positive impact of the restrictions on the severity of road accidents. The impact severity of death and serious accident, as well as bad injury and fatal accident, lacks significant relationships between the years 2020 and 2021. Restriction in mobility during the COVID-19 pandemic has led to stable traffic volumes and road accident rates, resulting in different distribution across both years; however, if the driver behavior or bad road conditions remained consistent, this could lead to similar impact severities throughout the pandemic. A clustering algorithm is one potential method for determining spatial distribution in the future work. An alternative statistical method or larger datasets should be considered as well in the future work to enhance the robustness of the analysis.

References

1. Prasetijo, J., Zhang, G., Mohd Jawi, Z., Mahyeddin, M.E., Zainal, Z.F., Isradi, M., Muthukrishnan, N.: Crash model based on integrated design consistency with low traffic volumes (due to health disaster (COVID-19)/movement control order). Innov. Infrastruct. Solut. **6**, 22 (2021)
2. Shankar, V., Venkatesan, M.: Pattern of injuries due to fatal road traffic accidents from rural Chennai. Onl. Int. J. Med. Soc. Sci. **1**, 30–36 (2015)
3. Supramaniam, P., Junus, S., Hashim, L., Chiew, S.C., Devesahayam, P.R.: Lost years, mortality burden: the impact of COVID-19 pandemic on premature death due to road traffic accidents in a northern state in Malaysia. BMC Public Health **24**, 1520 (2024)
4. Sameen, M.I., Pradhan, B.: Assessment of the effects of expressway geometric design features on the frequency of accident crash rates using high-resolution laser scanning data and GIS. Geomatics Nat. Hazards Risk 1–15 (2016)
5. World Health Organization: WHO Director-General's Opening Remarks at the Media Briefing on COVID-19, 11 Mar 2020. World Health Organization, Geneva (2020)
6. Gill, B.S., Jayaraj, V.J., Singh, S., Mohd Ghazali, S., Cheong, Y.L., Md Iderus, N.H., Sundram, B.M., Aris, T.B., Mohd Ibrahim, H., Hong, B.H., Labadin, J.: Modelling the effectiveness of epidemic control measures in preventing the transmission of COVID-19 in Malaysia. Int. J. Environ. Res. Public Health **17**, 5509 (2020)
7. Musa, K.I., Arifin, W.N., Mohd, M.H., Jamiluddin, M.S., Ahmad, N.A., Chen, X.W., Hanis, T.M., Bulgiba, A.: Measuring time-varying effective reproduction numbers for COVID-19 and their relationship with movement control order in Malaysia. Int. J. Environ. Res. Public Health **18**, 3273 (2021)
8. Godwin, P., Pang, R., Ong, A., Ong, N.: COVID-19: governance: impact of the movement control order on your business (Malaysia). In: Legal Briefings. Herbert Smith Freehills (2020)
9. Abdullah, L., Zamri, N.: Road accident models with two threshold levels of fuzzy linear regression. J. Emerg. Trends Comput. Inf. Sci. **3**, 225 (2012)
10. Retallack, A.E., Ostendorf, B.: Relationship between traffic volume and accident frequency at intersections. Int. J. Environ. Res. Public Health **17**, 1393 (2020)

11. Ayati, E., Abbasi, E.: Investigation on the role of traffic volume in crashes on urban highways. J. Saf. Res. **42**, 209 (2011)
12. Casado-Sanz, N., Guirao, B., Attard, M.: Analysis of the risk factors affecting the severity of traffic accidents on Spanish crosstown roads: the driver's perspective. Sustainability **12**, 2237 (2020)
13. Golob, T.F., Recker, W.W.: Relationships among urban freeway accidents, traffic flow, weather, and lighting conditions. J. Transp. Eng. **129**, 342 (2003)
14. Martin, J.L.: Relationship between crash rate and hourly traffic flow on interurban motorways. Accid. Anal. Prev. **34**, 619 (2002)
15. Fatoni, A., Adi, K., Widodo, A.P.: PIECES framework and importance performance analysis method to evaluate the implementation of information systems. E3S Web Conf. **202**, 15007 (2020)
16. Fadden, B.: Using the chi-square test to analyze voter behavior. In: Mathematics Senior Showcase 2020, pp. 1–10. Boise State University, ScholarWorks (2020)
17. Yang, E.J., Mohamed, A.A., Hoe, G.B., Lim, Y.K., Jaya, R.P., Ming, N.C., Ing, N.L.S., Mashros, N.: Effect of COVID-19 lockdown on mobility and traffic accidents in Malaysia. AIP Conf. Proc. **2688**, 020006 (2023)

The Problem of Recovering Convex Surfaces in a Semi-hyperbolic Space

Abdullaaziz Artykbaev and Gulnoza Kholmurodova

Abstract In three-dimensional Euclidean space, Alexandrov solved the problem of recovering a convex surface from a given extrinsic curvature of the vertices in the class of convex polyhedra and generalized it for the class of convex surfaces. The purpose of this article is to solve an analogue problem in semi-hyperbolic space. First we interpret the semi-hyperbolic space inside the sphere of isotropic space. Let us prove that convex surfaces of semi-hyperbolic space are represented by convex surfaces contained within the sphere of isotropic space. The sphere of isotropic space is a cylinder with rectilinear generatrices. Moreover, bounded convex surfaces in semi-hyperbolic space are expressed by convex surfaces strictly containing inside a sphere of isotropic space, and infinite surfaces using surfaces that have common points with a cylinder, that is, a sphere of isotropic space. For convex surfaces in semi-hyperbolic space, the extrinsic curvature is determined as a positive definite, additive function of the Borel set. A theorem for the existence and uniqueness of a surface with a given function of extrinsic curvature is proven which is equivalent to the existence and uniqueness of a solution to the Monge–Ampere equation.

Keywords Convex set · Convex surface · Cylindrical mapping · Geodesic mapping · Semi-hyperbolic space · Tangent space

1 Introduction

The study of the geometry of semi-hyperbolic space is difficult in that it is the geometry of the sphere of a four-dimensional semi-pseudo-Euclidean space. To present and identify the properties of surfaces in this option is difficult. Therefore, we propose an interpretation of a semi-hyperbolic space inside a unit sphere of an isotropic space. An isotropic space is an affine space where it is easy to represent a convex surface

A. Artykbaev · G. Kholmurodova (✉)
Tashkent State Transport University, Tashkent, Uzbekistan
e-mail: xolmurodova.gulnoza@tstu.uz

A. Artykbaev
e-mail: Artykbaev.abdullaaziz@tstu.uz

A. Shuib et al. (eds.), *Mathematical Modeling and Numerical Analysis*,
Springer Proceedings in Mathematics & Statistics 510,
https://doi.org/10.1007/978-981-96-9350-4_19

and its elements. Using the interpretation of semi-hyperbolic space, the extrinsic curvature of a convex surface is determined. The existence of a convex surface in a semi-hyperbolic space with a given extrinsic curvature is used to prove the existence of a solution to the Monge–Ampere equation in a non-simply connected domain.

2 Definition and Interpretation of Space ${}^{10}S_3^2$

The definition and basic properties of semi-Euclidean space are represented in the monograph of Rosenfeld [15]. Moreover, semi-Euclidean and semi-pseudo-Euclidean spaces are also studied by Artykbaev, Aydin, Yoon, Ismoilov, Sultanov, and other researchers [5, 7, 12, 13, 16–18]. The space ${}^{10}R_4^3$ is a four-dimensional affine space, and the scalar product of two vectors $X\{x_0, x_1, x_2, x_3\}$ and $Y\{y_0, y_1, y_2, y_3\}$ is defined by the following:

$$(X, Y) = \begin{cases} (X, Y)_1 = -x_0y_0 + x_1y_1 + x_2y_2 & (X, Y)_1 \neq 0 \\ (X, Y)_2 = x_3y_3 & (X, Y)_1 = 0. \end{cases} \tag{1}$$

The norm of a vector $|\bar{X}| = \sqrt{(\bar{X}, \bar{X})}$ and the distance between points is defined as the norm with ends at these points [10, 11]. If we define a sphere of imaginary radius $S \subset^{10} R_4^3$ as the locus of points equidistant from the origin of coordinates by imaginary unit, then its equation in affine coordinates has the form [9]:

$$-x_0^2 + x_1^2 + x_2^2 = -1. \tag{2}$$

Definition 1 A semi-hyperbolic space ${}^{10}S_3^2$ is a set isometric to the set of diametrically opposite points of a sphere of imaginary radius, the space ${}^{10}R_4^3$ [6]. Hence,

$${}^{10}S_3^2 = \{(x_0, x_1, x_2, x_3) \in^{10} R_4^3 : -x_0^2 + x_1^2 + x_2^2 = -1\}. \tag{3}$$

The distance between points in ${}^{10}S_3^2$ is defined as the angle between the corresponding vectors and is calculated by the following formula:

$$\cosh\delta = \frac{-x_0y_0 + x_1y_1 + x_2y_2}{\sqrt{-x_0^2 + x_1^2 + x_2^2}\sqrt{-y_0^2 + y_1^2 + y_2^2}} \tag{4}$$

if $\delta = 0$, then $d = |y_3 - x_3|$.

Plane $x_0 = 1$ of space ${}^{10}R_4^3$ is a tangent plane to a sphere of imaginary radius. The geometry of the plane $x_0 = 1$ is the geometry of the isotropic space R_3^2.

Applying the mapping

$$TX = \frac{X + e_0(e_0, X)}{(e_0, X)} \tag{5}$$

we interpret the space ${}^{10}S_3^2$ onto the tangent plane, which is a three-dimensional isotropic space R_3^2. Vector e_0—is the imaginary unit vector of basis vectors $\{e_0, e_1, e_2, e_3\}$ of space ${}^{10}R_4^3$.

Lemma 1 *The mapping TX is geodesic [4].*

Proof It is easy to notice that the map TX is the central projection of the sphere S—onto the tangent plane $x_0 = 1$. With this mapping, a point $\{x_0, x_1, x_2, x_3\} \in^{10} S_3^2$ corresponds to a point on the plane $x_0 = 1$ with coordinates $\{\frac{x_1}{x_0}, \frac{x_2}{x_0}, \frac{x_3}{x_0}\} \in R_3^2$. Since, $x_0 \neq 0$ is the one-valued mapping. With this mapping, straight lines correspond to straight lines and planes correspond to planes. This follows from the fact that straight lines and planes of space are defined as the intersection of a 2-dimensional and 3-dimensional planes of space with a sphere of imaginary radius. However, these planes also intersect with the plane $x_0 = 1$, by a straight line and a plane, respectively. The lemma is proved.

Lemma 2 *By the mapping TX, points of space ${}^{10}S_3^2$ correspond to points inside the sphere of space R_3^2.*

Proof Indeed, by the mapping TX the point $\{x_0, x_1, x_2, x_3\} \in^{10} S_3^2$ corresponds to point of R_3^2 with coordinates $\{\frac{x_1}{x_0}, \frac{x_2}{x_0}, \frac{x_3}{x_0}\}$. Considering that $-x_0^2 + x_1^2 + x_2^2 = -1$ we obtain

$$\frac{x_1^2}{x_0^2} + \frac{x_2^2}{x_0^2} = 1 - \frac{1}{x_0^2} < 1. \tag{6}$$

Consequently, points with coordinates $\{\frac{x_1}{x_0}, \frac{x_2}{x_0}, \frac{x_3}{x_0}\}$ correspond to the intrinsic point of the sphere of isotropic space R_3^2—which has the equation

$$X^2 + Y^2 = 1. \tag{7}$$

This means that the points of space ${}^{10}S_3^2$ are interpreted to the intrinsic points of the sphere in R_3^2, the straight lines correspond to the chords of this sphere and the planes are expressed by the part of the plane of space R_3^2 intersecting with the sphere of this space. The lemma is proved.

3 Results

It is easy to prove that at the intersection of the plane $z = 0$ of the isotropic space R_3^2 with the sphere, the Lobachevsky plane is realized. The interpretation of a semi-hyperbolic space ${}^{10}S_3^2$ inside a sphere of unit radius of space R_3^2 is necessary for the geometric representation of the convex surfaces of this space. Because, the following theorem is valid.

Theorem 1 *Convex surfaces of a semi-hyperbolic space* ${}^{10}S_3^2$ *are represented by convex surfaces inside a sphere of unit radius of an isotropic space* R_3^2*. Moreover, finite convex sets are convex sets strictly contained within the sphere, and infinite convex sets have common points with the sphere of space* R_3^2*.*

The proof of the theorem follows from Lemma 1 and Lemma 2 and the condition for the convexity of an affine space.

3.1 Recovering of a Convex Surface by a Given Extrinsic Curvature

Since convex surfaces in the space ${}^{10}S_3^2$ are interpreted as convex surfaces inside the sphere of the isotropic space R_3^2, to determine the extrinsic curvature in ${}^{10}S_3^2$ we use the properties of a convex surface of the isotropic space R_3^2.

Let the convex surface F of the space ${}^{10}S_3^2$ be given in the interpretation of this space inside the sphere by the equation $z = f(x, y)$ $(x, y) \in D$ $\{(x, y, 0) \in R_3^2, x^2 + y^2 < 1\}$. The space R_3^2 is defined in an affine space, therefore the support planes of the convex surfaces of the space R_3^2 are also the support planes for this surface considered in ${}^{10}S_3^2$. Therefore, to determine the extrinsic curvature of a convex surface, we use the cylindrical mapping of the convex surface of the isotropic space R_3^2. Each support plane $z - z_0 = f'_x(x - x_0) + f'_y(y - y_0)$ of a point $X(x_0, y_0)$ of a surface F, let us match the point with coordinates

$$\left\{\frac{f'_x}{\sqrt{f_x'^2 + f_y'^2}}, \frac{f'_y}{\sqrt{f_x'^2 + f_y'^2}}, \frac{1}{\sqrt{f_x'^2 + f_y'^2}}\right\}. \tag{8}$$

This point is a cylindrical image of the support plane on the sphere of isotropic space R_3^2. Using this construction, we determine the extrinsic curvature of a convex surface transferred to the set M in the domain D. When the set is $M \subset D$, then the extrinsic curvature is calculated by the formula:

$$\omega_F(M) = \iint\limits_M \frac{f''_{xx} f''_{yy} - f''^{\,2}_{xy}}{(f_x'^2 + f_y'^2)^{\frac{3}{2}}} \mathrm{d}x\mathrm{d}y. \tag{9}$$

Thus, the defined extrinsic curvature will be a positive definite function of Borel sets and is a monotonic function if the surface has no singular support planes [3]. The singular planes in ${}^{10}S_3^2$ when interpreted inside the sphere of space are the planes $z = \text{const}$. Because the cylindrical mapping corresponds to points at infinity on the sphere of space R_3^2. Obvious for singular planes $f'_x = f'_y = 0$. For this reason, in ${}^{10}S_3^2$—we will consider only surfaces that do not have singular support planes.

Lemma 3 *In* ${}^{10}S_3^2$ *there exist convex surfaces that do not have singular support planes.*

Proof According to Alexandrov, by a convex surface we mean a simply connected domain on the boundary of a convex body [1, 2]. The space we are considering is ${}^{10}S_3^2$—interpreted inside the sphere of space R_3^2. Consequently, convex surfaces inside the sphere will be convex surfaces of space. When we consider a bounded closed convex surface inside a sphere of space, it has two singular support planes. Excluding points with singular support planes, we obtain a set of surfaces that do not have singular support planes. Lemma is proved.

Let closed convex curves L_2 and L_1 be given on a singular plane $X\{x, y\}$ inside a sphere of isotropic space, and L_1 lies strictly inside L_2. Let us denote by G the domain enclosed between L_1 and L_2. Let L be a spatial closed curve, uniquely projected onto L_2—given in the space $z > 0$ by the equation $g(t), t \in L_2$. We denote the set of convex surfaces projecting one-to-one onto the domain G by m.

Theorem 2 *Dirichlet problem for the Monge–Ampere equation*

$$z''_{xx}z''_{yy} - z''^2_{xy} = \varphi(x, y, z, z_x, z_y) \tag{10}$$

in a non-simply connected domain $G \subset X\{x, y\}$ *at the boundary conditions* $z|_{t\in L_2} = f(t)$, $z|_{t\in L_1} = 0$ *and at the conditions that the curve* L_2 *has the specific curvature bounded from below by a positive number and if on the set* $G \times R \times R^2$ *the function* $\varphi(x, y, z, z_x, z_y)$ *is continuous and on the domain* $G \subset X\{x, y\}$ *satisfies the inequality*

$$0 \leq \varphi(x, y, z, z_x, z_y) \leq \beta(x, y)(z'^2_x + z'^2_y)^{\frac{3}{2}} \tag{11}$$

that is, the function $\beta(x, y) \geq 0$ *and summable in* G.

To prove the existence of a generalized solution, Pogorelov's extremal technique is used, which allows one to substantiate the existence of a convex polyhedron with a given value of extrinsic curvature [14]. This technique is based on the compactness of the set of convex polyhedra with bounded extrinsic curvature and the property of monotonicity of the extrinsic curvature of the polyhedron [3]. The connection of this problem with the Monge–Ampere equation was studied and the proof of Theorem 2 is based on the method of Bakelman [8].

4 Conclusion

The interpretation of a semi-hyperbolic space inside a unit sphere of an isotropic space made it possible to determine the extrinsic curvature of convex surfaces in this space. The resulting geometric construction of convex surfaces of semi-Euclidean space without singular support planes provided the basis for solving the problem of the existence of a solution to the Monge–Ampere equation in a non-simply connected domain.

Acknowledgements We express our sincere thanks to the Department of Higher Mathematics of the Tashkent State Transport University which has provided us their assistance.

Disclosure of Interests The authors have no competing interests to declare that are relevant to the content of this article.

References

1. Alexandrov, A.D.: Convex polyhedra. In: Springer Monographs in Mathematics (2005)
2. Alexandrov, A.D.: Intrinsic geometry of convex surfaces. In: Classics of Soviet Mathematics (2006)
3. Artykbaev, A., Ibodullaeva, N.: Generalized extrinsic curvature of a surface. AIP Conf. Proc. 1–5 (2024). AIP Publishing
4. Artykbaev, A., Mamadaliev, B.M.: Geometry of two-dimensional surfaces in a five-dimensional pseudo-Euclidian space of index two. AIP Conf. Proc. 1–4 (2023). AIP Publishing
5. Artykbaev, A., Mamadaliyev, B.M.: Geometry of two-dimensional surfaces in space ${}^2R_5$. Lobachevskii J. Math. **44**(4), 1243–1247 (2023)
6. Artykbaev, A., Saitova, S.S.: Interpretation of geometry on manifolds as a geometry in a space with projective metric. J. Math. Sci. **265**(1), 1–10 (2022)
7. Aydin, M.E.: A generalization of translation surfaces with constant curvature in the isotropic space. J. Geom. **107**(3), 603–615 (2016)
8. Bakelman, I.Y.: Convex Analysis and Nonlinear Geometric Elliptic Equations. Springer, Berlin (1994)
9. Chen, B.Y.: Pseudo-Riemannian Geometry, δ-Invariants and Applications. World Scientific Book (2011)
10. Ismoilov, S., Artykbaev, A.: Special mean and total curvature of a dual surface in isotropic spaces. Int. Electron. J. Geom. **15**(1), 1–10 (2022)
11. Ismoilov, S., Kholmurodova, G.N.: Equality of surfaces in Euclidian and semi-Euclidian spaces according to geometric characteristics, Monge–Ampere equation. Bull. Inst. Math. **6**(4), 39–46 (2023)
12. Kelleci, A.: Warped translation surfaces of finite type in simply isotropic 3-spaces. Fundam. J. Math. Appl. **3**(1), 137–143 (2020)
13. Medjahdi, B., Zoubir, H.: Affine factorable surfaces in the three-dimensional simply isotropic space. Int. J. Appl. Math. **51**(2), 2–14 (2021)
14. Pogorelov, A.V.: External Geometry of Convex Surfaces. Science, Moscow (1991)
15. Rosenfeld, B.A.: Non-Euclidian Spaces. Nauka, Moscow (1969)
16. Sipus, Z.M., Divjak, B.: Some special surfaces in the pseudo-Galilean space. Acta Math. Hung. **118**(3), 209–226 (2008)
17. Sultanov, B., Ismoilov, S.: Invariant geometric characteristics under the dual mapping of an isotropic space. Asia Pac. J. Math. **10**(20), 1–12 (2023)
18. Yoon, D.W.: Surfaces of revolution in the three dimensional pseudo-Galilean space. Glas. Mat. **48**(68), 415–428 (2013)

New Generalized Separation Axioms Structures in Fuzzy Bitopological Spaces

Ahlam Ahmed Alharbi and Adem Kilicman

Abstract This work aims to enhance several mathematical concepts within fuzzy bitopological spaces by analyzing various fuzzy separation axioms, such as T_h and T_{hw}, where $h = 0, 1, 2, 2_{\frac{1}{2}}$. The study employs different types of fuzzy generalized closed sets of (X, δ_i, δ_j), including fuzzy $(i, j) - g\alpha$-closed sets, fuzzy $(i, j) - gp$-closed sets, fuzzy $(i, j) - gs$-closed sets, and fuzzy $(i, j) - g\beta$-closed sets, with $i, j = \{1, 2\}$ and $i \neq j$. After that, we give some important theorems that explain how they relate to each other and what their main characteristics are. These are followed by some strong counterexamples that show the inverse connection is not valid. This study's conclusions are novel to the domain of fuzzy bitopological spaces. Furthermore, this work will serve as a valuable reference by introducing numerous fundamental concepts that remain unexplored.

Keywords Fuzzy bitopological spaces · Fuzzy generalized closed sets · Fuzzy generalized separation axioms

1 Introduction

This research begins by applying the concept of generalized closed sets in fuzzy bitopological spaces to the separation axioms. First, we will explain the sequence of studying this topic, which stems from fuzzy topological spaces, the main source of which was fuzzy topological spaces, introduced by Zadeh in 1965 [15]. Then,

A. A. Alharbi
Department of Mathematics, Faculty of Science, Taibah University, Madinah, Kingdom of Saudi Arabia
e-mail: aasehli@taibahu.edu.sa

Department of Mathematics, Faculty of Science, Universiti Putra Malaysia UPM, Serdang, Selangor, Malaysia

A. Kilicman (✉)
Faculty of Computer and Mathematical Sciences, Universiti Teknologi MARA, 40450 Shah Alam, Selangor, Malaysia
e-mail: kilicman@uitm.edu.my

A. Shuib et al. (eds.), *Mathematical Modeling and Numerical Analysis*,
Springer Proceedings in Mathematics & Statistics 510,
https://doi.org/10.1007/978-981-96-9350-4_20

Chang, in 1968, developed some fuzzy concepts [6]. Subsequently, several authors extended basic concepts to fuzzy sets and developed the theory of fuzzy topological spaces. For instance, Kandil introduced fuzzy bitopological space in 1989 [11]. Then, in 1997, Balasubramanian and Sundaram created a generalized fuzzy closed class in the field of fuzzy topology [5]. Furthermore, various experts have published multiple research articles on the generalized closed group of fuzzy spaces. Like, Zahran and El-Maghrabi discussed many circular operations in fuzzy space [16]. In addition, some scientists presented several studies on generalized closed groups in fuzzy space [8, 9, 12]. We want to continue developing the research on generalized closed sets and their applications in the field of fuzzy bitopology as mentioned in Refs. [1–4]. Also, studying separation axioms in fuzzy bitopological spaces is crucial for understanding their properties and behavior. It helps determine openness, compactness, and connectedness levels, essential for mathematics and computer science applications. Studying these axioms establishes a foundation for research in fuzzy set theory and topology, enabling informed decisions about the applicability of these spaces in mathematical models and theories. So, this paper intends to provide and investigate various generalized separation notions within the context of fuzzy bitopological spaces. Moreover, it examines their interrelations and attributes, while emphasizing significant theories and counterexamples. We have structured the research as follows: In Sect. 1, we provide an overview of the topic's historical background and its significance. In Sect. 2, we discuss some crucial foundational notions that hold significance in this investigation. Section 3 presents the concept of "generalized separation notions," focusing on generalized separation axioms in bitopological fuzzy spaces. We thoroughly explain the concept, covering important theorems, and specific properties. Finally, we summarize our results in Sect. 4.

2 Preliminaries

In this next section, we mention a few concepts which are fundamental to the research.

Definition 1 [13] Assume I stands for [0, 1] and X is not empty, thus:

(1) M is fuzzy set referred to a map with X domain and I range, $M(t) \in (0, 1]$ when $t \in M$, as $M(t) = 0$ if $t \notin M$.
(2) M in L as shown by $M \subseteq L$ if $M(t) \leq L(t)$, while $t \in X$.
(3) the combination $M \vee L$ is group defined as $(M \vee L)(t) = \text{upper}\{M(t), L(t)\}\ \forall\ t \in X$.
(4) the intersection $M \wedge L$ is defined as $(M \wedge L)(t) = \text{lower}\{M(t), L(t)\}\ \forall\, t \in X$.
(5) the completeness M^c is defined as $(M(t))^c = 1 - M(t), \forall\, t \in X$.

The concepts of fuzzy topology and fuzzy bitopology are explained as follows:

Definition 2 [13] (X, δ) is known as fuzzy topology when the next claims holds:

1. 0, 1 members of δ, since $0(t) = 0,\ 1(t) = 1$.
2. $M \wedge L$ member of δ, $\forall M, L \in \delta$.
3. $\vee_{i \in I} M_i$ member of δ, $\forall (M_{i \in I}) \in \delta$.

"Fuzzy topology space," or "fts" shortly is referred to (X, δ). Also, the parts of δ known as open fuzzy groups. If $F \in \delta$, then F^c is closed fuzzy group.

Definition 3 [11] A bitopology fuzzy spaces (X, δ_1, δ_2) as δ_1, δ_2 are fuzzy topological spaces on X and X not empty, denoted by "fbts."

Definition 4 [14] A fuzzy set of X is known as fuzzy point if it equals 0 for all $y \in X$, expect for one $x \in X$ and its value equals λ, where $(0 < \lambda \leq 1)$, denoted by x_λ.

The subsequent definitions elucidate the meanings of generalized fuzzy closed set and different types of separation axioms in fts.

Definition 5 [5] K is called generalized fuzzy closed when closure K is subgroup of R, and K is subgroup of R, R is fuzzy open.

i.e., K is generalized fuzzy closed if $cl(K) \leq R$, such that $K \leq R$, R is fuzzy open.

Definition 6 [10] A fts (X, δ) is known as:

(1) T_0 when $\forall x_r, y_s \in X, x_r \neq y_s$, $\exists\ U \in \delta$ such that $x_r \in U, y_s \notin U$, or $x_r \notin U, y_s \in U$.
(2) T_1 when $\forall x_r, y_s \in X, x_r \neq y_s, \exists U, V \in \delta$ such that $x_r \in U, y_s \notin U$, with $y_s \in V, x_r \notin V$.
(3) T_2 when $\forall x_r, y_s \in X, x_r \neq y_s, \exists U, V \in \delta$ such that $x_r \in U, y_s \in V$, with $U \wedge V = \phi$.
(4) $T_{2\frac{1}{2}}$ when for all fuzzy points x_r, y_t of A, $\exists$ fuzzy sets $B, C \subseteq A$ as $\mu_{x_r}(x) \leq \mu_B(x)$, $\mu_{y_t}(x) \leq \mu_C(x)$, with $cl(B)\, q\, cl(C)$.

Remark 1 [7] If (X, δ) is fuzzy $T_{2\frac{1}{2}}$, then it is T_2 and T_1, therefore T_0.
i.e., $T_{2\frac{1}{2}} \Rightarrow T_2 \Rightarrow T_1 \Rightarrow T_0$.

The concept of some kinds of generalized closed groups is the basis for constructing the definition of generalized separation axioms in fbts, and we introduced it in [3] as follows:

Definition 7 [3] Each fuzzy subgroup H of fbts (X, δ_i, δ_j) is called:

(1) Fuzzy (i, j)-generalized Φ-closed (in sum, $(i, j) - g\Phi - cld$) if $\delta_j - \Phi - cl(H) \leq W$, while $H \leq W$, $W \in \delta_i$, as Φ including ((α) alpha, (s) semi, (p) pre, and (β) beta).

(2) The $(i, j) - g\Phi$-open of X is a supplementary of fuzzy $(i, j) - g\Phi - cld$.

The following theorem shows the relations via each varieties of generalized closed groups in fbts.

Theorem 1 *[3] When (X, δ_i, δ_j) is fbts. Then, the next statements hold:*

(1) $\forall (i, j) - g - cld$ is $(i, j) - g\alpha - cld$.
(2) $\forall (i, j) - g\alpha - cld$ is $(i, j) - gp - cld$ and $(i, j) - gs - cld$.
(3) $\forall (i, j) - gp - cld$ or $(i, j) - gs - cld$ is $(i, j) - g\beta - cld$.

We have mentioned several important examples that show that the reverse of the previous relationships is not true in Ref. [3].

3 Types of Fuzzy Generalized Separation Axioms

This section explores the concept of generalized separation notions through the application of various forms of generalized closed groups in fbts, describing them in terms of significant theorems and features. Here, $h = 0, 1, 2, 2_{\frac{1}{2}}$ and Φ is made up of (α), (s), (p), and (β).

Definition 8 A fbts (X, δ_i, δ_j) is said to be:

(1) fuzzy generalized Φ T_0 (in sum, $(i, j) - g\Phi - T_0$) $\Longleftrightarrow$ $\forall$ different fuzzy points s, t of X, $\exists$ fuzzy $(i, j) - g\Phi$-open set A as $s \in A$, $t \cap A = 0$, or $t \in A$, $s \cap A = 0$.
(2) fuzzy generalized Φ T_{0w} (in sum, $(i, j) - g\Phi - T_{0w}$) $\Longleftrightarrow$ $\forall$ different fuzzy points s, t of X, $\exists$ fuzzy $(i, j) - g\Phi$-open set A as $s \in A \subseteq t^c$, or $t \in A \subseteq s^c$.
(3) fuzzy generalized Φ T_1 (in sum, $(i, j) - g\Phi - T_1$) $\Longleftrightarrow$ $\forall$ different fuzzy points s, t of X, $\exists$ fuzzy $(i, j) - g\Phi$-open sets A, B as $s \in A$, $t \cap A = 0$ and $t \in B$, $s \cap B = 0$.
(4) fuzzy generalized Φ T_{1w} (in sum, $(i, j) - g\Phi - T_{1w}$) $\Longleftrightarrow$ $\forall$ different fuzzy points s, t of X, $\exists$ fuzzy $(i, j) - g\Phi$-open sets A, B as $s \in A \leq t^c$ and $t \in B \leq s^c$.
(5) fuzzy generalized Φ T_2 (in sum, $(i, j) - g\Phi - T_2$) $\Longleftrightarrow$ $\forall$ different fuzzy points s, t of X, $\exists$ fuzzy $(i, j) - g\Phi$-open sets A, B as $s \in A, t \in B$, and $A \cap B = 0$.
(6) fuzzy generalized Φ T_{2w} (in sum, $(i, j) - g\Phi - T_{2w}$) $\Longleftrightarrow$ $\forall$ different fuzzy points s, t of X, $\exists$ fuzzy $(i, j) - g\Phi$-open sets A, B as $s \in A, t \in B$, and $A \subseteq B^c$.
(7) fuzzy generalized Φ $T_{2\frac{1}{2}}$ (in sum, $(i, j) - g\Phi - T_{2\frac{1}{2}}$) $\Longleftrightarrow$ $\forall$ different fuzzy points s, t of X, $\exists$ fuzzy $(i, j) - g\Phi$-open sets A, B as $s \in A, t \in B$, and $(i, j) - g\Phi - cl(A) \cap (i, j) - g\Phi - cl(B) = 0$.
(8) fuzzy generalized Φ $T_{2\frac{1}{2}w}$ (in sum, $(i, j) - g\Phi - T_{2\frac{1}{2}w}$) $\Longleftrightarrow$ $\forall$ different fuzzy points s, t of X, $\exists$ fuzzy $(i, j) - g\Phi$-open sets A, B as $s \in A, t \in B$, and $(i, j) - g\Phi - cl(A) \subseteq ((i, j) - g\Phi - cl(B))^c$.

The subsequent theorems are special characteristics for some generalized separation axioms.

The first theory gives equivalent meanings to the definition of $(i, j) - g\Phi - T_0$:

Theorem 2 *Let (X, δ_i, δ_j) be fbts. Then, the next claims are equivalent:*

(1) X is fuzzy $(i, j) - g\Phi - T_0$.
(2) $\forall$ points s, t of X, $\exists$ fuzzy $(i, j) - g\Phi$-open set A as $s \in A,\ t \cap A = 0$, or $t \in A,\ s \cap A = 0$.
(3) $\forall$ points s, t of X, $(i, j) - g\Phi - cl(s) \neq (i, j) - g\Phi - cl(t)$.

Proof (1) $\Rightarrow$ (2) It is clear from Definition 8.
(2) $\Rightarrow$ (3) Let s, t be distinct fuzzy points from X and A considers an $(i, j) - g\Phi$-open group as $s \in A,\ t \cap A = 0$.

So, $t \in A^c \Rightarrow (i, j) - g\Phi - cl(t) \subseteq (i, j) - g\Phi - cl(A^c) = A^c$.

Hence $s \in A$, thus $s \notin A^c$, after that $s \notin (i, j) - g\Phi - cl(t)$ and $s \in (i, j) - g\Phi - cl(s)$. As a result, $(i, j) - g\Phi - cl(s) \neq (i, j) - g\Phi - cl(t)$.
(3) $\Rightarrow$ (1) Let s, t be distinct fuzzy points of X with $(i, j) - g\Phi - cl(s) \neq (i, j) - g\Phi - cl(t)$.

Since $(i, j) - g\Phi - cl(s) \neq (i, j) - g\Phi - cl(t)$, thus $\exists$ fuzzy point q belongs to one and does not belong to the other. Say $q \in (i, j) - g\Phi - cl(s), q \notin (i, j) - g\Phi - cl(t)$, and hence $\exists\, U$ considers $(i, j) - g\Phi$-open set as $q \in U,\ U \cap (i, j) - g\Phi - cl(t) = 0$.

But $U \cap (i, j) - g\Phi - cl(s) \neq 0$ as $q \in (i, j) - g\Phi - cl(s)$.

Thus, the fuzzy space X will be $(i, j) - g\Phi - T_0$.

The following theory gives equivalent meanings to the definition of $(i, j) - g\Phi - T_1$:

Theorem 3 *Let (X, δ_i, δ_j) be fbts. Then, the next claims are equivalent:*

(1) X is fuzzy $(i, j) - g\Phi - T_1$
(2) $\forall$ point $s \in X$, $\{s\}$ is fuzzy $(i, j) - g\Phi - cld$ set.

Proof (1) $\Rightarrow$ (2) Suppose X be $(i, j) - g\Phi - T_1$, and $A = \{s\}$ be fuzzy subset of X. Since X is $(i, j) - g\Phi - T_1$, thus for all distinct points $s,\ t \exists U$ be $(i, j) - g\Phi$-open set as $t \in U,\ s \cap U = 0$. Hence, $U \cap A = 0$. Since $t \notin \{s\} = A \subseteq (i, j) - g\Phi - cl(A) = (i, j) - g\Phi - cl(\{s\}) = \{s\} = A$.

Thus, $t \notin (i, j) - g\Phi - cl(\{s\})$, and hence $(i, j) - g\Phi - cl(A) = \{s\} = A$.

Then, A is $(i, j) - g\Phi - cld$.
(2) $\Rightarrow$ (1) Suppose all fuzzy singleton set be $(i, j) - g\Phi - cld$, and s, t be distinct fuzzy points of X. Then, $\{s\}^c$ and $\{t\}^c$ will be $(i, j) - g\Phi$-open.

Also, $s \cap \{s\}^c = 0, t \in \{s\}^c$, and $t \cap \{t\}^c = 0, t \in \{s\}^c$.

As a result, X is $(i, j) - g\Phi - T_1$ space.

The following theory gives equivalent meanings to the definition of $(i, j) - g\Phi - T_2$:

Theorem 4 *A fbts* (X, δ_i, δ_j) *is fuzzy* $(i, j) - g\Phi - T_2$ *space* $\Longleftrightarrow$ $\forall$ *different points* s, t, $\exists$ *fuzzy* $(i, j) - g\Phi$*-open set* A *as* $s \in A$, $t \in ((i, j) - g\Phi - cl(A))^c$.

Proof Suppose X be $(i, j) - g\Phi - T_2$ space and $s, t \in X$, $s \neq t$. Thus, $\exists A, B$ known as $(i, j) - g\Phi$-open groups, such that $s \in A$, $t \in B$, $A \cap B = 0$, so $B \subseteq A^c$ or $A \subseteq B^c$, if we choose $A \subseteq B^c$. As B^c is $(i, j) - g\Phi - cld$ set, thus by taking $(i, j) - g\Phi$-closer on all sides we have, $(i, j) - g\Phi - cl(A) \subseteq B^c$ (from (4) of Theorem 5 in Ref. [1]), after that by taking the complement we have, $B \subseteq ((i, j) - g\Phi - cl(A))^c$. Since, $t \in B \subseteq ((i, j) - g\Phi - cl(A))^c$.

As a result, $t \in ((i, j) - g\Phi - cl(A))^c$.

Conversely, Let s, t be fuzzy distinct points, and A be $(i, j) - g\Phi$-open, $s \in A$, $t \in ((i, j) - g\Phi - cl(A))^c$. If we choose $B = ((i, j) - g\Phi - cl(A))^c$ which is fuzzy $(i, j) - g\Phi$-open set and including t, so we have

$$\begin{aligned} A \cap B &= A \cap ((i, j) - g\Phi - cl(A))^c \\ &= (A \cap X) - (A \cap (i, j) - g\Phi - cl(A)) \\ &= A - A \\ &= 0. \end{aligned}$$

Then, $\exists$ fuzzy $(i, j) - g\Phi$-open sets A, B such that $s \in A$, $t \in B$, and $A \cap B = 0$. As a result, X is fuzzy $(i, j) - g\Phi - T_2$ space.

The next theory needs to the definition of (i, j)-generalized Φ-closed neighborhood which defined as follows:

Definition 9 A fuzzy set A of (X, δ_i, δ_j) is known as fuzzy (i, j)-generalized Φ-closed neighborhood (in sum, $(i, j) - g\Phi - cld\, nbd$) of fuzzy point x_t, when A is fuzzy $(i, j) - g\,\Phi - cld$ set containing $(i, j) - g\,\Phi$-open set B and $x_t \in B \subseteq A$.

Theorem 5 *If* $\forall$ *different points* s, t *of* X, $\exists$ *two disjoint fuzzy* $(i, j) - g\Phi - cld\, nbd$ A, B *as* $s \in A$, *and* $t \in B$, *then a fbts* (X, δ_i, δ_j) *is fuzzy* $(i, j) - g\Phi - T_{2\frac{1}{2}}$.

Proof Let s, t be two fuzzy distinct points of X, and A, B be two fuzzy disjoint $(i, j) - g\Phi - cld\, nbd$ for s, t when $s \in A$, and $t \in B$. Then, from (4) of Theorem 5 in Ref. [1] we have, $A = (i, j) - g\Phi - cl(A)$, $B = (i, j) - g\Phi - cl(B)$, hence $(i, j) - g\Phi - cl(A) \cap (i, j) - g\Phi - cl(B) = 0$.

As a result, (X, δ_i, δ_j) is $(i, j) - g\Phi - T_{2\frac{1}{2}}$ space.

Remark 2 (i) In general, the converse of the given theorem is not true. This is because the process of intersection $(i, j) - g\Phi - cld$ classes need not to be as $(i, j) - g\Phi - cld$, as shown in Remark 3 in Ref. [2].

(ii) The converse of the above theorem holds when δ_i is an indiscrete fuzzy topology.

The subsequent theorems elucidate the interrelationships among all categories of generalized separation axioms in fbts.

Theorem 6 *Suppose (X, δ_i, δ_j) is fbts, then the next claims are true:*

(1) If X is fuzzy $(i, j) - g\Phi - T_h$ space, then its fuzzy $(i, j) - g\Phi - T_{hw}$ space.
(2) If (X, δ_j) is fuzzy T_h space (T_{hw} space), then (X, δ_i, δ_j) is fuzzy $(i, j) - g\Phi - T_h$ space ($(i, j) - g\Phi - T_{hw}$).

Theorem 7 *If (X, δ_i, δ_j) is fbts. Then the next statements are true:*

(1) Each fuzzy $(i, j) - g - T_h(T_h w)$ is fuzzy $(i, j) - g\alpha - T_h(T_h w)$.
(2) Each fuzzy $(i, j) - g\alpha - T_h(T_h w)$, is fuzzy $(i, j) - gs - T_1(T_h w)$ and $(i, j) - gp - T_h(T_h w)$.
(3) Each fuzzy $(i, j) - gs - T_h(T_h w)$, or fuzzy $(i, j) - gp - T_h(T_h w)$ is $(i, j) - g\beta - T_h(T_h w)$.

Theorem 8 $(i, j) - g\Phi - T_{2\frac{1}{2}} \Longrightarrow (i, j) - g\Phi - T_2 \Longrightarrow (i, j) - g\Phi - T_1 \Longrightarrow (i, j) - g\Phi - T_0$.

The subsequent examples illustrate the invalidity of inverse relationships. The first example shows that fuzzy $(i, j) - g\Phi - T_0$ space not fuzzy $(i, j) - g\Phi - T_1$ space.

Example 1 Assume $X = \{a, b\}, \delta_1 = \{0, 1, A\}$, $\delta_2 = \{0, 1, B, C\}$. Since $A(a, b) = \{0.7, 0.5\}$, $B(a, b) = \{0.7, 0.6\}$, $C(a, b) = \{0.2, 0.4\}$, then we notice that $\exists$ $D_1(a, b) = \{0.2, 0\}$, $D_2(a, b) = \{0.2, 0.5\}$ are fuzzy $(i, j) - g\Phi$-open, and $\exists$ $a = 0.1, b = 0.4$. So, $(a = 0.1, 0) \in D_1(a, b) = \{0.2, 0\} \cap b = 0$, but $b \in D_2$, $a \cap D_2 \neq 0$. Thus, X is fuzzy $(i, j) - g\Phi - T_0$ space not fuzzy $(i, j) - g\Phi - T_1$ space.

The next example shows that fuzzy $(i, j) - g\Phi - T_1$ space not fuzzy $(i, j) - g\Phi - T_2$ space.

Example 2 Assume $X = \{a, b, c\}$, $\delta_1 = \{0, 1, A\}$, $\delta_2 = \{0, 1, B, C\}$. Since $A(a, b, c) = \{0.8, 0.4, 0.3\}$, $B(a, b, c) = \{0.7, 0.6, 0.5\}$, $C(a, b, c) = \{0.2, 0.4, 0.3\}$, then we notice that $\exists$ $D_1(a, b, c) = \{0.1, 0, 0.2\}$, $D_2(a, b, c) = \{0, 0.5, 0.3\}$ are fuzzy $(i, j) - g\Phi$-open, and $\exists$ $a = 0.1, b = 0.4$.

Thus $(a = 0.1, 0, 0) \in D_1(a, b, c) = \{0.1, 0, 0.2\}$, $D_1 \cap b = 0$, and $b \in D_2, a \cap D_2 = 0$, but $D_1 \cap D_2 \neq 0$. Thus, X is fuzzy $(i, j) - g\Phi - T_1$ space not fuzzy $(i, j) - g\Phi - T_2$ space.

The next example shows that fuzzy $(i, j) - g\Phi - T_2$ space not fuzzy $(i, j) - g\Phi - T_{2\frac{1}{2}}$ space.

Example 3 Assume $X = \{a, b, c\}$, $\delta_1 = \{0, 1, A\}$, $\delta_2 = \{0, 1, B, C\}$.

Since $A(a, b, c) = \{0.8, 0.4, 0.3\}$, $B(a, b, c) = \{0.7, 0.6, 0.5\}$, $C(a, b, c) = \{0.2, 0.4, 0.3\}$, then we notice that $\exists$ $D_1(a, b, c) = \{0.1, 0, 0.2\}$, $D_2(a, b, c) = \{0, 0.5, 0\}$ are fuzzy $(i, j) - g\Phi$-open, and $\exists$ $a = 0.1, b = 0.4$.

Then $(a = 0.1, 0, 0) \in D_1(a, b, c) = \{0.1, 0, 0.2\}$, $D_1 \cap b = 0$, and $b \in D_2, a \cap D_2 = 0$, $D_1 \cap D_2 = 0$, but $(i, j) - g\Phi - cl(D_1) = B^c \cap (i, j) - g\Phi - cl(D_2) = C^c \neq 0 = B^c$.

Thus, X is fuzzy $(i, j) - g\Phi - T_2$ space not fuzzy $(i, j) - g\Phi - T_{2\frac{1}{2}}$ space.

Definition 10 [2] A fbts (X, δ_i, δ_j) is known as fuzzy $(i, j) - g\Phi$-compact when all fuzzy $(i, j) - g\Phi$-open cover of X has a finite subcover.

Theorem 9 *If K is fuzzy $(i, j) - g\Phi$-compact subset of fuzzy $(i, j) - g\Phi - T_2$ space and $p \notin K$, $\exists$ disjoint $(i, j) - g\Phi$-open sets A, B of X as $p \in A$, $K \subseteq B$.*

The diagram explaining the relations between all types of fuzzy $(i, j) - g\Phi - T_h$:

Remark 3 The reverse of the diagram in Fig. 1 is not generally true, as explained in Definition 8, Theorem 7, and from the connection via $(i, j) - g\Phi$-groups discussed in Theory 3.2.6 with Examples 3.2.10 through 3.2.14 in Ref. [2].

The diagram explaining the relations between all types of fuzzy $(i, j) - g\Phi - T_h$ and fuzzy $(i, j) - g\Phi - T_{hw}$:

Remark 4 The reverse of the diagram in Fig. 2 is not generally true, as explained in Definition 8, relation via fuzzy $(i, j) - g\Phi - T_h$ and fuzzy $(i, j) - g\Phi - T_{hw}$ discussed at Theorem 6, and Theory 3.2.6 with Examples 3.2.10 through 3.2.14 in Ref. [2].

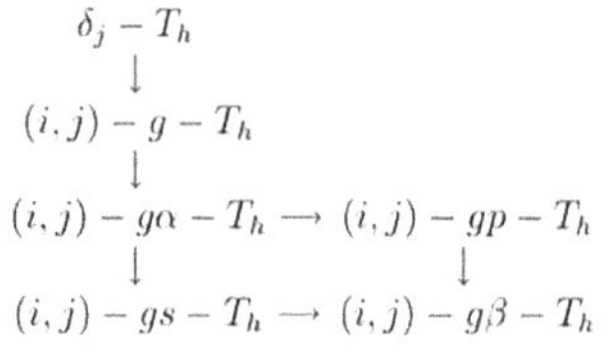

Fig. 1 Relationships of all fuzzy $(i, j) - g\Phi - T_h$

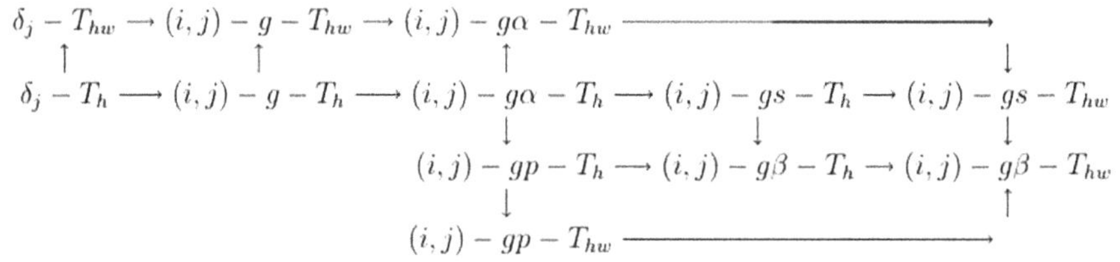

Fig. 2 Relationships of all fuzzy $(i, j) - g\Phi - T_h$ and $(i, j) - g\Phi - T_{hw}$

The following theory is predicated on the definitions of some functions previously studied in Ref. [3].

Theorem 10 *(1) The property of being fuzzy* $(i, j) - g\Phi - T_h$ (T_{hw}) *is hereditary.*
(2) If (X, δ_1, δ_2) *is fuzzy* $(i, j) - g\Phi - T_h(T_{hw})$ *space,* $f : (X, \delta_1, \delta_2) \rightarrow (Y, \sigma_1, \sigma_2)$ *is bijective* $(i, j) - g\Phi$*-open map, then* (Y, σ_1, σ_2) *is also fuzzy* $(i, j) - g\Phi - T_h(T_{hw})$ *space.*
(3) The property of being fuzzy $(i, j) - g\Phi - T_h(T_{hw})$ *is topological property.*

We demonstrate the validity of the paragraphs in this theory when $h = 0$.

Proof (1) Let (X, δ_i, δ_j) be an $(i, j) - g\Phi - T_0$ space, $Y \subseteq X$ be a subspace, $s, t \in Y, s \neq t$. So, $s, t \in X$, since X considered as $(i, j) - g\Phi - T_0$, thus $\exists A$ will be $(i, j) - g\Phi$-open as $s \in A, \{t\} \cap A = 0$ (or $t \in A, \{s\} \cap A = 0$). Thus, $s \in (Y \cap A), \{t\} \cap (Y \cap A) = 0$. Since $(Y \cap A)$ considers $(i, j) - g\Phi$-open group at the subspace Y. So, Y will be $(i, j) - g\Phi - T_0$. As a result, the property of being fuzzy $(i, j) - g\Phi - T_0$ is hereditary.
(2) Let $s^*, t^* \in Y, s^* \neq t^*$. Since f is bijective map, then $f^{-1}(t^*) = t, f^{-1}(s^*) = s$ as they be as two different fuzzy points of X, because the space X be as $(i, j) - g\Phi - T_0$, so $\exists$ an $(i, j) - g\Phi$-open set A such that $s \in A, \{t\} \cap A = 0$ (or $t \in A, \{s\} \cap A = 0$). But f is $(i, j) - g\Phi$-open map. So, $\exists f(A)$ will be $(i, j) - g\Phi$-open set of Y containing s^* and its intersection with t^* equal 0. Then, Y considers $(i, j) - g\Phi - T_0$ space. As a result, the property of being fuzzy $(i, j) - g\Phi - T_0$ is topological property.
(3) It is clear by using Corollary 6 in Ref. [4].

4 Conclusion and Future Studies

In this paper we presented several definitions of generalized separation axioms in fbts, including $T_0, T_1, T_2, T_{2\frac{1}{2}}, T_{0w}, T_{1w}, T_{2w}$, and $T_{2\frac{1}{2}w}$ with a study of a number of their properties and theories. Furthermore, we have discovered the relationship that implications all of these axioms, which are supported by counterexamples and illustrative figures.

This research also paves the way for future investigations into these separation axioms through the application of alternative fuzzy set types, such as gamma, theta, or regular sets. Furthermore, we can investigate these notions throughout diverse fuzzy spaces characterized by multiple topologies, as well as across other mathematical domains, including fuzzy metric spaces, fuzzy soft topologies, and fuzzy topological vector spaces. Additionally, we examine the appropriateness of these spaces in mathematical models and computer science applications.

Acknowledgements The authors express gratitude to the referee(s) for their insightful comments.

Disclosure of Interests The authors have no competing interests to declare that are relevant to the content of this article.

References

1. Alharbi, A., Kilicman, A.: Note on generalized neighborhoods structures in fuzzy bitopological spaces. Eur. J. Pure Appl. Math. 1980–1990 (2023)
2. Alharbi, A., Kilicman, A.: Generalized compactness in fuzzy bitopological spaces. Eur. J. Pure Appl. Math. 30–41 (2024)
3. Alharbi, A., Kilicman, A.: Generalized connectedness in fuzzy bitopological spaces. Appl. Math. Inf. Sci. 1019–1023 (2023)
4. Alharbi, A.A., Kilicman, A.: Generalized different types of mappings in fuzzy bitopological spaces. Eur. J. Pure Appl. Math. 2613–2631 (2023)
5. Balasubramanian, G., Sundaram, P.: On some generalizations of fuzzy continuous functions. Fuzzy Sets Syst. **86**(1), 93–100 (1997)
6. Chang, C.L.: Fuzzy topological spaces. J. Math. Anal. Appl. **24**(1), 182–190 (1968)
7. Cheng-Ming, H.: Fuzzy topological spaces. J. Math. Anal. Appl. **110**(1), 141–178 (1985)
8. El-Shafei, M.E.: Some applications of generalized closed sets in fuzzy topological spaces. Kyungpook Math. J. **45**(1), 13–19 (2005)
9. El-Shafei, M.E., Zakari, A.: Semi-generalized continuous mappings in fuzzy topological spaces. J. Fuzzy Math. **15**(1), 109 (2007)
10. Fora, A.A.: Separation axioms for fuzzy spaces. Fuzzy Sets Syst. **33**(1), 59–75 (1989)
11. Kandil, A., El-Shafee, M.E.: Biproximities and fuzzy bitopological spaces. Simon Stevin **63**(1), 45–66 (1989)
12. Nakajima, N.: Generalized fuzzy sets. Fuzzy Sets Syst. **32**(3), 307–314 (1989)
13. Palaniappan, N.: Fuzzy Topology, 2nd edn. Alpha Science International Ltd. (2002)
14. Wong, C.K.: Fuzzy points and local properties of fuzzy topology. J. Math. Anal. Appl. **46**(2), 316–328 (1974)
15. Zadeh, L.A.: Fuzzy sets. Inf. Control **8**(3), 338–353 (1965)
16. Zahran, A.M., El-Maghrabi, A.I.: Generalized ψ p-operations on fuzzy topological spaces. Abs. Appl. Anal. 1–12 (2011)

www.ingramcontent.com/pod-product-compliance
Ingram Content Group UK Ltd.
Pitfield, Milton Keynes, MK11 3LW, UK
UKHW020815121125
464966UK00001B/3
* 9 7 8 9 8 1 9 6 9 3 4 9 8 *